Multiphysics Modelling with
Finite Element Methods

SERIES ON STABILITY, VIBRATION AND CONTROL OF SYSTEMS

Founder and Editor: Ardéshir Guran
Co-Editors: M. Cloud & W. B. Zimmerman

About the Series

Rapid developments in system dynamics and control, areas related to many other topics in applied mathematics, call for comprehensive presentations of current topics. This series contains graduate level textbooks, monographs, and collection of thematically organized research or pedagogical articles addressing key topics in applied dynamics.

The material is ideal for a general scientific and engineering readership, and is also mathematically precise enough to be a useful reference for research specialists in mechanics and control, nonlinear dynamics, and in applied mathematics and physics.

Reporting on academic/industrial research from institutions around the world, the SVCS series reflects technological advances in mechanics and control. Particular emphasis is laid on emerging areas such as modeling of complex systems, bioengineering, mechatronics, structronics, fluidics, optoelectronic sensors, micromachining techniques, and intelligent system design.

Selected Volumes in Series A

Vol. 12 The Calculus of Variations and Functional Analysis: With Optimal Control and Applications in Mechanics
Authors: L. P. Lebedev and M. J. Cloud

Vol. 13 Multiparameter Stability Theory with Mechanical Applications
Authors: A. P. Seyranian and A. A. Mailybaev

Vol. 14 Stability of Stationary Sets in Control Systems with Discontinuous Nonlinearities
Authors: V. A. Yakubovich, G. A. Leonov and A. Kh. Gelig

Vol. 15 Process Modelling and Simulation with Finite Element Methods
Author: W. B. J. Zimmerman

Vol. 16 Design of Nonlinear Control Systems with the Highest Derivative in Feedback
Author: V. D. Yurkevich

Vol. 17 The Quantum World of Nuclear Physics
Author: Yu. A. Berezhnoy

Selected Volume in Series B

Vol. 14 Impact and Friction of Solids, Structures and Intelligent Machines
Editor: A. Guran

Vol. 15 Advances in Mechanics of Solids: In Memory of Prof. E. M. Haseganu
Editors: A. Guran, A. L. Smirnov, D. J. Steigmann and R. Vaillancourt

SERIES ON STABILITY, VIBRATION AND CONTROL OF SYSTEMS

Series A   **Volume 18**

Founder & Editor: **Ardéshir Guran**

Co-Editors: **M. Cloud & W. B. Zimmerman**

Multiphysics Modelling with Finite Element Methods

William B J Zimmerman

University of Sheffield, UK

World Scientific

NEW JERSEY · LONDON · SINGAPORE · BEIJING · SHANGHAI · HONG KONG · TAIPEI · CHENNAI

Published by

World Scientific Publishing Co. Pte. Ltd.

5 Toh Tuck Link, Singapore 596224

USA office: 27 Warren Street, Suite 401-402, Hackensack, NJ 07601

UK office: 57 Shelton Street, Covent Garden, London WC2H 9HE

Library of Congress Cataloging-in-Publication Data
Zimmerman, William B. J.
    Multiphysics modelling with finite element methods / William B. J. Zimmerman.
        p. cm. -- (Series on stability, vibration, and control of systems. Series A ; v. 18)
    Includes bibliographical references and index.
    ISBN-13 978-981-256-843-4
    ISBN-10 981-256-843-3
    1. Finite element method. 2. Information modeling.
    TA347.F5 Z55 2006
    511.42--dc22
                                                            2007298956

British Library Cataloguing-in-Publication Data
A catalogue record for this book is available from the British Library.

First published 2006
Reprinted 2007, 2008, 2010, 2012

Printed in Singapore.

ABOUT THE AUTHOR

William B. J. Zimmerman CEng FIChemE is the Professor of Biochemical Dynamical Systems in Chemical and Process Engineering at the University of Sheffield. His research interests are in fluid dynamics, reaction engineering, and microfluidic biotechnology. He is the author of Process Modelling and Simulation with Finite Element Methods (2004), the earlier version of this book, and the editor of Microfluidics: History, Theory and Applications (2005). He has previously created modules entitled Chemical Engineering Problem Solving with Mathematica, Modelling and Simulation in Chemical Processes, Numerical Analysis in Chemical Engineering, and FORTRAN programming. He has been modelling with finite element methods since 1986. He has authored over eighty scientific and scholarly works. He is a graduate of Princeton and Stanford Universities in Chemical Engineering, past Director of the MSc. in Environmental and Energy Engineering, originator of the MSc. in Process Fluid Dynamics, and a winner in US and UK national competitions of five prestigious fellowships:

2005-6 Royal Academy of Engineering and Leverhulme Trust Senior Research Fellow.

2000-5 EPSRC Advanced Research Fellow.

1994-99 Royal Academy of Engineering, Zeneca Young Academic Fellow.

1991-93 NATO postdoctoral fellow in science and engineering.

1988-91 National Science Foundation Research Fellow.

FOREWORD

I started running intensive modules training novice research students in modelling with COMSOL Multiphysics (then FEMLAB) in 2002. We just conducted the most recent, the seventh, in January 2006. I updated my course notes to FEMLAB 3.1 in early 2005 then to COMSOL Multiphysics 3.2 for the 2006 module. Many readers have contacted me saying that they have bought my previous book, but were disappointed that it was for FEMLAB 2.3. The intellectual content of the earlier book is still current, but I find it rather frustrating when a computer textbook does not match the latest version of the software. Given the speed with which COMSOL has been making updates, it is inevitable that textbooks that use it will suffer from version mismatch. Since the "look and feel" of FEMLAB changed with version 3, and now the name has changed, it seems appropriate to update my modelling textbook. The title has changed to reflect the now common usage of the jargon "multiphysics."

This book is intended to be accessible to all novice users of COMSOL Multiphysics and should be compatible with the next release, version 3.3, but is, of course, geared more towards chemical engineers. We have had civil, mechanical, electrical, and control systems engineers, applied mathematicians, physicists, chemists, and medical researchers take the intensive module, as well as industrial researchers and project engineers. I think the book is just as widely accessible to novice modellers.

I have no particular affiliation with COMSOL — we pay our annual maintenance for the classroom and research licenses just like other users. Over the years, however, I have built up good relations with Johan Sundqvist, Ed Fontes, Patrik Bosander, Jerome Long, Nikos Vasileides, Bertil Walden, Jukka Tarvo, Niklas Rom, Eric Favre, Mina Sierou, Anna Dzougoutov, Peter Georén, Jeff Hillier, and Lars Langemyr of COMSOL, who have been most supportive of my two projects, lending considerable resources for helping me to iron out difficulties in modeling, contributing to the intensive modules, and providing critiques of the draft chapters. The FEMLAB Users group at Yahoo! have been particularly supportive of my endeavours in the last few years. I heartily recommend that new users join and read the daily digest.

Some of the gurus have contributed case studies to this book (see http://groups.yahoo.com/group/FEMLAB_Users/). Many of the files produced with this book (.m, .mat, .mph) are available from my intensive module web site (http://eyrie.shef.ac.uk/femlab).

Many thanks to the team of collaborators and chapter co-authors who have encouraged this effort. Buddhi, Alex, Kiran, Jordan, Peter, George, Venkat, Dan, Ali, David, Jaime, Tanai and Julia have always had a kind word and a willingness to brainstorm and contribute.

CONTENTS

INTRODUCTION TO COMSOL MULTIPHYSICS

W.B.J. ZIMMERMAN

Department of Chemical and Process Engineering,
University of Sheffield,
Mappin Street, Sheffield S1 3JD, United Kingdom
E-mail: w.zimmerman@shef.ac.uk

COMSOL Multiphysics has superseded FEMLAB as the trade name for a powerful modelling package, based originally on the MATLAB programming language and integrated development environment. Perhaps a good fraction of the readers of this book were attracted by the title and the dust jacket description, so they might have little exposure to COMSOL Multiphysics previously. To them, I would heartily recommend attending a COMSOL seminar on their recurring academic roadshows. The experience of seeing COMSOL Multiphysics in action is more illustrative than the printed word and screen captures shown here. This Introduction provides an overview of why I wrote the book and developed an intensive training module for COMSOL Multiphysics modeling of chemical engineering applications — the unique features of COMSOL Multiphysics that the reader will want to assess for her own modeling objectives. The COMSOL Multiphysics User's Guide (available for download from the COMSOL web site) does a better job of familiarizing the reader with "What is COMSOL Multiphysics?" than the brief introduction in this chapter to the COMSOL Multiphysics graphical user interface (GUI). The point of the introduction to COMSOL Multiphysics here is to describe how completely determined models are set up in COMSOL Multiphysics, after which the methodology can be used in subsequent chapters without ambiguity. Nevertheless, I hope that this chapter whets your appetite for the cornucopia of modeling tools, along with an intellectual framework for using COMSOL Multiphysics for modeling, that is described in this book.

1. Overview of the book

Chapters 1–4 were taken as the text for the first intensive module "Chemical Engineering Modelling with FEMLAB." These chapters represent a personal odyssey with COMSOL Multiphysics. It was not originally my intention to write a book about COMSOL Multiphysics. For a long term project that I am still undertaking, I need a PDE engine that is readily customizable to additional terms and heterogeneous domains. Once I decided that COMSOL Multiphysics could fit the bill, I needed to become an expert on it. One nefarious way of doing that is to declare a course on it, rope graduate students and other interested external parties into attending, and then study like mad to produce a coherent set of lectures and computer laboratories. I already had several templates for this, having taught undergraduate and postgraduate modules on numerical analysis, modeling, and simulation. So I adapted the storyline of those modules with COMSOL Multiphysics models. Chapter One is the product of this adaptation. Chapter Two is an obvious outgrowth of my prior use of the PDE toolbox of MATLAB and a necessary explanation of finite element methods. Chapters 3–7 were far more deliberate attempts to exploit the powerful features of COMSOL Multiphysics by systematically exploring models that illustrate the feature of the theme of each chapter. I searched through my own repertoire of PDE modeling and sought out contributions from colleagues

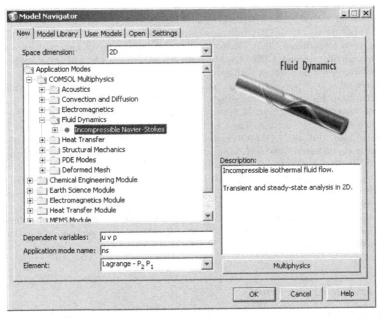

Figure 1.   The pre-built application modes are arranged in a tree structure on the Model Navigator. Here is the Incompressible Navier-Stokes mode under the Chemical Engineering Module. The Model Navigator specifies that this mode is 2-D, has three dependent variables, and uses a mixed type of element Lagrange p_2 for the velocities u and v, Lagrange p_1 for the pressure. Using mixed order discretization schemes is quite common in finite element methods for numerical stability of the Navier-Stokes solvers. The SIMPLE scheme [1] pioneered the approach. The Model Navigator allows the user to specify pre-built application modes or to customize a generic PDE mode (coefficient, general, weak) to build up their own model.

that would illustrate the features. Chapters 8 onwards are of a different type. These chapters would legitimately fit into the COMSOL Model Library as case studies of modeling with COMSOL Multiphysics, rather than organized along a particular programming theme. Nonetheless, the case studies highlight nonstandard aspects of COMSOL/MATLAB modeling, analysis, and postprocessing that are strikingly original. The Appendix represents an approach to COMSOL Multiphysics I would make if I were teaching a course on vector calculus and serves as a small primer to MATLAB programming.

Target audience

The book is aimed at graduate Chemical Engineers who use modelling tools and as a general introduction to COMSOL Multiphysics for scientists and engineers.

Attitude

The attitude of this book is to demonstrate particular features of COMSOL Multiphysics that make computational modelling easy to implement, and then emphasize those features that are advantages to modelling with COMSOL Multiphysics. This will be illustrated with reference to Chemical Engineering Modelling, which has a special history and well known applications. The features, however, are generally applicable in the sciences and engineering.

Bias

The book is slanted toward applications in fluid dynamics, transport phenomena, and heterogeneous reaction, which reflect some of the research interests of the author that routinely involve mathematical modelling by PDEs and solution by numerical methods.

1.1. *Modelling versus simulation*

This book is about modelling and programming. The first four chapters, the core of the taught module, focus completely on modelling. The remaining chapters are slanted towards the use of COMSOL Multiphysics for simulation. The distinction is that simulation has some stochastic and evolutionary elements. Simulations may have a PDE compute engine as an integral component, but generally involve much more "user defined programming." This book organizes case studies of modeling along the lines of a

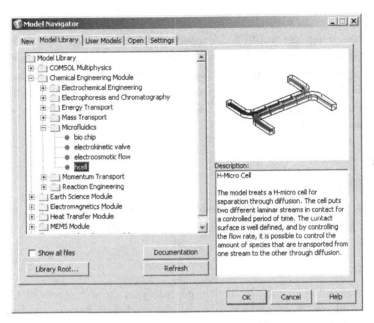

Figure 2. The Model Library contains already solved problems using existing application modes. The Model Library includes models created by COMSOL staff and donated by users. The growth in the content of the Model Library over the last twelve months has been phenomenal. Browsing the COMSOL Multiphysics models and the Model Library documentation of them is an excellent way of generating modeling ideas. The Model Library is organized by subject matter. Here, the microfluidic H-cell is highlighted, under the tree structure with branch Chemical Engineering Module and sub-branch Microfluidics. Microfluidics and MEMs are a frequent subject matter for multiphysics simulation [2].

cookbook — here are some models that are important in chemical engineering applications that are computable in MATLAB/COMSOL Multiphysics. What is lacking from this presentation style, however, are the philosophical and methodological aspects of modeling. This book is "How To," but not sufficiently "Why" and "How good?" are the models. There are two major classes of modeling activity — (1) rigorous physicochemical modeling, which takes the best understanding of physics and attempt to compute by numerical methods the exact value up to the limits of finite precision representation of numbers; (2) approximate modeling, which intends to approximate the true, rigorous dynamics with simpler relationships in order to estimate sizes of effects and features of the outcome, rather than exact, detailed accuracy. In this book, no attempt is made to systematically treat how to propose the equations and boundary conditions of modeling

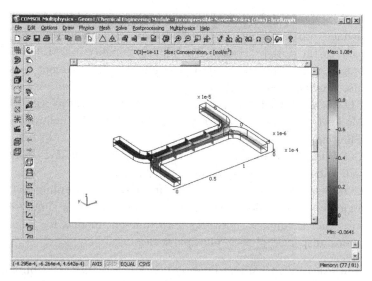

Figure 3. COMSOL Multiphysics' postprocessing screen. Here the solution for the last executed run of the microfluidic H-cell model is shown. COMSOL Multiphysics' GUI provides pull down menus and toolbars to initiate all building blocks of model construction — specifying analyzed geometries, meshing, specifying PDE equations and boundary conditions, analyzing and post processing the solutions found. Note that the status bar at the bottom shows the position of the cursor on the visualization window. The information window just above it echoes messages to the screen from the COMSOL/MATLAB commands executed in COMSOL Multiphysics' MATLAB workspace. The "Loading data from static_mixer.mat" message was the response to our request to load the model library entry for the turbulent static mixer.

— decisions about modeling objectives and acceptable approximations are presumed to have already been taken rationally. Yet, in most modeling conundrums and trouble shooting, whether or not the model itself is sensible is a key question, and what level of approximation and inaccuracy are acceptable, are part and parcel of the modeling activity. Numerics and scientific/engineering judgement about what should be modelled and how should not be separated.

1.2. Why should I use COMSOL Multiphysics for modelling?

(1) COMSOL Multiphysics has an integrated modelling environment.
(2) COMSOL Multiphysics takes a semi-analytic approach: You specify equations, COMSOL symbolically assembles FEM matrices and organizes the bookkeeping.

(3) COMSOL Multiphysics is fully compatible with MATLAB, so user defined programming for the modelling, organizing the computation, or the post-processing has full functionality. COMSOL Script is a MATLAB-like integrated programming environment that can also provide these facilities.

(4) COMSOL Multiphysics provides pre-built templates as Application Modes (see Figure 1) and in the Model Library for common modelling applications.

(5) COMSOL Multiphysics provides multiphysics modelling — linking well known "application modes" transparently.

(6) COMSOL innovated extended multiphysics — coupling between logically distinct domains and models that permits simultaneous solution. Examples: networks with different models for links and nodes, dispersed phases, multiple scales.

As we will learn in Chapters Four and Seven, extended multiphysics is very similar to the linkages provided by process simulation tools common for integrated flowsheets of process plant such as HYSYS and Aspen, or which can be developed in MATLAB's Simulink environment. COMSOL Multiphysics fully couples this functionality to a PDE engine that rivals CFD packages such as FLUENT and CFX or other commercial PDE engines such as ANSYS, but with competitive advantages listed above.

1.3. *Modelling strategies with COMSOL Multiphysics*

This book is about how I think about modelling and simulation. Perhaps my thoughts will serve as a guide to help you with the modelling problem that drew you to COMSOL Multiphysics. After posing myself the modelling problems in this book, I came up with a short list of guidelines for how to approach modelling with COMSOL Multiphysics:

(1) Don't re-invent the wheel. Read the Model Library and User's Guide/Web pages.

(2) Formulate a mathematical model. Compare with pre-built application modes.

(3) Can it all be done in the COMSOL Multiphysics GUI, or is the PDE engine only a subroutine?

COMSOL Multiphysics as an integrated modelling environment

COMSOL Multiphysics can be viewed two ways:

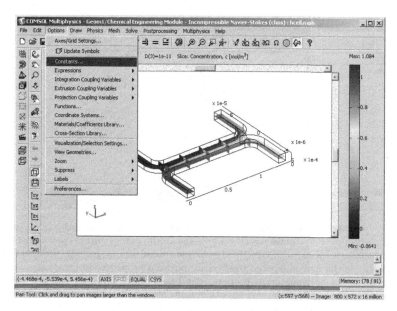

Figure 4. The Options Menu permits definition of many useful feature: constants, grids for drawing and visualization, and expressions used in entering the model equations are its most common uses.

(1) As an interactive, integrated GUI for setting up, solving, and post-processing a mathematical model — a PACKAGE.
(2) As a set of MATLAB subroutines for setting up, solving, and post-processing a mathematical model — a PROGRAMMING LANGUAGE.

This book intends to show how to implement models built both ways in an efficient way. The COMSOL Multiphysics GUI is so straightforward in setting up problems and trying "what if" scenarios that it must be the first port of call in "having a go." The great utility of a PACKAGE is that the barriers to entry are small, so the pay off is worth the investment of learning all the features of the tool.

1.4. *COMSOL Multiphysics' GUI allows sophisticated modularity in modelling*

Pre-built application modes provide templates for common calculations. The Model Library provides Case Studies. In setting up a model in COMSOL Multiphysics, there are conceptually five stages:

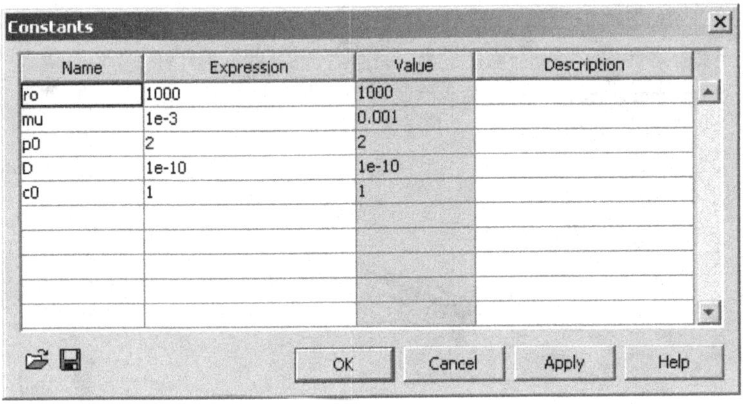

Name	Expression	Value	Description
ro	1000	1000	
mu	1e-3	0.001	
p0	2	2	
D	1e-10	1e-10	
c0	1	1	

Figure 5. COMSOL Multiphysics constants (density $ro = 100$, viscosity $mu = 10^{-3}$, pressure datum $p_0 = 2$, diffusivity $D = 1^{-10}$, and initial concentration $c_0 = 1$) defined for the microfludic H-cell. COMSOL Multiphysics does not, in general, presume any unit system, so the user is advised to enter all physical properties as constants. Some application modes have materials property databases, which have pre-assigned units. The onus is on the user to make sure that their units are consistent in mixing pre-defined application modes with user-defined equation application modes.

(1) Drawing. Specifying the domain(s) and constructing the geometries.

(2) Mathematical equations. Specifying the equations satisfied internally within the geometries (subdomain physics) and those on the boundaries or vertices (boundary and point physics).

(3) Meshing. COMSOL Multiphysics has powerful default mesh generator algorithms built-in, but also permits substantial user control over customization. One historic distinguishing feature of the finite element method is the permission of irregular and arbitrary meshes.

(4) Solving. The solver has robust default selections, but much of the black art of solving highly nonlinear systems comes from the ability to tailor the solving methodology, e.g. convergence criteria, active equations, initial guess, scaling of error estimates.

(5) Post-processing. What is the point of computing a solution if you cannot access the results in meaningful ways?

One of the long-standing criticisms of COMSOL Multiphysics is that it is not clear how to completely specify a model. However, practically a model can be set up by systematically traducing the Menu bar from left to right:

- The Model Navigator (Figure 1) accesses previously built application modes or existing models are loaded from the File Menu.
- The Options menu (see Figure 4) provides definition space for constants (see Figure 5), variables, and expressions used in either setup, solution, or post-processing phases.
- The Draw Menu (Figure 6) allows domain specifications in Draw Mode (Figure 7).
- The Physics Menu specifies the model equations. *The Point Mode* (Figure 8) provides entry for point constraints under Point Settings (Figure 9) dialogue box.
- The Boundary Mode in the Physics Menu (Figure 10) provides entry for boundary constraints through the Boundary Settings (Figure 11) dialogue box.
- The Subdomain Mode in the Physics Menu (Figure 12) permits PDE specifications (Figure 13) in the Subdomain Settings dialogue box.
- The Mesh Mode (Figure 14) shows the mesh and specifies it, which is generated by an elliptic mesh generator subject to constraints specified in the Mesh Parameters dialogue box. The Initialize Mesh menu entry generates the mesh, or the triangle button on the Toolbar.
- The Solve Menu specifies the type and parameters to be used in the solution scheme. The solution procedure is initiated by the Solve Button on the Solver Parameters dialogue box or the = button on the Toolbar. The solution is shown on the GUI main window with parameters defined in Post Mode. See Figure 3 again.
- The Postprocessing Menu provides various graphical and computational processing.
- The Multiphysics Menu allows switching between "active" modes for the specifications menus and permits additions and deletions of "application modes."
- The toolbar below the Menus provides helpful short-cuts. For instance, in 3-D, the draw mode, point mode, edge mode, boundary mode, and subdomain mode and mesh mode can be entered by a click.

The GUI makes the stages of computational modelling accessible in a much shorter time than traditional methods. Furthermore, the level of complexity in modelling is greater than any other PACKAGE. This has its advantages, as well as its own drawbacks.

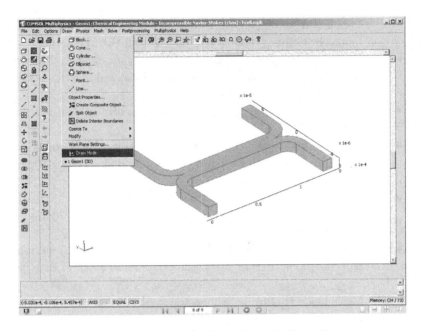

Figure 6. Draw Mode is selected from the Draw Menu.

1.5. *COMSOL (Script) as a programming language*

I have learned a seemingly ceaseless stream of programming languages — BASIC, Assembler (asm for 8088 & Cray Assembly Language), FORTRAN, PASCAL, LISP, APL, C, Mathematica, C++, MATLAB.

Programming is hard. The languages are full of commands and syntax with intricate details that need to be mastered before complicated problems can be tackled.

Engineers usually put problem solving first, and skills and techniques are acquired as necessary to solve problems. Programming strategies should reflect this. My FORTRAN programming strategy is simple — I find the "off-the-shelf" subprograms that do the integral steps of what I want to achieve, and then build a program "shell" around it to read in parameters, set up storage, call the essential subprograms, and then "post process" (also often with canned routines) and then write out the data to files. I treat COMSOL/MATLAB programming the same.

The key is to get the COMSOL Multiphysics GUI to do the work for you.

- The File Menu has the "Save Model m-file" and "Reset Mode m-file" options for you.

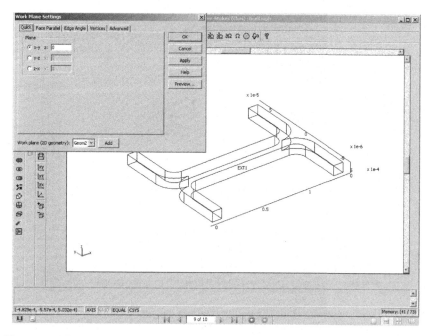

Figure 7. The single composite analyzed geometry (EXT1) of microfluidic H-cell. This geometry was drawn by geometry primitive commands (rectangles and arcs) and then merged together to form one contiguous domain in a 2-D work plane. Then it was extruded in the depth dimension to form a 3-D geometry.

- Set up the "workhorse" of your model in the GUI, and then export the model m-file, which provides most of the "program body" needed to use COMSOL Multiphysics as a programming language in MATLAB, thus providing all the subroutines and command syntax and logical structure, without the User needing to know the details.

- MATLAB m-files/m-file functions can then be set up to provide data entry, storage set up, post-processing, and output. Complicated programmes can be built up modularly without the user specifying, or even knowing all the details. MATLAB programming expertise is needed, but crucially NOT COMSOL Multiphysics programming expertise.

The book provides a wealth of examples of "user defined programming" with MATLAB m-file scripts and m-file functions calling COMSOL Multiphysics subprograms. In every case, however, I adapted models developed in the COMSOL Multiphysics GUI and read out as model m-files.

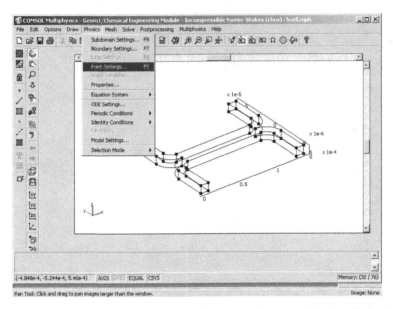

Figure 8.   Point mode shows all the points (vertices and specifically identified points) distinguished in the geometry model by black cubes.

I have yet to write a MATLAB program around COMSOL Multiphysics commands/functions "from scratch." The COMSOL Multiphysics Reference Manual provides a complete description of all the commands, so I have tried to get functionality out of the MATLAB programming that is not achievable through the GUI alone. Perhaps this is a good juncture to point out that each COMSOL Multiphysics GUI session has its own MATLAB workspace, separate from the one that launched it. So it is perfectly legitimate to write your own MATLAB m-file script and read it into the COMSOL Multiphysics GUI. The MATLAB workspace will not, however, execute all MATLAB commands, only those that are possible to do through the GUI alone, and the GUI responds by showing the intermediate steps — drawing the geometry, meshing, solving, and postprocessing. If you are writing your own user-defined m-file script for COMSOL Multiphysics/MATLAB, playing it back in the GUI shows you how far it gets before the program bugs (well, maybe you don't put them in your codes, but mine are usually infested to start) crash it. In this respect, MATLAB is a "macro" language for COMSOL Multiphysics. COMSOL Script provides a similar facility, now available since COMSOL Multiphysics 3.2. In this book, I will presume that the reader has COMSOL Multiphysics with

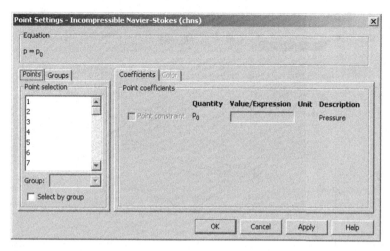

Figure 9. Point settings dialogue box. In the Incompressible Navier-Stokes Mode, only the pressure datum can be directly set in point settings.

MATLAB and the Chemical Engineering Module available. I will, however, try very hard to limit the examples where possible to the COMSOL Multiphysics GUI.

1.6. *Summary*

COMSOL Multiphysics has a powerful GUI that provides easy entry to try out "what if" scenarios and explore modelling methods/types without the investment of "programming time."

COMSOL Multiphysics has unique modelling advantages in "multiphysics" and "extended multiphysics" which may make COMSOL the only viable modelling tool for certain applications.

COMSOL Multiphysics provides a method of automatically creating MATLAB m-file source code that reduces the programming effort for setting up more complicated models. Exporting solutions to MATLAB also makes post solution analysis more flexible.

COMSOL/MATLAB programming provides automation opportunities, including running efficiently (least memory/processor overhead) as a background job. It is now possible with COMSOL Script to use the "comsol-batch" command from a DOS/UNIX/linux terminal window, running the .m file as a batch job. Together with COMSOL Script's capability of creating user interfaces through java, you can even create your own standalone software with a GUI.

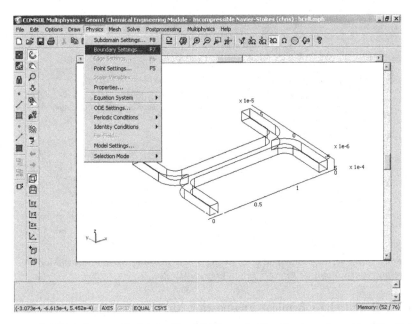

Figure 10. Boundary mode in the Physics Menu clarifies the boundary identifications and permits boundary data entry for the FEM model.

2. An a example from the Model Library

Figures 1–12 run through the major data entry points for PDE models with an example of the microfluidic H-cell from the Model Library, constructed on top of the Incompressible Navier-Stokes application mode. The figure captions tell the story, and the screen captures illustrate some of the key features of the COMSOL Multiphysics GUI. Rarely will the book show COMSOL Multiphysics GUI screen captures. From here on out, we will describe the information content for model specification in terms of the data entry needed for the dialogue boxes used in each model. This limitation to the printed word and graphical results of the models is a consequence of a desire to discuss many models, rather than to view menus and dialogue boxes with their content, limiting the number of models that can be effectively discussed.

2.1. *The microfluidic H-cell*

Figure 1 shows the Model Navigator, which permits selection among predetermined application modes, setting up user-specified "multiphysics"

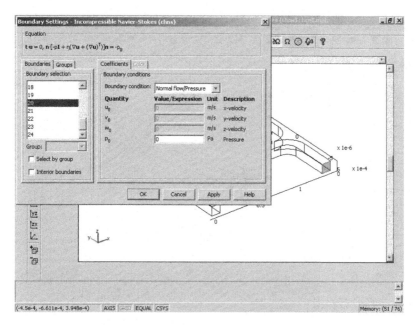

Figure 11. Boundary settings permit entry of boundary data for each boundary with a range of pre-built boundary conditions for the application mode. Here, pressure outflow conditions specified for boundary 20. The equation being solved is shown in the upper left of the boundary settings dialog box, and the boundary which is active is highlighted both in the scroll window and in the graphical display.

combinations of application modes, access to the models listed here by the user, and access to the Model Library, which houses many manyears of solved and explored models contributed by the FEMLAB user community and by the COMSOL development team. In this subsection, we will walk through the "microfluidic H-cell," which is modelled as a complicated 3-D geometry typical of microchannel operations [2]. This model can be found by selecting the Model Library tab in Figure 1, which brings up the Model Library dialogue page, whose menu tree is traversed in Figure 2. We arrive at the microfluidic H-cell model with illustration and short blurb description. Selecting the OK button brings up the COMSOL Multiphysics GUI with the geometry, model equations and boundary conditions completely specified. Figure 3 shows the postprocessing screen that the model was stored with in the file "hcell.mph." MPH files are the binary format for efficient disk storage of COMSOL Multiphysics variables, holding the complete state of the COMSOL Multiphysics environment. The postprocessing screen shows a color density plot of concentration of the last solution

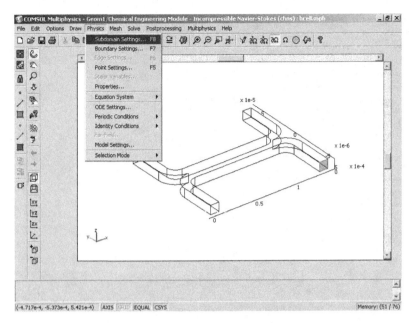

Figure 12.   In the microfluidic H-cell model, there is only one subdomain, exactly equivalent to the single composite geometry object specified and extruded in Draw Mode.

executed. Let's find out what situation that was, in terms of model equations and parameters.

The model is specified in a series of dialogue boxes. Traversing the pull down menus from left to right will show the pertinent specifications. Now pull down the Options menu, with the Constants choice highlighted as shown. The Options menu allows specification of a local database of constants, coupling variables, expressions, and differentiation rules, as well as specifying the display scales. Here, we only need to view the constants. Figure 5 shows the Constants dialogue box. We can see that $ro = 1$ and $mu = 1000$ are among the five constants specified. Note that these are pure numbers, i.e. no units are specified. Until recently. it was up to the user to employ a consistent set of units for his models, or to specify the model in dimensionless form with dimensionless control parameters. This is not necessarily a trivial task. COMSOL Multiphysics does make some unit systems available in the Materials/Coefficient Library and for pre-built application modes. You can specify the unit system (or none) in your model when creating it under the Physics menu, model settings. I am a creature of habit, so I make a habit of nondimensionalizing my equation systems.

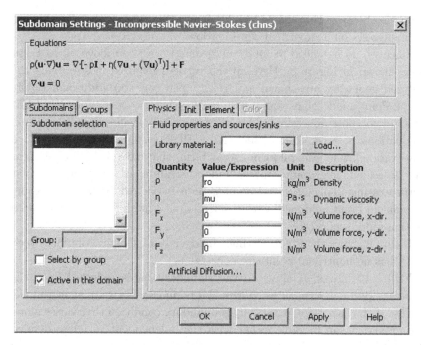

Figure 13. The subdomain settings dialogue box permits data entry for the PDE co-efficients defined in the equation line above the select tabs. Here the single subdomain 1 is selected, the same domain as shown in Figure 12. The constants defined in the (*ro* and *mu*) Constants dialogue box under the Options Menu are entered here. The init tab permits entry of initial conditions. The element tab permits customization of the element choice.

The Constants data entry permits mathematical expressions for the values in terms of other defined constants, which can ease the confusion of units with the description column.

The next pull down menu over is the Draw menu. Here we will only switch to Draw mode, which then takes over the display. Figure 7 shows the grey composite geometry EXT1 that was constructed when the model was originally created. Note that the Draw toolbar has replaced the postpro-cessing toolbar on the left. We could use these tools to enter new geometric primitives. EXT1 is a simply connected single domain, but we are not limited to either a single domain or to simply connected domains. COM-SOL Multiphysics accepts these graphic primitives, along with Boolean set theory operators (union and intersection) to construct analyzed geometries. Although the geometry specification can be done graphically in Draw mode, it can also be done through MATLAB functions, a power that is exploited in Chapter six on geometrical continuation.

Since we do not need to alter the geometry, we can move on to Point mode, shown in Figure 8. Here all the vertices required in specifying the analyzed geometry are shown as circles. You can add additional points within Point mode that you might need either for specifying the FEM model or for postprocessing. The FEM permits specification of a system of equations in weak form, which for a PDE system is equivalent to a conservation law in integral form. Weak terms that have no PDE equivalent may be added, like point sources and constraints. It may only be that postprocessing information is required at a particular point, so entering the point in Point mode will permit selection of a mesh to find the required solution more accurately.

Figure 9 shows the Point Settings dialogue box. The microfluidic H-cell model has no nontrivial pointwise constraints.

Figure 10 moves us along to the Boundary mode, selected from the Boundary pull down menu as shown. All boundary faces are shown in the display. If the user wants to specify a boundary condition that varies along a boundary, it can be done either with the independent variables defined when the model was created by the Model Navigator, typically x, y, and z for a typical 3-D geometry, or with intrinsic parameters such as the arc length s defined locally along an edge.

Figure 11 shows the Boundary Settings dialogue box. This application mode permits setting conditions on the variables among a predefined set of permitted conditions in the Incompressible Navier-Stokes mode. Again, the upper left corner shows the equation being satisfied on boundary 1.

Figure 12 shows us how to select Subdomain mode. Here there is exactly one subdomain (highlighted in the display). Subdomain mode is where the PDE system is usually specified. For simple PDEs, it is the equation(s) that is specified in subdomain mode. In pre-built application modes, however, the form of the equations is "hard-wired" in, and only the coefficients are specified in subdomain mode.

Figure 13 shows the Subdomain Settings dialogue box for domain 1. The upper left corner shows the equation(s) that are hard-wired into this application mode. The entry boxes are for the coefficients in the equations, which can be specified as constants, expressions involving other dependent or independent variables, or even MATLAB m-file functions or COMSOL Multiphysics interpolation or inline functions. The generality

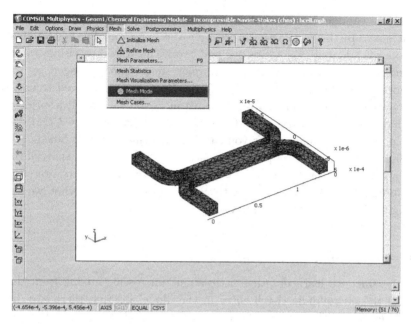

Figure 14. Mesh mode shows the existing mesh and permits specification of mesh parameters for the elliptic mesh generator routine. Mesh statistics gives some quantitative information about the features of the mesh.

of "user defined programming" for just coefficients in pre-built application modes is impressive.

Figure 14 shows the Mesh mode with the mesh set up for the saved solution here. Mesh mode, Solve mode, and Post mode are the places where the solution methodology are specified. But up to this point, we have specified a complete FEM model *analytically*. Mesh, Solve, and Post modes are about numerics, and to demonstrate these well takes a whole chapter, and is done simply in Chapter one.

Just hitting the Solve button (=) on the toolbar, however, gives us the solution with this mesh and the default numerical solution settings. Post mode (Figure 3) shows the color density plot of the surface velocity U for the conditions specified.

I doubt we are any the wiser about microfluidic mixing from this tour, but we now know the steps necessary to specify a model analytically. In subsequent chapters, these steps are referred to, and they are equivalent to specifying a PDE or FEM model completely. The microfluidic H-cell model and geometry specified here are both advanced models. Invariably, novice users wish to jump in at the deep end with the greatest model complexity all at once. In this book, we do precisely the opposite. The reductionist approach is adopted in chapters one and two with surgical precision, where we introduce the basics with even simpler steps than envisaged by the creators of COMSOL Multiphysics. Why? Because you do need to crawl before you can run, even if in other circumstances you are already a sprinter. If I have not convinced you in Chapter seven that with the COMSOL Multiphysics tools, figuratively you can fly, then this book has failed. The difficulty with complex computer packages is uncertainty on the part of the user about what the package does. So to remove the mystery, we start simple and build up capability with exact certainty about what we are asking COMSOL Multiphysics to do.

2.2. Why the tour of microfluidic H-cell model?

Clearly, since we learned rather little about microscale mixing from this tour of the microfluidic H-cell entry in the Model Library, there is a different reason for the tour itself. The rationale for showing these features of COMSOL Multiphysics is to give the non-COMSOL Multiphysics initiated reader some flavor of how the COMSOL Multiphysics GUI is laid out and how the data entry is organized. The actual intellectual content of models can be explained without the reader knowing the layout, but the reading experience would be more theoretically useful than practical. For this reason, if you are not already a COMSOL Multiphysics user, I would recommend requesting a demonstration license for both COMSOL and MATLAB. Mathworks [3] and COMSOL [4] will provide trial licenses for both products free of charge, with the software downloadable or available from CD-ROM shipped to you by request. The Users' Guide for COMSOL Multiphysics is very good, and you might want to read it after this Introduction and before Chapter One. I read all the documentation that comes with COMSOL Multiphysics cover to cover before designing and delivering my first intensive module on chemical engineering modeling with FEM-LAB [5] and highly recommend it. Nevertheless, I felt there was something missing in the COMSOL references, even though the Model Library and Chemical Engineering Module references have a wealth of fascinating case studies. I think it is the perspective of an expert user that is missing, but forgive my hubris in thinking it is my perspective!

By now, you must be thinking that this book is a thinly veiled sales pitch for COMSOL/MATLAB. I would be dishonest if I did not make my preference for modeling with COMSOL/MATLAB clear at the outset of this book. There are many packages for modeling available on the market, but COMSOL Multiphysics is the first I have seen for general purpose modeling that is equation based in generating the PDE engine. Equations are the language of mathematical modeling and mathematical physics, and COMSOL Multiphysics aims to speak the language of its target user community. So this book represents my personal odyssey in learning how to adapt COMSOL Multiphysics to modeling of chemical engineering processes, especially but not exclusively PDE based. In the next section, I give a synopsis of the themes treated in each chapter. As an experienced programmer with nearly two decades of computational modeling and FEM experience, I could not have achieved these results in the six months spent writing this book by any other package in my arsenal, nor even by adapting research codes written by myself and other expert numerical analysts with which I am proficient. This is also the last endorsement for COMSOL Multiphysics you will read in a book which only rarely makes use of other tools. Some readers might notice Mathematica, MATLAB and gnuplot graphics.

In the first version of this book, I made a list of negatives collected from FEMLAB users in general and my students in particular. Systematically, in the new version releases since 2002, the negatives on my list have disappeared. I have only five substantive complaints about COMSOL Multiphysics:

(1) The error messages are obscure.
(2) Backwards compatibility is problematic.
(3) The GUI does not give as many handles over the mesh as it used to.
(4) Some solver features in older versions have not survived.
(5) There has been an explosion of modules, model library examples, and new features which are hard for even experienced users to keep up with!

COMSOL Multiphysics users previously complained that many interesting post-processing manipulations required MATLAB programming and exporting of results to the MATLAB workspace, that the COMSOL Multiphysics graphics were poor, and the COMSOL Multiphysics error messages are obtuse and cryptic. Progress has been made on all these fronts. Ferreting out errors in syntax is more difficult than with C/FORTRAN compilers, although MATLAB m-file scripts are generally more informative when they crash about the nature of the problem than the same m-file run in the GUI. In part this comes from the ability to interrogate the variables in the MATLAB workspace much as one uses a debugger to tease out post-crash information from C. Perhaps a future advance in COMSOL will include access to the COMSOL GUI workspace. There are a lot of positives, however, in that COMSOL Multiphysics has regularly expanded its capabilities. Features that we used to have to code manually in home grown research codes are now routine, such as the moving boundary (arbitrary Langrange Eulerian transformations) mode. The multigrid solver is now a standard option, especially important for capturing complicated, multiple scale dynamics.

Modelling or conceptual errors, however, are notoriously difficult to identify. We can lay those at the user's door morally, but since COMSOL Multiphysics is not "idiot-proof," we are free to specify "badly conditioned or inconsistent" models (politely, wrong). COMSOL Multiphysics may never generate an error message at all. With just about every novice user who has sought my advice, I have shown them where they have specified an inconsistent boundary condition like $0 = 1$ in General PDE mode. Yet, in many cases, COMSOL Multiphysics generates output that is not superficially wrong, but certainly not satisfactory in the sense of modeling, conceptual, or syntactical errors. At this point, the "tough love" approach is all that can be advised — there is no substitute for experience. This book encapsulates many of my experiences. I haven't tried to sugar-coat my chapters so that all models are magically perfect. Think of this as a cookbook that shows both good recipes and bad ones, but each labeled and the latter coming with a health warning. For instance, in Chapter Seven, I tried four attempts at modeling the population balance equations before the last came good. So you will learn from my mistakes that I own up to as well as from my triumphs. For better or worse, every modeling attempt I made during the six months of writing this book has been included. I will pat myself on the back for persistence, because in the end they all worked, but at many points I had my doubts and frustrations. I am pleased not to have cherry picked the models. Of course I have not shown every single computation nor "what if" line that I pursued in each model.

3. Chapter synopsis

Chapter One treats the basics of numerical analysis with COMSOL. No doubt many of the example models are artificial in that if you were handed the modeling problems in Chapter One, COMSOL would not be the obvious choice of computational platform. The topics of root finding, numerical integration by marching, numerical integration of ordinary differential equations, and linear system analysis are universal to numerical analysis. They form the basis of my previous lecture courses in FORTRAN programming and chemical engineering problem solving with Mathematica. For pedagogical purposes, Chapter One provides a firm basis for understanding what COMSOL does. The common applications in chemical engineering that are treated as examples — flash distillation, tubular reactor design, diffusive-reactive systems, and heat conduction in solids — are understandable to the nonchemical engineer as well. Perhaps the single most important modeling feature introduced here, however, is the use of a conceptual 0-dimensional model. Consisting of a single element, the 0-D construct introduces a variable which is a scalar for which an ODE in time or an algebraic equation can be specified. This construct is important for describing equations or systems of equations that are mixed (partial) differential-algebraic, and is utilized with the extended multiphysics feature of COMSOL in the more complicated models presented later.

Chapter Two might be thought the normal point of departure for a textbook on finite element methods (FEM). In my opinion, COMSOL is not so much a tool about FEM, but a modeling tool that happens to use FEM in its automated methodology. The key actions of COMSOL that reduce the drudgery of modeling are (1) the translation of systems of equations in symbolic form to an algorithm that can be computed numerically, (2) the provision of a wide array of numerical solver, analysis, and post-processing tools at either the "touch of a button" or (3) through a powerful "scripting language" can be programmed in MATLAB as subroutines (function calls) and automated. So much of modeling of partial differential equations in the past has been devoted to the computer implementation of algorithms that the modeler did not get the chance to properly consider modeling alternatives. Who would consider a different modeling scheme if it meant spending three graduate student years building the tools before the scheme could be tested? COMSOL is a paradigm shift for modelers — it frees them to ask those "what if" questions without the price of coding a new computer program. Nonetheless, COMSOL uses FEM as the powerhouse of its PDE engine. Chapter Two gives an overview of how FEM is implemented in COMSOL. For experienced FEM users, the takeaway message is that COMSOL translates PDEs specified symbolically into the assembly of the FEM augmented stiffness matrix — the stiffness matrix, the load vector, and auxiliary equations for Lagrange multipliers representing boundary conditions and auxiliary conditions. Chapter Two illustrates these points about partial differential equations and the finite element method through treatment of canonical types of linear, second order PDEs: elliptic, parabolic, and hyperbolic and gives an overview of FEM, with particular emphasis on the treatment of boundary and auxiliary conditions by the method of Lagrange multipliers.

Chapter Three is about multiphysics modeling. What is it? How does COMSOL do it so well? There are applications: thermoconvection, non-isothermal chemical reactors, heterogeneous reaction in a porous pellet. Furthermore, the workhorse methodology for nonlinear solving, parametric continuation, is explained. I won't steal the thunder of Chapter Three here by explaining multiphysics modeling in detail. Suffice to say that multiphysics modeling means the ability to treat many PDE equations simultaneously, and the provision of pre-built PDE equations that can be mixed and matched in the specification of a model so that the symbolic translation to a FEM assembly is transparent to the user.

Chapter Four is about extended multiphysics: the central role of coupling variables and the use of Lagrange multipliers. Example applications are: a heterogeneous reactor; reactor-separator-recycle; buffer tank modelling; and an immobilized cell bioreactor model.

Chapter five starts the advanced concepts in modeling — nonlinear dynamics and simulation. Chapter six deals with geometric continuation, and Chapter seven treats integral equations and inverse problems. All three chapters are largely drawn from my own research portfolio, but there are also newly developed treatments or extended studies from previous works. Rather than systematically exploring the features of COMSOL as in Chapters 1–4, Chapters 5–7 pose the question "Can COMSOL be bent to solve the problems that interest me in stability theory (five), complex geometries and modulating domains (six) or inverse problems (seven), where I know the questions and desired forms of the answers, but can COMSOL provide the solution tools?" These chapters will have their own audiences for the direct questions they treat, but should provide many users with fertile proving grounds and a basketful of "tricks of the trade." Getting information into and out of the COMSOL GUI is one of the weaknesses of the package. Many of my tricks are how to use the MATLAB interfaces to do intricate I/O.

Chapters eight and beyond are purely about applications and case studies and are not all co-authored by me. To a large degree, Chapters 5–7 are about my applications and their generalizations, used to demonstrate

COMSOL functionality. Chapters 8 and beyond are the applications of colleagues for which we thought COMSOL and the concepts of Chapters 1–4 should be exploitable. My co-authors of these chapters have other agendas and that is evident in the narrative voice adopted in these chapters.

Chapters eight and nine are about the level set method for modeling two phase flows that are dominated by interfacial dynamics and transport. The subject matter was mastered and modelled in record time for one of my doctoral students. The simulations are a reflection of the need for researchers to be able to run numerical experiments in complex systems dynamics to augment understanding of laboratory experiments. Such *"in silico"* experiments are more flexible than laboratory experiments, provide a much greater wealth of detailed knowledge, but at the expense of modeling errors of all varieties.

Chapter ten is a purely fluid dynamics case study, with applications to microfluidic mixing. Chapters 11–14 focus on multiphysics modelling with electrokinetic, plasma, electrochemical, and MHD phenomena mixed with transport equations. A substantial fraction of COMSOL users are numbered in the microfluidics community, especially with biotech end-uses. Rather early on, we targeted COMSOL as a potentially useful modeling tool for microfluidic reactor networks for the "chemical-factory-on-a-chip" community. The extended multiphysics capabilities of COMSOL for designing such factories are an explosive growth area which should benefit the community. Microfluidics 2003 [6] was sponsored by COMSOL for just this reason.

The appendix, a MATLAB/COMSOL Multiphysics primer for vector calculus, is a compromise between the recurrent suggestion of students taking the module for more MATLAB instruction and my desire for the students to grasp vector calculus more intuitively. I am actually a late convert to MATLAB, with apologies to Cleve Moler, its creator. I was one of the graduate students gifted with the beta test edition of MATLAB 1.0 while he was developing it. At the time, computational power was expensive and there was a bias against interpreted environments for scientific computing. To programmers, the same matrix utilities were available as library subroutines, and the final product, a compiled executable, was more efficient. MATLAB has come a long way since version 1.0 beta, and the number of man years and breadth of applications in the toolboxes, as well as judicious use of compilation within the environment, simply invalidates my early prejudices. I cannot access programming libraries with anywhere near the functionality of the MATLAB toolboxes. The GUIs for the toolboxes make manmonths of programming effort evaporate at the touch of a button (OK, the click of a mouse). And if speed is still an issue, the MATLAB C compiler is available. Or just my favorite trick of running MATLAB as a background job (no GUIs to clutter the memory) is usually sufficient for big jobs. So to get the most functionality out of COMSOL Multiphysics, MAT-LAB programming ability is valuable. But anything other than a primer is outside the scope of this book. I presume a modest MATLAB familiarity of the reader which is readily achieved from over-the-counter books. So to add more MATLAB support, I decided to write a short primer about vector calculus representations and computations in MATLAB/COMSOL Multiphysics for the appendix. This project could easily get out of hand, so I apologize for abridging it for convenience. MATLAB was never intended for vector calculus directly, but vector calculus is fundamental to PDEs and therefore to COMSOL Multiphysics.

Enjoy the journey through this book. As it is an odyssey, the destination is not the focus. Certainly the reader, however, has a concrete objective in modeling for wanting to use COMSOL Multiphysics. Perhaps somewhere in this odyssey you will find tools to bring to bear on your problem and will find useful in reaching your modelling objectives.

References

[1] S. V. Patankar, *Numerical Heat Transfer and Fluid Flow* (Hemisphere Publishing Corporation, New York, 1980).

[2] W. B. Zimmerman, *Microfluidics: History, Theory and Applications*, CISM Lecture Series, Vol. 466 (Springer-Verlag-Wien, 2005).

[3] MATLAB demonstration version can be found at
http://www.mathworks.com.

[4] COMSOL Multiphysics demonstration CD-ROM can be requested at
http://www.comsol.com.

[5] W. B. J. Zimmerman, http://eyrie.shef.ac.uk/femlab.

[6] W. B. J. Zimmerman, http://eyrie.shef.ac.uk/fluidics.

Chapter One

COMSOL MULTIPHYSICS AND THE BASICS OF NUMERICAL ANALYSIS

W.B.J. ZIMMERMAN

Department of Chemical and Process Engineering,
University of Sheffield,
Mappin Street, Sheffield S1 3JD, United Kingdom
E-mail: w.zimmerman@shef.ac.uk

In this chapter, several key elements of numerical analysis are profiled in COMSOL Multiphysics with 0-D and 1-D models. These elements are root finding, numerical integration by marching, numerical integration of ordinary differential equations, and linear system analysis. These methods underly nearly all problem solving techniques by numerical analysis for chemical engineering applications. The use of these methods in COMSOL Multiphysics is illustrated with reference to some common applications in chemical engineering: flash distillation, tubular reactor design, diffusive-reactive systems, and heat conduction in solids.

1. Introduction

This chapter is rather busy, as it must accomplish several different goals. Primarily, it is intended to introduce key features of how COMSOL Multiphysics works. Secondarily, it is to illustrate how these key features can be used to analyze simple enough chemical engineering problems that 0-D and 1-D spatial or spatial-temporal systems can describe them. The chapter is also intended to whet your interest to investigate modeling and simulation with COMSOL Multiphysics by presenting at least a glimpse of the power of the COMSOL Multiphysics and MATLAB tools when applied to chemical engineering analysis.

Because COMSOL Multiphysics is not intended to be a general tool for problem solving, some of these goals are achieved in a roundabout fashion. The author has previously taught courses in chemical engineering problem solving by numerical analysis using FORTRAN, $Mathematica^{TM}$, and $MATLAB^{TM}$, and used all the examples implemented here with those tools. Furthermore, the most extensive compilations of chemical engineering problem solving by numerical analysis have been done in POLYMATH [1], which only seems to be used by the chemical engineering community through the CACHE program.

In this book's predecessor, at this point we introduced the concept of a zero-dimensional domain to solve nonlinear algebraic equations and time-dependent ordinary differential equations. Conceptually, the concept of a zero-dimensional domain is simply a single finite element. Understanding the finite element method from the perspective of what happens in a single finite element is pedagogically very useful. COMSOL Multiphysics, however, has made it much simpler to solve 0-D algebraic and time-dependent ODEs by creating a separate dialogue box to specify them. So in this chapter, we will solve some examples both ways.

2. Method 1: Root finding

Typically, courses in numerical analysis go into great detail in the description of the algorithm classes used for root finding. From experience, there are only two algorithms that are really useful — the bisection method and Newton's method. Instead of presenting all the methods, here we will consider why root finding is one of the most useful numerical analysis tools. Finding roots in linear systems is fairly easy. Nonlinear systems are the challenge, and nearly all interesting dynamics stem from nonlinear systems. The interest in root finding in nonlinear systems results from its utility in describing inverse functions. Why? Because with most nonlinear functions, the "forward direction," $y = f(u)$, is well described, but the inverse function of $u = f^{-1}(y)$ may be analytically indescribable, multi-valued (nonunique), or even nonexistent. But if it exists, then the numerical description of an inverse function is identical to a root finding problem — find u such that $F(u) = 0$ is equivalent to $F(u) = f(u) - y = 0$. Since the goal of most analysis is to find a solution of a set of constraints on a system, this is equivalent to inverting the set of constraints. COMSOL Multiphysics has a core function for solving nonlinear systems, femnlin, and in this section its use to solve 0-D root finding problems will be illustrated.

femnlin uses Newton's method which with only one variable u uses the first derivative $F'(u)$ which is used iteratively to drive toward the root. The method takes a local estimate of the slope of the function and projects to the root. The slope can be computed either analytically (Newton-Raphson Method) or numerically (the secant method). If the slope can be computed either way, you can use Taylor's theorem to project to the root. The basic idea is to use a Taylor expansion about the current guess u_0:

$$f(u) = f(u_0) + (u - u_0)f'(u_0) + \cdots, \qquad (1)$$

which can be re-arranged, ignoring higher order terms in $(u - u_0)$ to estimate the root as

$$u = u_0 - \frac{f(u_0)}{f'(u_0)}. \qquad (2)$$

This methodology is readily extendable to a multiple dimension solution space, i.e. u is a vector of unknowns, and division by $f'(u_0)$ represents multiplication by the inverse of the Jacobian of f. The next subsection illustrates root finding in COMSOL Multiphysics.

2.1. *Root finding: A simple application of the COMSOL Multiphysics nonlinear solver*

As implied in the previous section, root finding is a "0-D" activity, at least in terms of the spatial-temporal dependence of the solution vector of unknowns, u, which can be a multi-dimensional vector. COMSOL Multiphysics does not have a "0-D" application mode, so we improvise in 1-D. This has the undesirable feature that we will unnecessarily solve the problem redundantly at several points in space. Given the small size of the problem, the efficiency of COMSOL Multiphysics coding, and the speed of modern microprocessors, this causes no guilt whatsoever!

Start up MATLAB and type COMSOL Multiphysics in the command window. After several splash screens, you should be facing the Model Navigator window. Follow the steps in Table 1 to set up a 0-D application mode to solve the nonlinear polynomial equation:

$$u^3 + u^2 - 4u + 2 = 0. \qquad (3)$$

Physics: Subdomain settings specifies the equation to be satisfied in each subdomain in Table 1. Notice the equation in the upper left given in vector notation. In 1-D, this equation can be simplified to

$$d_a \frac{\partial u}{\partial t} - \frac{\partial}{\partial x}\left(c\frac{\partial u}{\partial x} + \alpha u\gamma\right) + au + \beta\frac{\partial u}{\partial x} = f. \qquad (4)$$

Clearly, $\alpha\gamma$ and β are redundant with the simplification to 1-D. Since we want to find roots in 0-D, however, all the coefficients on the LHS of (4) can be set to zero. By rearranging the polynomial, we can readily see that $a = 4$ and $f = u^3 + u^2 + 2$. Note that we discretize the domain with a single element by specifying the maximum element size to be one, giving us 0-D!

Table 1.   Root finding example in coefficient mode. File name: rootfinder.mph.

Model Navigator	Select 1-D. COMSOL Multiphysics:PDE modes:PDE, coefficient form
Draw Menu	Specify objects: Line. Coordinates pop-up menu. $x : 0\ 1$ name: interval OK
Physics Menu: Boundary settings	Select domains: 1 and 2 (hold down Ctrl key) Select Neumann boundary conditions Leave defaults $q = 0\ g = 0$ OK
Physics Menu: Sub-domain settings	Select domain: 1 Set $c = 0$; $a = 4$; $f = u\char`^3 + u\char`^2 + 2$; $d_a = 0$ Apply. Select init tab: set $u(t_0) = -2$ OK
Mesh menu: mesh parameters	Set maximum element size 1 Hit remesh. OK
Solve menu: solver parameters	Stationary nonlinear. Solve. OK
Post-processing: Point Evaluation	Boundary selection: 1. Expression: u. OK

By specifying the initial guess of as $u(t_0) = -2$, we find the root nearest to this value. If you are wondering why $a = 4$ was set, rather than all of the dependence put into f, it is so that the finite element discretization of the RHS of (4) does not result in a singular stiffness matrix.

The post-processing stage shows the result in the output window:

$$\text{Value} : -2.732051, \text{Expression} : u, \text{Boundary} : 1.$$

The analytically determined root nearest to this is $-1 - \sqrt{3}$, showing the numerical solution in good agreement. According to the structure of the quadratic formula of algebra, clearly another root is $-1 + \sqrt{3}$, and by inspection, the third root is 1. Returning to the subdomain settings, set the initial guess to $u(t_0) = -0.5$ and COMSOL Multiphysics converges to $u = 0.732051$, again a good approximation. $u(t_0) = 1.2$ as an initial guess converges to $u = 1$.

This exercise clarifies two features of nonlinear solvers and problems — (i) nonlinear problems can have multiple solutions; (ii) the initial guess is key to convergence to a particular solution. With Newton's method, it is usually the case that convergence is to the nearest solution, but overshoots in highly nonlinear problems may override this stereotype. These features persist in higher dimensional solution spaces and with spatial-temporal dependence.

The COMSOL Multiphysics model mph-file rootfinder.mph contains all the MATLAB source code with FEMLAB extensions to reproduce the current state of the FEMLAB GUI. This file is available from the website

Table 2. Root finding example in general mode. File name: rtfindgen.mph.

Model Navigator	1-D, COMSOL Multiphysics:PDE modes, general form
Options	Set Axes/Grid to [0,1]
Draw	Name: Interval; Start = 0; Stop = 1
Physics Menu/ Boundary Settings	Set both endpoints (domains) to Neumann BCs
Physics Menu/Subdomain Settings	set $\Gamma = 0$; $d_a = 0$; $F = u^3 + u^2 - 4*u + 2$
Mesh mode	Set Max element size, general = 1; Remesh
Solve	Use default settings (nonlinear solver, **exact** Jacobian)
Post-process	After five iterations, the solution is found. Click on the graph to read out $u = 0.732051$. Play with the initial conditions to find the other two roots

http://eyrie.shef.ac.uk/femlab. Just pull down the file menu, select Open model m-file, and use the Open file dialog window to locate it. You can rapidly place your nonlinear function in the Subdomain settings, specify an initial guess, and use the stationary nonlinear solver to converge to a solution. But what if your function does not have a linear component to put on the LHS of (4)? For instance, $\tanh(u) - u^2 + 5 = 0$ results in a singular stiffness matrix when FEMLAB assembles the LHS of (4). The suggestion is to set the coefficient of the second derivative of u, $c = 1$ in the Subdomain settings. Coupled with the Neumann boundary conditions, this artificial diffusion cannot change the fact that the solution must be constant over the single element, yet it prevents the stiffness matrix from becoming singular.

Root finding in General Mode

The difficulty with a singular stiffness matrix assembly for $\tanh(u) - u^2 + 5 = 0$ can be averted by using General Mode, which solves

$$e_a \frac{\partial^2 u}{\partial t^2} + d_a \frac{\partial u}{\partial t} + \frac{\partial \Gamma}{\partial x} = F, \tag{5}$$

Table 3. Root finding in ODE settings.

Physics Menu: ODE settings	Name: v. Equation: $\tanh(v) - v\hat{\ }2 + 5$ OK
Solve menu: solver parameters	Stationary nonlinear. Solve. OK
Post-processing	Point evaluation. Boundary 2. Expression: v
Report window	Value: -2.008819, Expression: v, Boundary: 2

where $\Gamma(u, ux)$ is in principle the same functionality as the coefficient form (4), but is treated differently by the Solver routines. In Coefficient Mode, the coefficients are treated as independent of u unless the numerical Jacobian is used, which brings out some of the nonlinear dependency — iteration does the rest. The exact Jacobian in General Mode differentiates both Γ and F with respect to u symbolically in assembling the stiffness matrix. Typically, General Mode requires fewer iterations for convergence than Coefficient Mode with the numerical Jacobian. The use of the exact Jacobian below does not require any special treatment to avoid a singular stiffness matrix in the treatment of the linear terms as the coefficient mode did. In general, General Mode is more robust at solving nonlinear problems than Coefficient Mode. It is my opinion that Coefficient Mode is a "legacy" feature of COMSOL Multiphysics — the PDE Toolbox of MATLAB, in many ways a precursor to FEMLAB and COMSOL Multiphysics, uses coefficient representations extensively. Further, the coefficient formulation with numerical Jacobian is a long standing FEM methodology, so for benchmarking against other codes, it is a useful formulation.

Table 2 holds the recipe for General Mode — a minor modification of what we just did.

Although setting up this template (rtfindgen.mph) for root finding of simple functions of one variable was rather involved, and in fact MATLAB has a simpler procedure for root finding using the built-in function fzero and inline declarations of functions, the COMSOL Multiphysics GUI now provides the utility to solve algebraic constraints as auxiliary equations in auxiliary variables, termed state variables. As an adjunct to our general mode root finder model, follow the steps in Table 3 to solve a nonlinear equation for a state variable v.

The next subsection applies our newly constructed nonlinear root finding scheme to a common chemical engineering application, flash distillation, which clarifies a few more features of the COMSOL Multiphysics GUI.

2.2. *Root finding: Application to flash distillation*

Chemical thermodynamics harbors many common applications of root finding, since the constraints of chemical equilibrium and mass conservation are frequently sufficient, along with constitutive models like equations of state, to provide the same number of constraints as unknowns in the problem. In this subsection, we will take flash distillation as an example of simple root

Table 4. Initial composition to the flash unit and partition coefficients K at equilibrium.

Component	X_I	K_i at 65°C and 3.4 bar
Ethane	0.0079	16.2
Propane	0.1281	5.2
i-Butane	0.0849	2.6
n-Butane	0.2690	1.98
i-Pentane	0.0589	0.91
n-Pentane	0.1361	0.72
Hexane	0.3151	0.28

finding for one degree of freedom of the system, which is conveniently taken as the phase fraction ϕ.

A liquid hydrocarbon mixture undergoes a flash to 3.4 bar and 65°C. The composition of the liquid feed stream and the "K" value of each component for the flash condition are given in the table. We want to determine composition of the vapor and liquid product streams in a flash distillation process and the fraction of feed leaving the flash as liquid. Table 4 gives the initial composition of the batch.

A material balance for component i gives the relation

$$X_i = (1 - \phi)y_i + \phi x_i \,, \tag{6}$$

where X_i is the mole fraction in the feed (liquid), x_i is the mole fraction in the liquid product stream, y_i is the mole fraction in the vapour product, and ϕ is the ratio of liquid product to feed molar flow rate. The definition of the equilibrium coefficient is $K_i = y_i/x_i$. Using this to eliminate x_i from the balance relation results in a single equation between y_i and X_i:

$$y_i = \frac{X_i}{1 - \phi(1 - \frac{1}{K_i})} \,. \tag{7}$$

Since the y_i must sum to 1, we have a nonlinear equation for ϕ:

$$f(\phi) = 1 - \sum_{i=1}^{n} \frac{X_i}{1 - \phi(1 - \frac{1}{K_i})} = 0 \,, \tag{8}$$

where n is the number of components. This function $f(\phi)$ can be solved for the root(s) ϕ, which allows back-substitution to find all the mole fractions in the product stream. The Newton-Raphson method requires the derivative $f'(\phi_k)$ at the current estimate to determine the improved estimate, and COMSOL Multiphysics will compute this analytically as an option. It is

Table 5.   Flash distillation example.

Model Navigator	Select 1-D. COMSOL Multiphysics:PDE modes:PDE, general form
Draw Menu	Specify objects: Line. Coordinates pop-up menu. x: 0 1 name: interval OK
Options: Constants	Enter the data from Table 4 X_1, 0.0079, etc.
Options: Scalar Expressions	Define expression for the terms in the RHS of (8) $t_1 - X_1/(1 - u*(1 - 1/K_1))$ $t_2 - X_2/(1 - u*(1 - 1/K_2))$ etc.
Physics Menu: Boundary settings	Select domains: 1 and 2 (hold down Ctrl key) Select Neumann boundary conditions Leave defaults $q = 0$ $g = 0$ OK
Physics Menu: Sub-domain settings	Select domain: 1 Set $F = 1 + t_1 + t_2 + \cdots + t_7$; $d_a = 0$ Select Init tab; set $u(t_0) = 0.5$
Mesh menu: mesh parameters	Set maximum element size 1 Hit remesh. OK
Solve menu: solver parameters	Stationary nonlinear. Solve. OK
Post-processing: Point Evaluation	Boundary selection: 1. Expression: u. OK Report window: Value : 0.458509, Expression: u

fairly straightforward to arrive at the Newton-Raphson iterate as

$$f'(\phi) = \sum_{i=1}^{n} \frac{X_i(1 - \frac{1}{K_i})}{[1 - \phi(1 - \frac{1}{K_i})]^2} \cdot \qquad (9)$$

Now onto the COMSOL Multiphysics solution for root finding. As an exercise, we will set up the solution using the general PDE mode. We could just load rootfinder.mph or rtfindgen.mph and customize it, but of course becoming familiar with COMSOL Multiphysics' features is an important goal.

Start up COMSOL Multiphysics and await the Model Navigator window. If you already have a COMSOL Multiphysics session started, save your workspace as a model MPH-file or the commands as a model m-file, and the pull down the file menu and select New. Follow the steps as arranged in Table 5 to set up the flash distillation example. Note that we have two additional stages in this example — Options: Constants and Options: Expressions. Constants can be defined and then used wherever a pure number might be legally used in a COMSOL Multiphysics data entry field. Expressions are similar to constants in that they can also be used wherever a COMSOL Multiphysics data entry field permits, but have the additional feature that they depend on the dependent variable(s). They are also available for post-processing. Try putting t_1 and t_2 into the post-processing data display as expressions:

$$\text{Value} : -0.013865, \text{Expression} : t_1, \text{Boundary} : 1,$$
$$\text{Value} : -0.203441, \text{Expression} : t_2, \text{Boundary} : 1.$$

Another useful set of information comes from the solver log. Pull down the solver menu to the bottom and select View Log. The Solver Log dialogue window pops up. This shows when the solver ran, what solver command was executed (here femnlin) and how circuitous the path to the solution from the initial condition was. Here it took three iterations to achieve absolute error of 10^{-9}. This information comes in particularly handy if your solution does not converge or is slowly convergent.

Exercises.

1.1. Find the roots of the equation $f(u) = u^3 - 3u^2 + \frac{5}{2}u - \frac{1}{2} = 0$. As this function is a cubic polynomial, there is an analytic solution in the irrational numbers, $u = 1$, $u = 1 - \frac{1}{\sqrt{2}}$, $u = 1 + \frac{1}{\sqrt{2}}$.

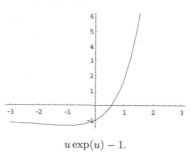

$u \exp(u) - 1$.

1.2. Find the root of the equation $f(u) = ue^u - 1 = 0$. This function is transcendental, which means that it has no analytic solution in the rational numbers. If you use Coefficient Mode, put $c = 1$ to aid convergence.

3. Method 2: Numerical integration by marching

Numerical integration is the mainstay of numerical analysis. The first duty of scientific computing before there were digital computers were to fill the handbooks with tables of special functions, nearly all of which were solutions to special classes of ordinary differential equations. And the computational methodology? One-dimensional numerical integration.

There are two classes of 1-D integration: initial value problems (IVP) and boundary value problems (BVP). The latter will be considered in the next section. The easiest to integrate are IVPs, as if all the initial conditions are all specified at a point, it is straightforward to step along by small increments according to the local first derivative. Clearly, if the ODE is first order, i.e.

$$\frac{dy}{dt} = f(t), \qquad y|_{t+\Delta t} = y|_t + \Delta t f(t). \tag{10}$$

The second statement in (10) is true exactly in the limit of $\Delta t \to 0$. It is termed the Euler method and is the most straight-forward way of integrating a first order ODE. In one dimension, you simply step forward according to the local value of the derivative of f at the point (x_n, y_n), where n refers to the n-th discretization step of the interval upon which you are integrating. Thus,

$$y_{n+1} = y_n + h f(x_n, y_n),$$
$$x_{n+1} = x_n + h. \tag{11}$$

This assumes that the derivative does not change over the step of size h, which is only actually true for a linear function. For any function with curvature, this is a lousy assumption. Consider, for instance, how far wrong we go with a large step size in Figure 1. So clearly, one important point in improving on Euler's Method is to be able to use big steps, since it requires small steps for good accuracy. Euler's method is called "first order" accurate, as the error only decreases as the first power of h.

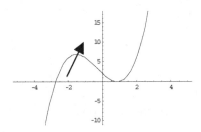

Figure 1.   Euler step in numerical integration.

Runge-Kutta methods

So if we want to use big step sizes, we need a "higher order method," one that reduces the error faster as step size decreases. A k-th order method has error which diminishes as h^k. Given that it is curvature that we know we are neglecting, we can estimate the curvature of the graph $y(x)$ by evaluating the slope $f(x)$ at several intermediate points between x_n and x_{n+1}. Second order accuracy is obtained by using the initial derivative to estimate a point halfway across the interval, then using the midpoint derivative across the full width of the interval.

$$k_1 = hf(x_n, y_n),$$

$$k_2 = hf\left(x_n + \frac{1}{2}h, y_n + \frac{1}{2}k_1\right), \tag{12}$$

$$y_{n+1} = y_n + k_2 + O(h^3).$$

The upshot is that by making two function evaluations, we have saved a whole order in accuracy. So, for instance, with a first order method, N calculations gives us an error $O(1/N)$, but for a second order method, $2N$ calculations gives us error $O(1/4N^2)$. It would take N^2 calculations to do so well with a first order method.

Higher order Runge-Kutta methods

Can we do better? Clearly, we can use a three midpoint method to achieve third order accuracy, a four midpoint method for fourth order accuracy, etc. When should we stop? Well, there is more programming work for higher order methods, so our time is a consideration. But intrinsically, functions may not be very smooth in their k-th derivative that we are estimating. It is possible that in increasing the "*accuracy* of the approximation," the round-off error of higher derivative terms so estimated becomes appreciable. If that is the case, with each successive step, the error may grow rapidly. This implies that higher order methods are less *stable* than lower order methods.

The common choice for integrating ODEs is to use a fourth order Runge-Kutta method. This is fairly compact to programme, gives good accuracy, and typically has good stability character.

Other methods

There are two other famous problems in numerical integration that need particular programming attention:

Numerical Instability. Suppose your integration diverges to be very far from known test-cases, even with a high order accuracy method. Then it is likely that your method is numerically unstable. You can cut down your step size and eventually achieve numerical stability. However, this means a longer calculation. If you are computing a great many such integrations and the slowness really bothers you, try a semi-implicit method like predictor-corrector schemes.

Stiff Systems. Stiff systems usually have two widely disparate length or time scales on which physical mechanisms occur. Stiff systems may have "numerical instability" of the explosive sort mentioned above, or they may have nonphysical oscillations. Try the book of Gear [2] for a recipe to treat stiff systems.

3.1. *Numerical integration: A simple example*

Higher order ODEs are treatable by marching methods by reduction of order. Suppose you have an ODE:

$$\frac{d^2y}{dx^2} + q(x)\frac{dy}{dx} = r(x). \tag{13}$$

Unless $q(x)$ and $r(x)$ are constants, then you are out of luck with most textbook analytic methods for finding a solution. There are special cases of $q(x)$ and $r(x)$ that lead to analytic solutions, but these days you are better off computing the numerical solution in nearly all cases anyway. Why? Because you need to plot the graph of the solution $y(x)$ to make sense of it, so you will need to harness some computing horse power for the graphics. How? First let's reduce the order of the second order system above to two first order systems:

$$\frac{dy}{dx} = z(x),$$

$$\frac{dz}{dx} = r(x) - q(x)z(x).$$

Each of these ODEs can be numerically integrated by time marching methods as in (11) or (12), *simultaneously*. A simple example is

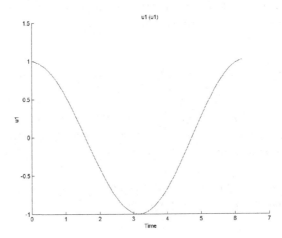

Figure 2. $u_1(t)$ over one period.

$$\frac{d^2 u}{dt^2} + u = 0\,. \tag{14}$$

Reduction of order yields two first order ODEs:

$$\frac{du_1}{dt} = -u_2\,,$$
$$\frac{du_2}{dt} = u_1\,. \tag{15}$$

Taking the initial condition to be $u_1 = 1$ and $u_2 = 0$, we can now set up a 0-D spatial system to integrate this coupled set of ODEs.

Start up COMSOL Multiphysics and await the Model Navigator window. If you already have a COMSOL Multiphysics session started, save your workspace as a model MPH-file or the commands as a model m-file, and then pull down the File menu and select New. Follow the set up information in Table 6.

This application mode gives us two dependent variables u_1 and u_2 and one space coordinate x. Notice the equation in the upper left in Subdomain Settings is given in vector notation. Because we have a vectorial set of variables, all the data entry tabs are for vectorial (F) or matrix (d_a) input.

linspace(0,2*pi,50) is the MATLAB command to create a vector of length 50 which uniformly goes from 0 to 2π Data display gives $u_1(t = 2\pi) = 1.004414$. Given that the analytic solution is $u_1(t = 2\pi) = 1$, this is

Table 6. Time integration by marching of a simple example.

Model Navigator	Select 1-D. COMSOL Multiphysics:PDE modes:PDE, general form Insert dependant variables: $u_1 u_2$ Select Element: Lagrange-Linear
Draw Menu	Specify objects: Line. Coordinates pop-up menu. x: 0 1 name: interval OK
Physics Menu: Boundary settings	Select domains: 1 and 2 (hold down Ctrl key) Select Neumann boundary conditions Leave defaults $q = 0$ $g = 0$ OK
Physics Menu: Subdomain settings	Select domain: 1 Set $F_1 = -u_2$, $F_2 = u_1$; Select Init tab; set $u_1(t_w) = 1$
Mesh menu: mesh parameters	Set maximum element size 1 Hit remesh. OK
Solve menu: solver parameters	Time-dependent solve. Enter on General tab Times: linspace (0, 2*pi, 50) Solve. OK
Post-processing: Cross-section plot parameters	Point tab. Accept the default of u_1 General tab. Accept the default of all times OK

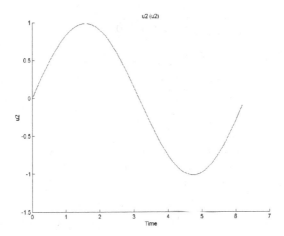

Figure 3. $u_2(t)$ over one period.

rather inaccurate (0.4%). Previously, FEMLAB permitted the user to select the time-integration scheme among several built-in solvers for MATLAB and FEMLAB. COMSOL Multiphysics does not permit this flexibility, but rather uses internal algorithm selection. It does, however, permit the user to adjust the local error tolerances (relative and absolute) on the General Tab of the Solver Parameters dialogue box. I changed the relative tolerance to 0.001 and absolute to 0.0001 to produce a somewhat better endpoint calculation of $u_1(t = 2\pi) = 0.998027$. Note that cumulative global error is of the order of the accuracy of the method 0.001.

These two figures (2 and 3) clarify that FEMLAB can reproduce the numerical integration of the cosine and sine functions with high fidelity if given a small enough time step. Although we think of sine and cosine as "analytic functions," when tabulated this way, it is clear that the distinction between analytic functions and those that require numerical integration is specious — they are no more analytic than Bessel functions, elliptic functions, etc.

Exercise.

1.3. Try integrating equations (15) from the same initial conditions using the ODE Settings on the Physics Menu and naming v_1 and v_2 as state variables. The only particular difference for time dependent ODEs from algebraic equations is that you must use the notation for dv_1/dt, v_1t, etc.

4. Numerical integration: Tubular reactor design

In this section, a coupled set of first order nonlinear ODEs are solved simultaneously for the design of a tubular reactor undergoing a homogeneous chemical reaction. Typically, the key element in the design of a tubular reactor is the estimate of the length of the reactor.

A tubular reactor is used to dehydrate gaseous ethyl alcohol at 2 bar and 150°C. The formula for this chemical reaction is

$$C_2H_5OH \rightarrow C_2H_4 + H_2O .$$

Some experiments on this reaction have suggested the reaction rate expression at 2 bar pressure and 150°C, where C_A is the concentration of ethyl alcohol (mol/litre) and R is the rate of consumption of ethyl alcohol (mol/s/m^3):

$$R = \frac{52.7C_A^2}{1 + \frac{0.013}{C_A}} .$$

The reactor is to have a 0.05 m diameter and the alcohol inlet flowrate is to be 10 g/s. The objective is to determine the reactor length to achieve various degrees of alcohol conversion. We wish to determine reactor length for the outlet alcohol mole fractions 0.5, 0.4, 0.3, 0.2, and 0.1.

Chemical engineering design theory

Assuming small heat of reaction, plug flow and ideal gas behaviour, it can be shown that the reacting flow is described by four ordinary differential equations in terms of the dependent variables C_A, C_W (the water concentration), V (the velocity) and x (the distance along the reactor from the inlet):

$$\begin{aligned}
\frac{dC_A}{dt} &= -R\left(1 + \frac{C_A}{C}\right) , \\
\frac{dC_W}{dt} &= R\left(1 - \frac{C_W}{C}\right) , \\
\frac{dV}{dt} &= \frac{RV}{C} , \\
\frac{dx}{dt} &= V .
\end{aligned} \tag{16}$$

The last equation states that the superficial velocity creates an equivalence between distance along the reactor and the residence time t that a fluid element has to react. These equations are subject to the initial condition of the flow at the inlet $(t = 0)$:

$$C_A(0) = C \quad V(0) = V_0 \,,$$
$$C_W(0) = 0 \quad x(0) = 0 \,. \tag{17}$$

Approach

Clearly from the initial condition and stoichiometry, $C_W = C_E$ (the concentration of ethyl alcohol, and the value of C is constant as temperature and pressure are assumed constant. C can be found from the ideal gas law, with

$$C = \frac{p}{T(8314 \; \frac{J}{\text{kmolK}})} \,. \tag{18}$$

And the initial flow velocity V_0 can be determined from the flowrate given, the inlet density (the molecular weight of ethyl alcohol is 46 kg/kmol), and the tube cross-sectional area. The equations will need to be integrated numerically in space-time t until the required alcohol mole fractions have been reached. Use either simple Euler or Runge-Kutta numerical integration.

You may note that it is possible to solve for C_A without recourse to the other variables, but C_W, V, and x depend explicitly on t. But since the requirement is to find positions x where specific mole fractions occur, it is best to solve for all four variables simultaneously.

Partial results

A resolved numerical solution gives

$$\frac{C_A}{C} = 0.1 \,,$$
$$t = 5.65225 \,, \tag{19}$$
$$x = 18.5435 \,,$$

with a profile for C_A/C as in Figure 4.

COMSOL Multiphysics implementation

We wish to create our pseudo-0-D simulation environment yet again, this time with four dependent variables. Start up COMSOL Multiphysics and await the Model Navigator window. Follow the steps in Table 7 to set up the tubular reactor design model. This application mode gives us four dependent variables $u_1 u_2 u_3 u_4$ and one space coordinate x. Since there are several parameters, it is useful to specify them with named constants. Furthermore, the rate law expression recurs, so it is convenient to define it as a scalar expression.

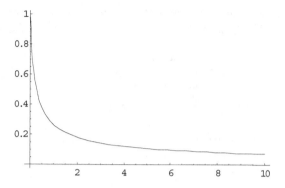

Figure 4. Profile of normalized alcohol concentration versus space time t.

Try plotting point plots of u_1, u_2, u_3 and u_4 for the whole range of times. How good is the qualitative agreement with Figure 4? Does it agree numerically with the fully resolved solution?

Exercises.

1.4. Find the value of $y'(x = 1)$ from the system of equations below. Plot y' for x between 0 and 3.

$$y'' + y' + y^2 = 0 \,,$$
$$y(x = 0) = 1 \,,$$
$$y'(x = 0) = 0 \,.$$

1.5. Linear systems of ODEs result from first order reversible reaction systems in a continuously stirred tank reactor. For instance, consider the isomerization reactions

$$A \leftrightarrow B \leftrightarrow C$$

with forward reaction rates k_1 and k_3, respectively, as written; reverse reaction rates k_2 and k_4, as written. First order kinetics leads to the following system of ODEs:

$$\frac{dc_A}{dt} = -k_1 c_A + k_2 c_B \,,$$
$$\frac{dc_B}{dt} = k_1 c_A - k_2 c_B - k_3 c_B + k_4 c_C \,, \tag{20}$$
$$\frac{dc_C}{dt} = k_3 c_B - k_4 c_C \,.$$

Table 7. Tubular reactor design modelling steps in COMSOL Multiphysics.

Model Navigator	Select 1-D. COMSOL Multiphysics:PDE modes:PDE, general form Set dependent variables: $u_1 u_2 u_3 u_4$ Select Element: Lagrange-Linear. OK
Draw Menu	Specify objects: Line. Coordinates pop-up menu. x: 0 1 name: interval OK
Options Menu: Constants	Fill out the table as below Name Expression P 200000 T 423 R 8314 MM 46 Flowrate 0.01 Dia 0.05 C P/(RT) area pi*Dia^2/4 rho MM*C vel Flowrate/rho/area OK
Options Menu: Scalar Expressions	Define rate $= 52.7 * u_1 \hat{} 2/(1 + 0.013/u_1)$
Physics Menu: Boundary settings	Select domains: 1 and 2 (hold down Ctrl key) Select Neumann boundary conditions Leave defaults $q = 0$ $g = 0$ OK
Physics Menu: Sub-domain settings	Select domain 1 F tab; set $F_1 = -\text{rate} * (1 + u_1/C)$; $F_2 = \text{rate} * (1 - u_2/C)$; $F_3 = \text{rate} * u_3/C$; $F_4 = u_3/C$ Init tab; set $u_1(t_0) = C$; $u_3(t_0) = $ vel. OK
Mesh menu: mesh parameters	Set maximum element size 1 Hit remesh. OK
Solve menu: solver parameters	Time-dependent solve. Enter on General tab Times: linspace (0, 10, 100) Solve. OK
Post-processing: Cross-section plot parameters	Point tab. Accept the default of u_1 General tab. Accept the default of all times OK

It may surprise you, but because the above system is linear, it has a general, analytic solution. Though general, it lends little insight into the dynamics of the system. Plot the graph of concentrations versus time for the initial value problem. Start with pure $C_A = 1$ with parametric values $k_1 = 1$ Hz, $k_2 = 0$ Hz, $k_3 = 2$ Hz, $k_4 = 3$ Hz and plot the graph versus time of concentrations.

5. Method 3: Numerical integration of ordinary differential equations

In the previous section, numerical integration was treated by marching methods, commonly referred to as "time-stepping," although in the reactor design application, it was clearly spatial integration. In marching methods, the unknowns are found *sequentially*. The other common method for integration is to approximate the ODE and solve *simultaneously* for the unknown dependent variables at the grid points. With marching methods, all solutions must be initial value problems (IVP). The number of initial conditions must match the order of the ODE system. But for second order and higher systems, a second type of boundary condition is possible — the boundary value problem (BVP), where in 1-D, there are conditions at the initial and final points of the domain. Hence, these are two point boundary value problems. Marching methods can laboriously treat BVPs by shooting — artificially prescribing an IVP and guessing the initial conditions that satisfy the actual BVP by trial and error. In higher dimensional PDEs, a BVP specifies conditions on the boundaries of the domain.

One of the major advantages of the finite element method is that it naturally solves two-point BVPs. As an example, the reaction and diffusion equation in 1-D is

$$\frac{D}{L^2}\frac{\partial^2 u}{\partial x^2} = R(u)\,, \qquad (21)$$

where u is the concentration of the species, D is the diffusivity, L is the length of the domain, $R(u)$ is the disappearance rate by reaction, and x is the dimensionless spatial coordinate. If the unknown function $u(x)$ is approximated by discrete values $u_j = u(x_j)$ at the grid points $x = x_j = j\Delta x$, then with central differences, the system of equations becomes

$$\sum_{j=1}^{N} M_{ij} u_j = \frac{L^2 \overline{\Delta x}^2}{D} R_i\,, \qquad (22)$$

where M_{ij} is a tridiagonal matrix with the diagonal element -2, and 1 on the super and subdiagonals:

$$M = \begin{bmatrix} -2 & 1 & 0 & 0 & \cdots \\ 1 & -2 & 1 & 0 & \cdots \\ 0 & 1 & \ddots & \ddots & \ddots \\ 0 & 0 & \ddots & \ddots & \ddots \\ \vdots & \ddots & \ddots & \ddots & \ddots \end{bmatrix}\,, \qquad (23)$$

and $R_j = R(u_j)$. This system can be solved by iteration for u_i^n by matrix inversion, where n refers to the n-th guess:

$$u_i^n = \frac{L^2 \overline{\Delta x}^2}{D} \sum_{j=1}^{N} M_{ij}^{-1} R_j\,, \qquad (24)$$

and $R_j = R(u_i^{n-1})$. For either IVP or BVP, the appropriate rows of the matrix M in (23) can be altered to accommodate the boundary conditions. As written, (23) supposes $u = 0$ at both $x = 0$ and $x = 1$. This is a Dirichlet type boundary condition, and is the natural boundary condition for finite difference methods — natural because it occurs if no effort is made to overwrite rows of (23) with specified boundary conditions.

We will now illustrate the solution of (21) with COMSOL Multiphysics on a small 1-D domain with first order reaction $R(u) = ku$ and representative values for the resulting dimensionless parameter, the Damkohler number:

$$Du = \frac{kL^2}{D} = \frac{(10^{-3} \text{ s}^{-1})(10^{-3} \text{ m}^{-1})^2}{1.2 \times 10^{-9} \text{ m}^2/\text{s}} = 0.833, \qquad (25)$$

and with boundary conditions $u = 1$ at $x = 0$ and no flux at $u = 1$.

This exercise interacts with MATLAB to explore the structure of a COMSOL Multiphysics representation of solution data and model layout. In windows, COMSOL Multiphysics with MATLAB is a desktop icon option, if you have a MATLAB licence. In UNIX/linux, the equivalent functionality can be launched with

comsol matlab path-ml nodesktop-ml nosplash

from linux command line. The "matlab" argument tells femlab to launch a matlab command window. The "path" argument sets up the matlab command window with access to the COMSOL library of commands.

First launch COMSOL Multiphysics and enter the Model Navigator. Follow the steps in Table 8. This application mode gives us one dependent variable u, but in a 1-D space with coordinate x. h and r are the two handles on Dirichlet BCs in Coefficient Mode. If you want to set u to a given value U_0 on a boundary, then it is accomplished with setting $h = 1$ and $r = U_0$. Clicking on the triangle symbol creates a default mesh (15 elements) and the triangle with the embedded upside-down triangle to refines the mesh (30 elements).

You should get a graph with the information as in Figure 5. Clearly the desired boundary conditions are met: $u = 1$ at $x = 0$ and the slope vanishes at $x = 1$. But did COMSOL Multiphysics solve the problem we thought we posed?

Now resolve with the stationary nonlinear solver. First note by View Log on the Solver Menu that COMSOL Multiphysics takes thirteen iterations to converge. Do you notice that the final value has dropped from 0.86 to 0.69? One might wonder why there is a difference. The linear (coefficient

Table 8. ODE example of a two-point boundary value problem in a reaction-diffusion system.

Model Navigator	Select 1-D. COMSOL Multiphysics:PDE modes:PDE, *coefficient* form Set dependent variable: u Select Element: Lagrange-Linear. OK
Draw Menu	Specify objects: Line. Coordinates pop-up menu. x: 0 1 name: interval OK
Physics Menu: Boundary settings	Select domains 1 Check Dirichlet and set $h = 1$; $r = 1$ Select domain 2 Select Neumann boundary conditions. OK
Physics Menu: Subdomain settings	Select domain 1 Set $C = -1$; $f = 0.833 * u$; $d_a = 0$ Select Init tab; set $u(t_0) = 1 - x$ OK
Meshing	Click on triangle symbol to mesh
Solve menu: solver parameters	Note stationary linear default. OK General tab. Set solution form to "Coefficient" Solve with button ($=$)
Post-processing: Point evaluation	Select Boundary 2 and expression u. Reports: Value: 0.861167, Expression: u, Boundary: 2

form) solver only evaluates $R(u)$ once at the initial condition $u(t_0) = 1 - x$ and thus only needs one iteration of (24). The nonlinear solver evaluates $R(u)$ for each iteration at the old estimate for u. Thus, the nonlinear solver might "forget" the initial guess completely after a number of iterations as it homes in on a converged solution.

Let's test this explanation. Changing the initial condition should change the stationary linear solution. Return now to Subdomain settings and try the initial condition $u(t_0) = 1$. What final value do you get for $u(x = 1)$? Now try the stationary nonlinear solver. Do you get the same solution as with the other initial condition?

This example should illustrate the importance of selecting the right solver for your equations. If there is any dependence of f on the dependent variables, then the stationary nonlinear solver should be used. The linear solver is faster, but it also presumes that the coefficients of the PDE do not depend on the dependent variable u (else the problem would be nonlinear). When in doubt, use the nonlinear solver. After all, (21) with $R(u) = ku$, is a *linear* problem, but COMSOL Multiphysics only finds the correct steady state solution with the nonlinear solver! The slow convergence rate is also the consequence of the form of the model — general mode with the exact Jacobian solver option for the nonlinear solver converges in two iterations to the correct profile.

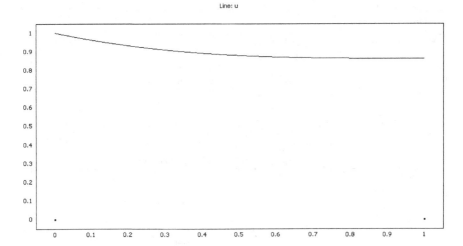

Figure 5.   Concentration profile at steady state.

We argued that (22) is the finite difference matrix equation for this problem, yet later applied the argument that (24) should describe the COMSOL Multiphysics finite element problem. Because we used Lagrange linear elements, in this special case the finite element and finite difference matrix operators coincide, up to the boundary conditions. To see this, we will take a foray into the MATLAB representation of COMSOL Multiphysics problems.

Pull down the File Menu and select Export FEM structure as "fem." This puts the current solution as a MATLAB data structure in the MATLAB workspace. We can then manipulate it using the built-in MATLAB functions and commands, as well as the special function set of COMSOL Multiphysics.

In your MATLAB workspace, try the commands

```
>>x=fem.mesh.p;
>>u=fem.sol.u;
>>plot(x,u)
```

This should pop up a MATLAB Figure plotting the solution u versus the array of mesh points. No doubt your plot looks scrambled. This is because COMSOL Multiphysics stores the mesh points and the associated solution variables so as to make the specification of the matrix equations sparse and compact. We can make sense of the solution by ordering the mesh points using the MATLAB sort command and the solution.

In your MATLAB workspace, try the commands

```
>>[xx, idx]=sort(x);
>>plot (xx, u(idx))
```

The final MATLAB manipulation we will consider here is interrogation of the finite element matrix. The fem structure does not hold the finite element stiffness matrix, but rather contains the information necessary for FEMLAB functions to construct it. This activity is a vital part of the finite element method, and the FEMLAB function that does it is called assemble. Type in the command below:

```
>>[K,L,M,N]= assemble(fem);
>>K'/15
```

This plot should resemble Figure 5, with the exception that it represents your last COMSOL Multiphysics solution. In fact, we can only make sense of the solution format of the *fem* structure so readily because this is a single dependent variable, one-dimensional problem. Otherwise, multiple variables and dimensions leave a mesh and solution structure that only COMSOL Multiphysics tools/functions can readily decode. Figure 6 shows the sparsity structure generated by the MATLAB command spy (K'). It is instructive to compare this with (23), since it is clear that 1-D Lagrange-linear elements with uniform grid are comparable to finite difference methods in the formulation of the matrix equations.

You should now see a MATLAB sparse representation of a matrix, all of the elements of which are 1, -2, and 1, arranged on different diagonals. This is the stiffness matrix of the finite element method, and up to the ordering of the unknowns, is equivalent to (23). If you return to the Subdomain Settings, element tab, and select Lagrange quadratic elements, and repeat the solution, exporting FEM and assemble K as above, you will note that although sparse, the matrix is distinctly different from the Lagrange linear elements.

Exercise.

1.6. The coefficient form has a PDE term αu. Repeat the implementation of the reaction-diffusion example, but this time entering $\alpha = 0.833$ and $f = 0$ for the subdomain settings. Now compare the stationary linear and nonlinear solver solutions. Can you explain why this formulation leads to this result? What effect does this formulation of the problem have on the stiffness matrix K. Can you think of a difficulty that

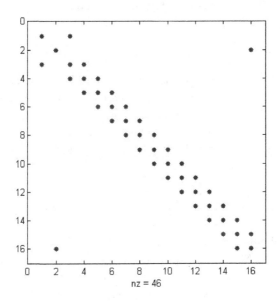

Figure 6.   Sparsity structure of K' from the matlab command spy (K').

might occur if the Da is chosen so that the diagonal element is nearly zero in magnitude, i.e. $Da\Delta x^2 = 2$?

6. Method 4: Linear systems analysis

Central to MATLAB, and hence to COMSOL Multiphysics, is linear systems analysis. In this section, we will briefly review the concepts of linear operator theory — typically lumped as "matrix equations" in undergraduate engineering mathematics modules. The good news is that it is not necessary to do any matrix manipulations yourself. That was the *raison d' être* for MATLAB: to serve as a user interface to libraries of subroutines for engineering matrix computations. It should be noted that COMSOL Script would equally well serve this purpose in the examples in this chapter. Much of the history of scientific computing is encapsulated in efficient and sparse methods for matrix computations. An excellent guide to matrix computations, but surely for experts, is the book of Golub and Van Loan [3]. However, at the introductory level to MATLAB, a good and readable survey can be found in the up-to-date book by Hanselman and Littlefield [4].

Briefly, the standard matrix equations look like this:

$$a_{11}x_1 + a_{12}x_2 + a_{13}x_3 + \cdots + a_{1N}x_N = b_1,$$
$$a_{21}x_1 + a_{22}x_2 + a_{23}x_3 + \cdots + a_{2N}x_N = b_2,$$
$$a_{31}x_1 + a_{32}x_2 + a_{33}x_3 + \cdots + a_{3N}x_N = b_3, \tag{26}$$
$$\vdots$$
$$a_{M1}x_1 + a_{M2}x_2 + a_{M3}x_3 + \cdots + a_{MN}x_N = b_M.$$

Here there are N unknowns x_j which are related by M equations. The coefficients a_{ij} are known numbers, as are the constant terms on the right hand side, b_i. In engineering, models are frequently derived that satisfy such linear systems of equations. Mass and energy balances, for instance, commonly generate such sets of linear equations.

Solvability

When $N = M$, there are as many constraints as there are unknowns, so there is a good chance of solving the system for a unique solution set of x_j's. There can fail to be a unique solution if one or more of the equations is a linear combination of the others (row degeneracy) or if all the equations contain only certain combinations of the variables (column degeneracy). For square matrices, row and column degeneracy are equivalent. A set of degenerate equations are termed **singular**. Numerically, however, at least two additional things can go wrong:

- While not exactly linear combinations of each other, some of the equations may be so close to linearly dependent that within round-off errors on the computer they are.
- Accumulated round-off errors in the solution process can swamp the true solution. This frequently occurs for large N. The procedure does not fail, but the computed solution does not satisfy the original equations all that well.

Guidelines for linear systems

There is no "typical" linear system of equations, but a rough idea is that round-off error becomes appreciable:

- N as large as 20–50 can be solved by normal methods in single precision without recourse to specialist correction of the two numerical pathologies.
- N as large as several hundred can be solved by double precision.

- N as large as several thousand can be solved when the coefficients are sparse (i.e. most are zero) by methods that take advantage of sparseness. MATLAB has a special data type for sparse matrices, and a suite of functions that exploit the sparseness.

However, in engineering and physical sciences, there are problems that by their very nature are singular or nearly singular. You might find difficulty with $N = 10$. Singular value decomposition is a technique which can sometimes treat singular problems by projecting onto nonsingular ones.

Common tasks in numerical linear algebra

Equation (26) can be succinctly written as a matrix equation (cf. equation (22)).

$$\mathbf{A} \cdot \mathbf{x} = \mathbf{b} . \tag{27}$$

- Solution for the unknown vector $\mathbf{x}$, where $\mathbf{A}$ is a square matrix of coefficients, and $\mathbf{b}$ is a known vector.
- Solution with more than one $\mathbf{b}$ vector with the matrix $\mathbf{A}$ held constant.
- Calculation of the matrix $\mathbf{A}^{-1}$, which is the inverse of a square matrix $\mathbf{A}$.
- Calculation of the determinant of a square matrix $\mathbf{A}$.
- If $M < N$, or if $M = N$ but the equations are degenerate, then there are effectively fewer equations than unknowns — an underdetermined system. In this case, either there can be no solution, or there is more than one solution vector $\mathbf{x}$. The solution space consists of a particular solution $\mathbf{x}_p$ plus any linear combination of typically N–M vectors called the nullspace of $\mathbf{A}$. The task of finding this solution space is called *singularvalue decomposition.*
- If $M > N$, there is, in general, no solution vector x to (26). This overdetermined system happens frequently, and the best compromise solution that comes closest to satisfying the equations is sought. Usually, the closeness is "least-squares" difference between the right and left hand sides of (26).

Matrix computations in MATLAB

Matrix inversion is easily entered using the inv(matrix) command. Solution of matrix equations is represented by the matrix division\operator as here:

```
>> A=[ 3 -1 0; -1 6 -2; 0 -2 10];
>> B=[1; 5; 26];
```

```
>> X=A\B
X =
1.0000
2.0000
3.0000
```

Determinants

Determinants are used in stability theory and in assessing the degree of singularity of a matrix. Why do you need to know the determinant? Most of the time, you want to know when a determinant is zero. However, when the determinant is zero, or numerically close to zero, it is numerically difficult to compute due to "round-off" swamping effects mentioned earlier. This is yet another application for singular value decomposition.

MATLAB computes determinants by the simple function $\det(A)$. Either enter by hand the matrix below at the MATLAB command line, or cut and paste from the file matrix2.dat:

```
>> A=[0.45, -0.244111, -0.0193373, 0.323972, -0.118829;
-0.244111, 0.684036, -0.103427, 0.205569, 0.00292382;
-0.0193373,-0.103427,0.8295, 0.0189674, -0.011169;
0.323972, 0.205569,0.0189674, 0.659479, 0.197388;
-0.118829,0.00292382,-0.011169, 0.197388, 0.776985]
```

The determinant is found from the det command

```
>> det(A)

ans =
-1.9682e-008
```

Principal axis theorem: Eigenvalues and eigenvectors

MATLAB has built-in functions for computing the eigenvalues and eigenvectors of a matrix:

```
>> eig(A)

ans = -0.0000
0.7000
0.8000
0.9000
1.0000
```

The eig() function can also return the eigenvectors as the columns of the matrix V when called as below:

```
>> [V,D]=eig(A)
```

V =

-0.6836	0.0000	-0.5469	-0.4785	-0.0684
-0.4181	0.6162	0.1831	0.4530	-0.4547
-0.0837	0.4003	0.6189	-0.6232	0.2479
0.5409	0.2582	-0.2415	-0.4042	-0.6474
-0.2416	-0.6272	0.4755	-0.1190	-0.5550

D =

-0.0000	0	0	0	0
0	0.7000	0	0	0
0	0	0.8000	0	0
0	0	0	0.9000	0
0	0	0	0	1.0000

The eigs() function is a variant of eig() which computes a specific number of eigenvalues/eigenvector pairs for sparse matrices. Its use will be demonstrated in the next subsection in conjuction with COMSOL Multiphysics.

The matrix A has a determinant that is little different from zero and a single eigenvalue that is effectively zero. The eigenvector associated with it is effectively the null space of A — the direction that gets mapped to zero:

```
>> A*V(:,1)

ans =
1.0e-007 *
0.2669
0.1633
0.0327
-0.2112
0.0943
```

All the other eigenvectors can be verified by the property that they map onto themselves, scaled by the eigenvalue, for instance:

```
>> A*V(:,2)./ V(:,2)

ans =
0.7000
0.7000
0.7000
0.7000
0.7000
```

In MATLAB, the ./ division operator is element-by-element division. The colon above refers to the whole of the column.

Because the system is nearly singular, we should not be surprised that the solutions to any matrix equation involving it are poorly conditioned. For instance,

```
>> B=[0; 1; 0; 1; 0];
>> A\B

ans =
1.0e+006 *
2.1487
1.3142
0.2631
-1.7001
0.7593
```

Since the elements of A are of order one, the forcing vector B is of order one, one would expect the solution to (27) to be order one, not order one million. For chemical engineers, this is like being told that a mass balance involves input flow rates of about 1 kg/hr, constraints on mass balances with appreciable fractions in the splitters (order one), and that the solution mass flow rates are about one million kg/hr for internal streams. Not likely. Yet this is the solution proposed by a nearly singular matrix.

Singular value decomposition (SVD)

SVD offers a better solution in many respects. All matrices have a unique decomposition, similar to the principal axis theorem for eigenvalues and eigenvectors

$$A = U \cdot diag \cdot V^T , \qquad (28)$$

where U and V are square real and orthogonal. $diag$ is a diagonal matrix which contains the singular values. In terms of U, V, and $diag$, the system (27) is readily solved

$$A^{-1} = V \cdot [1/diag_j] \cdot U^T , \qquad (29)$$

U and V being orthogonal means that their transposes are also their inverses. The inverse of a diagonal matrix is just the reciprocal of the diagonal elements. So the only time we have a problem solving the system is when one or more of the singular values ($diag_j$), relative to the largest, is close to zero. It follows that ($1/diag_j$) is a very large number, which distorts our numerical solution, sending it off to infinity along a direction which is spurious. A good approximation is to throw these spurious directions away completely by setting $(1/diag_j)$ for the offending singular values to zero! The vector,

$$x = V \cdot [1/diag_j] \cdot U^T b \tag{30}$$

with this substitution for nearly zero elements, should be the smallest in magnitude to approximately satisfy the equations.

In the case of our example matrix A, the MATLAB command svd() gives the singular values if called with one output, and the three matrices U, $diag$, V if called with three:

```
>>[U,D,V]=svd(A)

U =
```

-0.0684	-0.4785	0.5469	0.0000	-0.6836
-0.4547	0.4530	-0.1831	-0.6162	-0.4181
0.2479	-0.6232	-0.6189	-0.4003	-0.0837
-0.6474	-0.4042	0.2415	-0.2582	0.5409
-0.5550	-0.1190	-0.4755	0.6272	-0.2416

```
D =
```

1.0000	0	0	0	0
0	0.9000	0	0	0
0	0	0.8000	0	0
0	0	0	0.7000	0
0	0	0	-0.0000	1.0000

```
V =
```

-0.0684	-0.4785	0.5469	0.0000	-0.6836
-0.4547	0.4530	-0.1831	-0.6162	-0.4181
0.2479	-0.6232	-0.6189	-0.4003	-0.0837
-0.6474	-0.4042	0.2415	-0.2582	0.5409
-0.5550	-0.1190	-0.4755	0.6272	-0.2416

Reassuringly, $U = V$ since the matrix A is symmetric.

The SVD prescription for solution with smallest magnitude is implemented as follows:

```
>> ss=[1. 1./0.9 1./0.8 1./0.7 0];
>> dinv=diag(ss);
>> V*dinv*U'*B

ans =
```

```
0.0893
1.2820
0.1479
1.0317
-0.2130
```

This is a far more physically acceptable solution, for instance, for internal mass flow rates in the hypothetical mass balance discussed above.

This excursion into linear systems theory is important for modeling with COMSOL Multiphysics because finite element methods are matrix based. When the generalized stiffness matrix becomes nearly singular, COMSOL Multiphysics may not be providing a satisfactory solution. These matrix computations and their sparse implementations in MATLAB can readily serve as diagnostics for the health of the COMSOL Multiphysics solution. They also provide an insight into the natural dynamics of the system through the eigen analysis of the operator. These ideas will be made concrete with an example computed as a COMSOL Multiphysics model in the next subsection.

6.1. *Heat transfer in a nonuniform medium*

The steady state heat transfer equation is commonly met in engineering studies as the simplest PDE that is analytically solvable: Poisson's equation. Nevertheless, series solutions for complicated geometries may be intractable. The author has recently shown that some series so derived are purely asymptotic and poorly convergent [5]. Consequently, numerical solutions are likely to be better behaved than series expansions. Furthermore, any variation on the processes of heat transfer may destroy the analytic structure. In this section, we will consider the typical one-dimensional heat transfer problem in a slab of nonuniform conductivity and a distributed source that is differentially heated on the ends:

$$-\frac{d}{dx}\left(k\frac{dT}{dx}\right) = f(x),$$

$$T|_{x=0} = 1 \quad T|_{x=1} = 0.$$

(31)

Launch COMSOL Multiphysics with MATLAB and enter the Model Navigator: Follow Table 9 for the instruction set to set up the heat transfer problem in a medium with a distributed source. The solution should be found fairly quickly resulting in a nearly linear profile with almost a slope of -1. The Solver Log shows two step solution, which, since the problem

Table 9.   Heat transfer in a nonuniform medium — Distributed source.

Model Navigator	Select 1-D. COMSOL Multiphysics: Heat Transfer:Conduction: Steady state analysis. Set dependent variable: u Select Element: Lagrange-Linear. OK
Draw Menu	Specify objects: Line. Coordinates pop-up menu. x: 0 1 name: interval OK
Physics Menu: Boundary settings	Select boundary selection 1 Set boundary condition: temperature; $T_0 = 1$ Select boundary selection 2 Set boundary condition; $T_1 = 0$. OK
Physics Menu: Subdomain settings	Select domain 1 Set $k = 1$; $Q = -x * (1 - x)$ Select Init tab; set $T(t_0) = 1 - x$. OK
Meshing	Click on triangle symbol to mesh
Solver	Click on the solve (=) button to solve
Post-processing: Data display	Specify $x = 0.5$ Value: 0.474097 $[K]$, Expression: T, Position: (0.5)

is *linear*, is guaranteed. Verify that $T|_{x=0.5} = 0.474097$. This problem has an analytic solution with $T|_{x=0.5} \sim 0.475$:

$$T = 1 - \frac{13}{12}x + \frac{x^3}{6} - \frac{x^4}{12}. \tag{32}$$

Now try $k = 1 - x/2$. There is also an analytic solution in this case, but in the complex numbers requiring logarithms in the complex plane and a branch cut. The analytic solution gives $T|_{x=0.5} \sim 0.550$. How good is your solution?

Now for the linear systems theory. Pull down the File menu and select Export FEM structure as "fem." This is the second time we have exported the FEM structure, so it might be useful to explain a bit more about it. As with any MATLAB variable, if you type the variable name on the command line, MATLAB will either provide the value of the variable or show its data structure. Try

```
>> fem

fem =

      version: [1x1 struct]
         appl: {[1x1 struct]}
         geom: [1x1 geom1]
         mesh: [1x1 femmesh]
       border: 1
      outform: 'general'
         form: 'general'
```

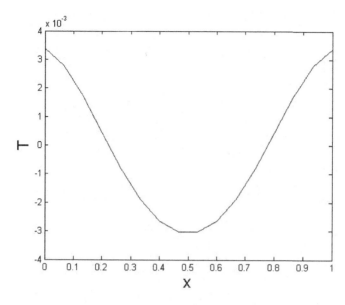

Figure 7. Projection solution for the purely Neumann solution to the nonuniform conductivity and distributed source heat transfer problem.

```
units: 'SI'
  equ: [1x1 struct]
  bnd: [1x1 struct]
 draw: [1x1 struct]
xmesh: [1x1 com.femlab.xmesh.Xmesh]
  sol: [1x1 femsol]
```

Of course, you can go further down the FEM structure and investigate branches, twigs, and leaves on the tree. We have already pruned

$$\text{fem.sol.u},$$
$$\text{fem.mesh.p1}.$$

The different branches contain a complete COMSOL Multiphysics model — specifying equations (fem.appl{1}.equ) and boundary conditions (fem.appl{1}.equ) for the geometry held in fem.geom. Some of these fields are writable by MATLAB assignment statements, like any other variable. Interrogate the boundary conditions, for instance, with

```
>> fem.appl{1}.bnd
```

```
ans =
```

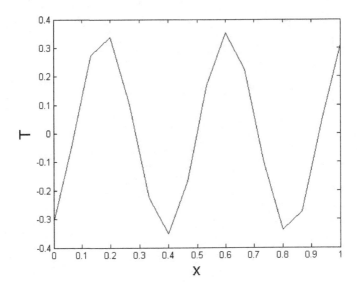

Figure 8.   Smallest nonzero eigenvalue/eigenvector pair for the nonuniform conductivity and distributed source heat transfer problem.

```
   type: 'T'
     TO: {[0]   [1]}
    ind: [2 1]

>> fem.appl{1}.bnd.T0{2}

ans =

     1

>> fem.appl{1}.bnd.T0{2}=10

ans =

    10
```

It is now possible to upload this FEM structure to COMSOL Multiphysics using the FILE menu Import field. If you do this, check under Physics Menu — Boundary Settings, you should find that the boundary condition on boundary 1 is now $T = 10$. Clearly, we could make this change to the model far more readily by using the pull down menus in the

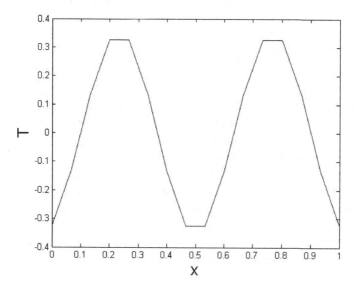

Figure 9. Next smallest nonzero eigenvalue/eigenvector pair for the nonuniform conductivity and distributed source heat transfer problem.

GUI. Nevertheless, altering FEM structures in MATLAB is a very powerful feature. In later chapters, we will use this feature to good effect. Given the complexity of the FEM data structure, however, it is best to keep our alterations to simple features.

You can now manipulate the solution in MATLAB. As in the last section, you can assemble the stiffness matrix and analyze it with eigs. As K is sparse, you can find the smallest six eigenvalues in magnitude with the eigs command.:

```
>>[K,L,M,N] = assemble(fem);
>>dd=eig(K)
>>dd=eigs(K, 6,-0.1)
>>[V,D]= eigs(K, 6, -0.1)
```

Note that K has one zero eigenvalue, and that all its eigenvalues are negative otherwise. This should not worry you, as COMSOL Multiphysics implements its boundary conditions through the block matrix N and auxiliary forcing vector M. It could replace rows of K and elements in L to approximate boundary conditions, but the structure of boundary conditions in COMSOL Multiphysics allows for more general types of boundary conditions when augmenting the matrix equations with N and M. The fact that K is singular as a block matrix is a consequence of the natural boundary conditions for finite element methods being Neumann conditions (no flux). There are an infinity of solutions to the pure Neumann boundary conditions, as an arbitrary value can be added to any solution and it is still a solution. That K is singular naturally tripped up the author when he first used finite element methods as an undergraduate. Purely Neumann boundary conditions (zero heat flux in this example) are ill-posed. For instance, if you change the boundary conditions in our example to purely Neumann conditions, you should find that the steady state solution is not solvable — leads to a singular matrix. Some reference value (Dirichlet condition) must be specified for there to be an unique solution. Yet MATLAB can solve such a problem by SVD or by the principal axis theorem. Since the matrix K is positive-semi-definite, all its eigenvalues are real. So pseudo-inversion to eliminate the zero eigenvalue of K follows by following the recipe for the pseudo-inverse of the previous section:

```
>>ss=1./dd;
>>ss(6)=0;
>>dinv=diag(ss);
>>uneumann=V*dinv*V'*L
```

Finally, interpreting this solution must be done remembering that the structure of a FEMLAB mesh is not monotonic. These commands plot the solution:

```
>>[xs, idx]=sort(fem.mesh.p);
>>plot(xs, fem.sol.u(idx)) ;
```

Similarly, the approximate Neumann solution found from the projection onto the first five eigenvectors with smallest magnitude nonzero eigenvalues is found from

```
>>plot(xs, uneumann(idx));
```

Figure 7 shows the projection of the solution onto just the first six modes using the pseudo-inverse method. You can think of this plot as the deviation from the linear function satisfying the boundary conditions, i.e. $1 - x$. Furthermore, the eigenvectors can be interpreted the same way:

```
>>col1=V(:,1);
>>plot(xs,col1(idx));
>>col2=V(:,2);
>>plot(xs,col2(idx));
```

It is one of my pet complaints that eigenvalues and eigenvectors are not interpreted in mathematics classes, so the student does not learn why they are taught and thus dismisses them as "esoteric." The eigenvectors in this problem represent the decaying "modes" as the solution approaches steady state at long times. The eigenvalues are the (exponential) decay rates, in this case, since the COMSOL Multiphysics sign is reversed from common practise. *Negative* eigenvalues would represent growth rates of unstable modes. Clearly, all the modes here are dissipative of energy. That not all analysis share this feature — some are unstable, can lead to physically interesting phenomena (pattern formation, explosion), but also to numerically difficult modelling — nonconvergent computational models.

Please note that in the case of the Neumann solution, any constant value can be added to the solution and it will remain a solution. The eigenvectors are not normalized, so they can be multipled by any number and still be eigenvectors. Figures 8 and 9 show the two eigenpairs with smallest eigenvalues in magnitude. These are the slowest decaying modes, and therefore the pattern of the expected "standing waves" that disappear last as the steady-state is formed.

References

[1] M. B. Cutlip and M. Shacham, *Problem Solving in Chemical Engineering with Numerical Methods* (Prentice-Hall, Upper Saddle River, NJ, 1999).

[2] W. C. Gear, *Numerical Initial Value Problems in Ordinary Differential Equations* (1971).

[3] G. H. Golub and C. F. Van Loan, *Matrix Computations*, 3rd edn. (Johns Hopkins University Press, Baltimore, London, 1996).

[4] D. Hanselman and B. Littlefield, *Mastering MATLAB 7: A Comprehensive Tutorial and Reference* (Prentice-Hall, Saddle River, NJ, 2005).

[5] W. B. Zimmerman, On the resistance of a spherical particle settling in a tube of viscous fluid, *International Journal of Engineering Science* **42**(17–18) (2004) 1753–1778.

Chapter Two

ANALYZING EVOLUTION EQUATIONS BY THE FINITE ELEMENT METHOD

W.B.J. ZIMMERMAN and B.N. HEWAKANDAMBY

Department of Chemical and Process Engineering,
University of Sheffield,
Mappin Street, Sheffield S1 3JD, United Kingdom
E-mail: w.zimmerman@shef.ac.uk

Partial differential equations (PDEs) arise naturally in science and engineering from complex balance equations. Commonplace PDEs are derived from conservation laws for transport of mass, momentum, species and energy. Because these conservation laws are integral equations over the domain, the PDEs that arise from the continuum hypothesis have a structure that is readily represented by the finite element method as an approximation. In this chapter, the three different classes of differential equations that arise in spatial-temporal systems — elliptic, parabolic, and hyperbolic — are defined and representative cases are treated by COMSOL Multiphysics computations. An overview of the finite element method is given, but greater depth of detail will await later chapters where the applications particularly exploit features of finite element methods that intrinsically permit elegant and accurate computation.

1. Introduction

Partial differential equations are usually found in science and engineering applications as the local, infinitesimal constraint imposed by conservation laws that are typically expressed as integral equations. The whole class of transport phenomena due to conservation of mass, momentum, species and energy lead to PDEs in the continuum approximation. Chemical engineers are well acquainted with shell balances in transport phenomena studies for heat, mass and momentum transfer.

In contrast to the previous chapter, where 0-D and 1-D spatial systems were treated by COMSOL Multiphysics with example applications in chemical engineering, the chemical engineering curriculum is not overflowing with 2-D and 3-D example computations of the solutions to PDEs. A rare example is found in [1]. In fact, historically, many of the common chemical engineering models and design formula are simplifications of higher spatial dimension dynamics that are treated phenomologically. Resistance coefficients in fluid dynamics, mass transfer and heat transfer coefficients, Thiele moduli in heterogeneous catalysis, McCabe-Thiele diagrams for distillation column design, and many more common techniques are convenient semi-empiricisms that mask an underlying transport or nonequilibrium thermodynamics higher spatial dimension system, possibly expressible as a PDE system, but traditionally thought too difficult to solve given the complexity of the fundamental physical chemistry. These simpler methodologies are still preferable for quick estimates desired for preliminary design calculations, but may be insufficient for detailed design, retrofit, or process analysis and optimization purposes. For fundamental science, these methods are still migrating from chemical engineers to biotechnologists or material scientists in the first approach to multidisciplinary work. Nevertheless, computational fluid dynamics (CFD) has forever changed the paradigm for what is considered the state-of-the-art in transport modeling. Phenomenological methods may still have a niche, and a particularly important one in interpreting distributed system models, yet CFD has a unique role for visualization and quantification of transport phenomena.

COMSOL Multiphysics is not a "commercial CFD code," but it will do some CFD. There are several general purpose CFD packages available, with their own advantages in supporting certain applications. By CFD, most process engineers would envisage support for many turbulence and combustion models. COMSOL Multiphysics, however, has a different niche in the area of multiphysics. In addition to the traditional transport phenomena that CFD treats, COMSOL Multiphysics includes application modes for electrodynamics, magnetodynamics, and structural mechanics, permitting simultaneous treatment of these and transport phenomena. But its greatest strengths are actually least trumpeted — first, the ease of "user defined programming," which is the ability to implement the user's own model or parametric variation of coefficients, boundary conditions, initial conditions and to link to simultaneous physics, even on other domains; second, that it is built on MATLAB (or COMSOL Script) so that all the programming functionality needed to set up greater complexity of models or simulations is available, treating COMSOL Multiphysics as a convenient suite of subroutines for high-level finite element programming and analysis. In the last chapter, we saw some of the power of user defined programming and analysis. In this chapter, we introduce COMSOL Multiphysics' core strength of finite element modeling of higher dimensional PDE systems.

The greater functionality of multiphysics, extended physics, and treatment of non-PDE constraints will be left for later chapters.

2. Partial differential equations

PDEs are classified according to their order, boundary condition type, and degree of linearity (yes, no or quasi). Amazingly, most PDEs encountered in science and engineering are second order, i.e. the highest derivative term is a second partial derivative. Is this a coincidence? Lip service is usually paid at this point to variational principles underlying most of physics. Yet, recently Frieden [2] has demonstrated that all known laws of physics can be derived from the principle of minimum Fisher information, which naturally introduces a second order operator of a field quantity as the highest order term in the associated law of physics — from the wavefunction in Schrodinger's equation to classical electrodynamics. Thus, classification and solution of second order spatial temporal systems in 2-D and 3-D are of wide applicability and importance in the sciences and engineering. For this reason, and that finite element methods (FEM) are intrinsically well-suited to treating second order systems, FEM are techniques with wide applicability.

In this chapter, we focus on second order systems in 2-D and 3-D. There are three canonical exemplar systems that are nearly uniformly treated in the standard textbooks. They are:

$$\text{Laplace's equation (elliptic)}: \quad \frac{\partial^2 u}{\partial x^2} + \frac{\partial^2 u}{\partial y^2} = 0,$$

$$\text{Diffusion equation (parabolic)}: \quad \frac{\partial u}{\partial t} = \frac{\partial^2 u}{\partial x^2},$$

$$\text{Wave equation (hyperbolic)}: \quad \frac{\partial^2 u}{\partial t^2} = \frac{\partial^2 u}{\partial x^2}.$$

We will treat these classes of PDEs with COMSOL Multiphysics modelling, but in each case adding the spice of introducing a new or special feature of COMSOL Multiphysics. The terms elliptic, parabolic and hyperbolic are traditional guides to the features of a PDE system from characterization of the linear terms by reference to the general linear, second order partial differential equations in one dependent and two independent variables:

$$a\frac{\partial^2 u}{\partial x^2} + 2b\frac{\partial^2 u}{\partial x \partial y} + c\frac{\partial^2 u}{\partial y^2} + d\frac{\partial u}{\partial x} + e\frac{\partial u}{\partial y} + fu + g = 0, \tag{1}$$

where the coefficients are functions of the independent variables x and y only, or constant. The three canonical forms are determined by the following criterion:

$$\text{elliptic}: \quad b^2 - ac < 0\,,$$
$$\text{parabolic}: \quad b^2 - ac = 0\,,$$
$$\text{hyperbolic}: \quad b^2 - ac > 0\,.$$

These classifications serve as a rough guide to the information flow in the domain. For instance, in elliptic equations, information from the boundaries is propagated instantaneously to all interior points. Thus, elliptic equations are termed "nonlocal," meaning that information from far away influences the given position, versus "local," where only information from nearby influences the field variable. In parabolic systems, information "diffuses," i.e. it spreads out in all directions. In hyperbolic systems, information "propagates," i.e. there is a demarcation between regions that have already received the information, regions that will receive the information, and possibly regions that will never receive the information. If the system is linear or quasi-linear (i.e. some coefficient depends on the dependent variable or a lower order partial derivative than that it multiplies), this classification system and the intuition about how information is transported serves as a robust guide to second order systems. For nonlinear systems, however, nonlinearity can destroy the information transport structure. In nonlinear systems, information may be "bound," i.e. never transferred, beyond given attractors, or it may be created from noise (one view) or lost (a different view) by forgetting initial conditions in a given window in time.

2.1. *Poisson's equation: An elliptic PDE*

A modest variant on Laplace's equation is the Poisson equation:

$$\nabla^2 u = f(x)\,. \tag{2}$$

We saw this equation in 1-D form in (21) of Chapter 1 which described heat transfer in a nonuniform medium with a distributed source. Here, the thermal conductivity is uniform. In order to give a different spin on (2), one should note that it is the equation satisfied by the streamfunction with an imposed vorticity profile:

$$\nabla^2 \psi = -\omega(x)\,. \tag{3}$$

There are two common types of vortices that are easy to characterize — the Rankine vortex, where vorticity is constant within a region, and the

Table 1. Poisson's equation for unit distributed source.

Model Navigator	Select 2-D. COMSOL Multiphysics|PDE Modes|Classical PDEs|Poisson's Equation Set Element: Lagrange-Quadratic OK
Draw Menu	Specify objects: Circle Set defaults (unit radius; centre at origin). OK
Physics Menu: Boundary settings	Accept defaults for Dirichlet and $h = 1$, $r = 0$ $(u = 0)$ OK
Physics Menu: Sub-domain settings	Select domain 1 Accept defaults $c = 1$; $f = 1$ OK
Meshing	Click on triangle symbol to mesh
Solver	Click on the solve (=) button to solve
Post-processing: Plot parameters	Unselect Surface on General tab Contour tab: tick box for contour plot. OK

point-source vortex, where vorticity falls off rapidly and thus is idealized as point vortex. One might be curious about the streamlines generated by these two vortex types.

We will investigate these streamlines computationally. Start up COM-SOL Multiphysics and enter the Model Navigator. Follow the steps outlined in Table 1 to set up the Poisson's solution with unit distributed source (constant vorticity vortex). This application mode gives us one dependent variable u, in a 2-D space with coordinates x and y. We draw a solid circular disc in the domain, but COMSOL Multiphysics generates four segments of the boundary. It is informative to note that the default mesh produces 762 elements. The contour plots for the streamfunction ψ are shown in Figure 1.

The streamlines are viewed by a contour plot. Figure 1 was generated by the "Export: Image" option on the File Menu.

Clearly, the streamlines are concentric circles. As the boundary is a streamline ($\psi = 0$) and the maximum occurs in the center ($\psi = 0.25$), the volumetric flow rate induced by the constant vorticity is 0.25. Refining the mesh yields 3045 elements, but no apparent change in the solution (still concentric circles). Since the difference between contour values of streamfunction is the volumetric flowrate between the contourlines, and the tangent to the streamline is the direction of flow, the meaning of Figure 1 is clear — constant vorticity leads to recirculating flow in the rotating cavity. Diffusion of vorticity everywhere from the walls leads to the fastest flow near the wall, and the slowest in the center of the circular cavity.

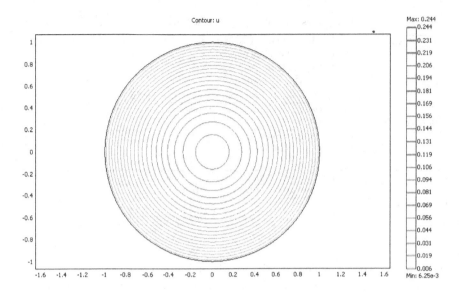

Figure 1.   Streamlines for constant vorticity in a circular cavity with slip.

Now, for the other case: point source at the origin. We could conceivably approximate a point source by drawing a small circle centered at the origin as a second domain, and in that domain, have $f = 1/\pi R^2$, where R is the smaller circle's radius, and $f = 0$ outside. Then f integrates to unity, and in the limit $R \to 0$, f approaches the Dirac delta function. However, the limit can be approached more elegantly with the power of finite elements and weak terms.

Pull down the Draw menu and specify a point at the origin. By explicitly creating a distinguished point in the geometry, two things happen — it is possible (but not required) to specify point contributions to the equation system or point constraints on it, as well as the point itself becomes a node in the mesh generated and constrains the mesh thus. Node points of the mesh are easier (and perhaps more accurately estimated) to find the dependent variable values for since no intra-element interpolation is required. It would also be possible to require a finer (or coarser) mesh in the vicinity of a distinguished point in the geometry. Follow the steps in Table 2.

Figure 2 shows the streamlines for the weak point source term defined in Table 2. It should be apparent from the value of the streamfunction at the origin that the volumetric flow rate is higher than in the distributed source

Table 2. Streamlines for a point vortex at the center of a circular cavity.

Draw Menu	Specify objects: Point
	Set point at $x = 0$, $y = 0$. OK
Physics:	Point selection: Set to 3 (origin)
Point settings	Under week term enter u_test
	OK
Physics Menu: Subdo-	Select domain 1
main settings	Set $c = 1$; $f = 1$
	OK
Meshing	Click on triangle symbol to mesh
Solve Menu	Advanced tab: Change solution form to weak
Solver parameters	OK
Post-processing	Click on the point at the origin. Report window:
	Value: 0.805017, Expression: u, Position: (0, 0)

example. There is substantial bunching of the contour lines at the origin, and the circular symmetry is not quite respected. Certainly the concentric circles are lost in the approach to the origin. The reader might certainly wish to know what "u_test" is, and how "weak point terms" translate into point sources. The discussion of weak terms and weak constraints is delayed until later in this chapter when the finite element method is more fully explained. It should suffice, however, to mention here that one of the great strengths of the finite element method, illustrated here, is the treatment

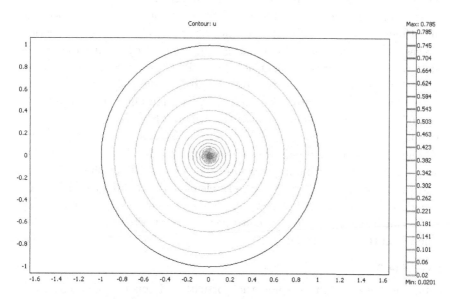

Figure 2. Streamlines for a point vortex at the center of a circular cavity.

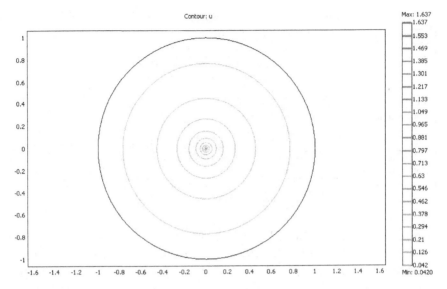

Figure 3.   Streamlines for a point vortex at the center of a circular cavity, adapted mesh.

semi-analytically of discontinuities such as the Dirac delta function here and step functions. These arise naturally in engineering dynamics due to composite materials or material property changes at interfaces.

Refining the mesh to 2986 elements does not visually improve the smoothness of the circular contours, however, the maximum streamfunction increases from 0.805 to 0.915. Greater improvement comes from adapting the mesh, with closer refinement near the origin. Now pull down the Mesh menu and select the Parameters option. Under the point tab, enter for vertex 3 maximum element size 0.001. Upon remeshing, this should give ca. 1658, but with high density near the origin. This gives $\psi = 1.57$ at the origin.

Figure 2 has a maximum streamfunction of 1.57. Remeshing to 6632 elements achieves maximum streamfunction of 1.68. Remeshing again to 26528 elements results in 1.79. 106112 element remeshing gives 1.90. Although it is not clear that grid convergence is ever achieved, the qualitative arrangement of streamlines has converged as the swirling falls off with distance from the source rapidly.

Exercise 2.1.

Solve for the streamlines when the vorticity falls off exponentially with radius, i.e.

$$\omega = \omega_0 \exp(-\sqrt{x^2 + y^2})\,.$$

2.2. *The diffusion equation: A parabolic PDE*

The 1-D unsteady diffusion equation can refer equally well to any of three common transport phenomena:

$$\text{Mass diffusion :} \quad \frac{\partial c}{\partial t} = D \frac{\partial^2 c}{\partial x^2}\,,$$

$$\text{Thermal conduction :} \quad \frac{\partial T}{\partial t} = \frac{k}{\rho c_p} \frac{\partial^2 T}{\partial x^2}\,,$$

$$\text{2-D Vorticity transport :} \quad \frac{\partial \omega}{\partial t} = \nu \frac{\partial^2 \omega}{\partial x^2}\,,$$

where c, T, and ω are concentration, temperature, and the z-component of vorticity in a 2-D flow, respectively, and their corresponding diffusivities are D, α, and ν. This equation is thoroughly studied in the undergraduate curriculum. It has solutions by Fourier and Laplace transforms, and similarity solutions for initial and boundary conditions that collapse on the variable $\eta = \frac{x}{\sqrt{Dt}}$. That doesn't leave much room for finite element methods — just another technique for a tired old problem, right? Wrong. COMSOL Multiphysics can still give this problem a boost which is not commonly considered. COMSOL Multiphysics solutions are well suited to nonconstant coefficients, i.e. transport properties that depend on the field variable. For instance, for suitably low pressures and high temperatures, a gas must satisfy the ideal gas law:

$$\rho = \frac{nM}{V} = \frac{PM}{RT}\,, \tag{4}$$

where R is the gas constant and M is the relative molecular mass of the species. Under these conditions, it is rare to find a gas that has a constant heat capacity. For instance, over a range of temperatures, the heat capacity of CO_2 gas is well approximated by a quadratic in temperature (in MJ/kg-mol°C), with T in °C:

$$c_p = 36.11 + 0.04233T - 2.887 \times 10^{-5}T^2\,. \tag{5}$$

It follows that

$$\frac{k}{\rho c_p} = \frac{kR}{36.11PM} f(T)\,,$$

$$f(T) = \frac{(T + 273)}{1. + 1.172 \times 10^{-3}T - 7.995 \times 10^{-7}T^2}\,. \tag{6}$$

Table 3. Heat transfer in a nonlinear medium — Diffusivity varies with temperature. Steady state analysis.

Model Navigator	Select 1-D. COMSOL Multiphysics|PDE Modes|Classical PDEs|Heat Equation Keep defaults for dependant variables and element type. OK
Draw Menu	Specify objects: Line. Set $x : 0\ 1$ Set Name: interval. OK
Physics Menu: Boundary settings	Select domain 1 Check Dirichlet and $h = 1$; $r = 500$ ($T = 500$) Select domain 2 Check Dirichlet and $h = 1$; $r = 400$ ($T = 400$) OK
Options Menu: Constants	Name Expression $a1$ 1.172E−3 $a2$ 7.995E−7 F400 421.5 OK
Options Menu: Scalar Expressions	Name Expression FofT $(u + 273)/(1 + a1 * u + a2 * u^2)/F400$ OK
Physics Menu: Sub- domain settings	Select domain 1 Set $f = 0$; $c = 1$; $d_a = 1/\text{FofT}$ Select Init tab; set $u(tw) = 500 - 100 * x$. OK
Meshing	Click on the triangle on the button bar to mesh Click twice on the "Refine mesh" button
Solver	Click on the solve (=) button to solve (stationary linear solver)
Post-processing: Click on 0.5	Value: Value: 450, Expression: u, Position: (0.5)

Suitable scalings for time and position,

$$\tau = t\frac{kR}{36.11PML^2}\,, \qquad X = \frac{x}{L}\,, \tag{7}$$

substituted into the thermal conduction equation yields this simple form:

$$\frac{\partial T}{\partial \tau} = f(T)\frac{\partial^2 T}{\partial X^2}\,. \tag{8}$$

Now, let's consider conduction across a stagnant CO_2 gas layer of length L where horizontal boundaries are held constant at $400°C$ and $500°C$.

Start up COMSOL Multiphysics and enter the Model Navigator. Follow Table 3. This application mode gives us one dependent variable u in a 1-D space with coordinate x. Click on the mesh triangle on the toolbar, and then refine the mesh twice to 60 elements. Click on the = button on the toolbar to solve (stationary linear solver). The steady state solution is unchanged from the initial condition, since for long enough times, the accumulation term is immaterial. It certainly is for the stationary solver. So we can only

Table 4. Heat transfer in a nonlinear medium — Diffusivity varies with temperature. Transient analysis.

Physics Menu: Sub-domain settings	Init tab; set $u(t_0) = 400$ OK
Solve Menu Solver parameters	General tab. Select time-dependent solver Set out put times: 0: 0.01: 0.2 Advanced tab. Set solution form: General. OK Solve (=)
Post-processing: Domain plot	Accept defaults. OK
Post-processing: Cross-section plot parameters	Point tab. Set x: 0.5. OK

influence the transient solution with a temperature dependent diffusivity. So change the initial condition to $u(t_0) = 400$, i.e. the left boundary jumps to $u = 500$ to define time $t = 0$. Now pull down the Solve menu and select the Parameters option. This pops up the Solver Parameters dialog window. Fill in the changes in Table 4 for the transient analysis. The two post-processing steps are used to create Figures 4 and 5 which show the approach to the steady state (nearly linear) profile and the rise of the mid-point temperature toward the average.

Since Figures 4 and 5 show the rate of advance of the diffusive front, the important question is the role of the nonlinear medium. In particular, since diffusivity increases with temperature, we find that the profile reaches steady state more rapidly than with constant diffusivity. The self-similarity with $\eta = \frac{x}{\sqrt{Dt}}$ is not apparent in Figure 4, as the higher temperatures home in on the steady-state linear profile faster than the lower temperatures. Figure 5 shows the rise in temperature to the steady state value at the midpoint of the domain, which has the expected s-shape, but again rises faster than expected at short times.

Exercise 2.2.

Solve for the same plots as Figures 4 and 5 with constant diffusivity $f(T) = 1$. How do the profiles differ?

2.3. *Wave equations: Hyperbolic PDEs*

The canonical 1-D wave equation has also been studied to death. Nor does it particularly turn up in chemical engineering applications. The obvious place is the study of sound waves, which receives little attention in the

u

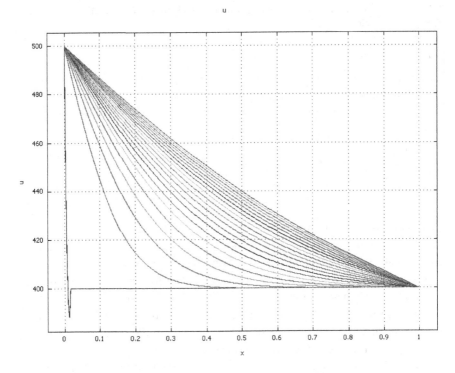

Figure 4.   Temperature profiles from $t = 0$ to $t = 0.2$.

chemical engineering curriculum. Does this mean that the wave equation is unimportant in the chemical and process industries? More complicated wave equations or nonlinear evolution equations turn up all the time. For instance, reactors are known to exhibit chemical waves, waves on interfaces in condensers, swirl atomizers, and distillation columns effect mass transfer, and acoustics, power ultrasound, and sonochemistry are receiving much attention on the research front. It just so happens that chemical engineers are taught little about waves, and thus it is difficult to find classical textbook analyses of chemical engineering unit operations in which waves play any role.

We will leave the canonical 1-D wave equation for an exercise. As an example here, let's introduce the Korteweg-de Vries equation for wave propagation instead:

$$\frac{\partial u}{\partial t} + 6u\frac{\partial u}{\partial x} + \frac{\partial^3 u}{\partial x^3} = 0\,. \tag{9}$$

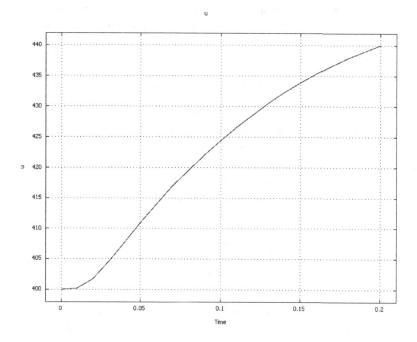

Figure 5.   Temperature at position $x = 0.5$.

This is the wave equation for surface waves on shallow layers — typically what you get from skipping stones — and is ubiquitous in physics since it can arise in any wave propagation media that does not dissipate (much) and is "shallow" in some sense. The authors have been involved in recent derivations of similar wave equations in geophysical [3] and surface coating applications [4]. In both of these works, the numerical analysis was conducted with purpose-built simulations using the method of lines (MOL). MOL discretizes the spatial derivative terms using finite differences, thereby creating a set of ordinary differential equations in time to be solved simultaneously for the values of u at the grid points. As we shall learn later in this chapter, COMSOL Multiphysics implements the finite element method for time dependent partial differential equation systems in a similar way. Rather than ODEs for the values of u at the grid points, COMSOL Multiphysics solves ODEs for the generalized "degrees of freedom" for the problem, from which it is usually a simple task to construct the value of the dependent variable u anywhere. We did not consider FEM-LAB as a sufficient tool for modelling the wave equations like (9) derived in

[3] and [4] for technical reasons — we wanted precise control over the numerical integration procedure, adopting either stiff system or fully implicit approaches for the time integration, since this class of equations can readily become numerically unstable. COMSOL Multiphysics, however, does have robust solvers for time integration and automatically selects stiff solvers. So out of curiosity, we decided to give COMSOL Multiphysics a crack at the Korteweg-de Vries equation for soliton propagation.

One big reason that we thought of using COMSOL Multiphysics where FEMLAB was previously rejected is the treatment of higher order spatial derivatives. $\frac{\partial^3 u}{\partial x^3}$ is unavailable in FEMLAB, to our knowledge. It u_{xxx} is not defined in COMSOL Multiphysics either, but u_{xxt} is, which got us thinking about how to construct u_{xxx}. Our idea is to introduce an auxiliary variable v, creating a system equivalent to (9).

$$
\frac{\partial u}{\partial t} + \frac{\partial}{\partial x}[3u^2 + v] = 0\,,
$$
$$
v = \frac{\partial^2 u}{\partial x^2}\,.
$$

(10)

The first equation is in general, conservative form. The second equation is an elliptic equation for v. So the overall system is a differential-algebraic equation in that not all variables are evolutionary. Nevertheless, the original system is hyperbolic in character.

Since waves are intended to propagate hypothetically indefinitely, ideally they should be simulated in an infinite media. Since no computer has infinite memory, it is more sensible to make the medium large enough the boundary effects are negligible, and then introduce periodic boundaries. Here we will simulate a single soliton propagating according to (9). This isolated or localized structure is compact enough that if the domain is long enough, it does not interact with its periodic image. We'll see this by animating the solution. The single soliton solution to (9) is known analytically to be:

$$
u(x, t) = 2a^2 \text{sech}^2(ax - 4a^3 t)\,.
$$

Here, a is the amplitude parameter. Note that the propagation rate (ratio of coefficients of t to x) depends on a^2. Consequently, larger waves propagate faster. This is a consequence of the nonlinearity of the Korteweg-de Vries equation. It is necessary to know the initial value of this soliton to simulate it, as well as its second spatial derivative to define v initially:

$$
v(x, t = 0) = 4a^4(\cosh(2ax) - 2)\text{sech}^4(ax)\,.
$$

Table 5. Nonlinear wave evolution example with the Korteweg-de Vries soliton propagation.

Model Navigator	Select 2-D. COMSOL Multiphysics|PDE Modes|General Mode. Set two dependent variables: u v Set Element: Lagrange-*Cubic* OK
Draw Menu	Specify objects:Line x: -10 10. Name: interval OK
Options Menu Constants	Define a constant $a = 1$. We might want to change this to study amplitude effects
Options Menu Scalar Expressions	Define expressions for initial$u$2 $* \hat{a}2 * (\text{sech}(a*x))$ʃ2 initial$v$4 $* \hat{a}4 * (-2 + \cosh(2*a*x)) * (\text{sech}(a*x))$ʃ4
Physics Menu: Boundary settings	Make both endpoints have Neumann conditions with $G_1 = G_2 = 0$
Physics Menu Periodic conditions	Select boundary 1. Enter expression: u. Click in the constraint name box to automatically generate constraint name pconstr1. Destination tab. Check box for boundary 2. Check box for "Use selected boundaries as destination." Enter expression u. Source vertices tab. Select boundary 1 and add across with $>>$. Destination tab. Select boundary 2 and add across with $>>$. Return to Source tab and repeat all steps substituting "v" for "u" for pconstr2
Physics Menu: Sub-domain settings	Select domain 1 Set $\Gamma_1 = 3 * \hat{u}2 + v$; $\Gamma_2 = ux$; $F_1 = 0$; $F_2 = v$; da (v_2 entry) $= 0$; Init tab. $U(t_0) = $ initial$u$; $v(t_0) = $ initial$v$. OK
Mesh Menu Mesh parameters	Global tab. Set maximum element size 0.02 Remesh. OK
Solve Menu Solver parameters	General tab. Set output times to 0:0.01:3 Advanced tab. Set contraint handling to Lagrange Multipliers. Set Null space function to orthonormal. Set solution form to weak. OK Click on the Solve (=) button
Post-processing: Plot parameters	Select line tab. Specify either u (default) or v as the plotted variable. Animate tab. Start animation

Start up COMSOL Multiphysics and enter the Model Navigator. Table 5 contains the set up information for the soliton propagation model.

Figures 6 and 7 show the initial states of the soliton solution and its Laplacian for the evolution of the solution to (10). The animation described in Table 5 provides a clear example of periodic boundary conditions. It should be noted that *positive* amplitude waveforms travel rightward in the Korteweg-de Vries system (9), but *negative* amplitude waves travel leftward. Since there is only a first order derivative in time, it is not possible for the

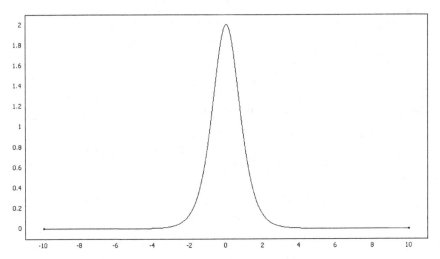

Figure 6.   Initial waveform u for the soliton propagation solution to the KdV equation.

Korteweg-de Vries equation to admit travelling waves in both directions, such as the canonical wave equation.

Exercise 2.3.

The canonical wave equation, $\frac{\partial^2 u}{\partial t^2} = \frac{\partial^2 u}{\partial x^2}$, has its own application mode under Classical PDEs. This application mode treats second spatial and temporal derivatives. Alter the boundary condition (Neumann BCs) and implement periodic boundary conditions as in Table 5 for the canonical wave equation. Try the initial conditions $u(t_0) = \sin(10 * Pi * x)$ and $u\_t(t_0) = -10 * Pi * \cos(10 * Pi * x)$. What do you expect to see in the animation for $u(t)$? $u\_t(t)$? Did anything unexpected occur? Note that COMSOL Multiphysics has a built-in constant pi.

Exercise 2.4.

Try solving the Korteweg-de Vries model for $a = 2$. The amplitude factor results in an a^2 factor for the increase in magnitude and a further a^2 factor increase in propagation speed, theoretically. What happens to your simulation? Any suggestion about how to overcome this difficulty?

3. The finite element method

By now, you must be wondering how COMSOL Multiphysics actually accomplishes this magic of solving PDE systems. Finite element analysis has

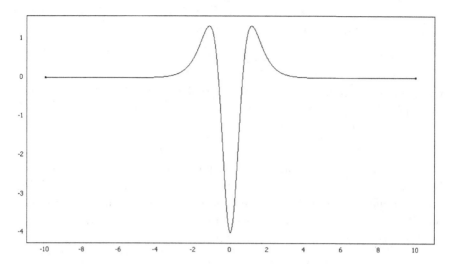

Figure 7. Initial waveform v for the Laplacian of the soliton propagation solution to the KdV equation.

been around for several decades, and has had commercial packages available since the 1980s. A good introduction can be found in the book of Reddy [5]. It is not the intention here to describe FEM in any great detail, nor to describe the full COMSOL Multiphysics implementation, but rather to give an impression of the type of calculations that occur in FEM, and an understanding of why COMSOL Multiphysics is a particularly convenient tool for implementing FEM.

The essence of the finite element method is to state any constraints on the field variables in *weak form*. To understand what a weak form is (and why mathematicians termed it weak), it should be understood that the *strong form* of a system of constraints is the partial differential equation system and appropriate boundary conditions. Why is it strong? Because the field variables are required to be continuous and have continuous partial derivatives up through the order of the equation. That is a strong requirement. The weak form places a weaker restriction on the functions that could satisfy the constraints — discontinuities must be integrable.

To see the equivalence between a PDE and its weak form, consider a stationary PDE for a single dependent variable u in three spatial dimensions in a domain Ω, e.g. the general form:

$$\nabla \cdot \Gamma(u) = F(u) \,. \tag{11}$$

Let's suppose that v, called a test function, is any arbitrary function defined on the domain Ω and restricted to a class of functions $v \in V$. Multiplying (11) by v and integrating over the domain results in

$$\int_\Omega v\nabla \cdot \Gamma(u)dx = \int_\Omega vF(u)dx\,, \tag{12}$$

where dx is the volume element. Upon applying the divergence theorem, we achieve

$$\int_{\partial\Omega} v\Gamma \cdot ndS - \int_\Omega \nabla v \cdot \Gamma(u)dx = \int_\Omega vF(u)dx\,. \tag{13}$$

When the PDE is constrained by Neumann boundary conditions, the boundary term on (13) vanishes. This is one of the reasons that FEM have Neumann natural boundary conditions. Recall in Chapter 1 we observed that finite difference methods have natural Dirichlet conditions. This results in the condition on the volume integral:

$$\int_\Omega [vF(u) - \nabla v \cdot \Gamma(u)]dx = 0\,. \tag{14}$$

This must hold for every $v \in V$. Now for the magic: finite elements and basis functions. Let's suppose that u is decomposed onto a series of basis functions:

$$u(x) = \sum_i u_i\phi_i(x)\,. \tag{15}$$

For instance, if the ϕ_i are sines and cosines with the fundamental and progressive harmonics, then (15) is a Fourier series. Instead, in FEM, the basis functions are chosen to be functions that only have support within a single element, i.e. they are zero in every element but one.

Figure 8 gives an example of two Lagrange linear basis functions in 1-D. Clearly, any function $u(x)$ can be approximated to arbitrary accuracy with piecewise linear basis functions and sufficiently small elements. The basis functions can be taken to be higher order, in which case more than one unknown u_i is needed per element. So the number of unknowns rises with the order of the basis functions. The number of basis function rises with order as well. For Lagrange linear basis functions, the representation of the function in any element is through two basis functions:

$$\phi = \xi\,, \qquad \phi = 1 - \xi\,, \tag{16}$$

where ξ is the local coordinate in the element. So for N elements, there are $2N$ basis functions ϕ_i. Lagrange quadratic elements have three basis

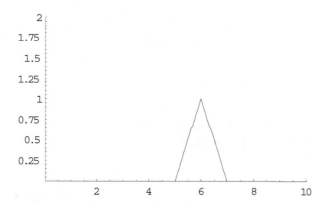

Figure 8. Two piecewise linear basis functions in 1-D on adjacent elements.

functions:

$$\phi = (1 - \xi)(1 - 2\xi), \quad \phi = 4\xi(1 - \xi), \quad \phi = \xi(2\xi - 1). \tag{17}$$

Thus, for Lagrange quadratic elements there are $3N$ basis functions for N elements.

Recall that (14) must be satisfied for all $v \in V$, which we now take to be the function space of all functions that are linear combinations of the basis functions ϕ_i, i.e.

$$v(x) = \sum_i v_i \phi_i(x). \tag{18}$$

But because v enters (14) linearly, it suffices to show that if (14) is satisfied for each of the basis functions ϕ_i playing the role of v, then it is satisfied for all linear combinations of the basis functions (18), and thus for all $v \in V$. Thus, the condition (18) is equivalent to a system of $(k + 1)N$ equations [(14) for each ϕ_i] in $(k + 1)N$ unknowns (the u_i), where k is the order of the element ($k = 1$ linear, $k = 2$ quadratic, etc.).

Then there is the reason why FEM with COMSOL Multiphysics has such utility. COMSOL Multiphysics automates the assembly of the $(k+1)N$ equations (14). First, we note that $\Gamma(u)$ and $F(u)$ are general, potentially nonlinear, functions of u. So, in general, the solution is not achievable in closed form. In Chapter One, we showed that COMSOL Multiphysics has a built-in nonlinear solver for 0-D problems, i.e. $f(u) = 0$, where u was a single unknown value. The nonlinear solver was based on Newton's Method. The N-dimensional analogue of Newton's Method for the vector equation

$$L(U) = 0,\qquad(19)$$

where U is the vector of unknowns u_i and $L(U)$ is the system of equations found by substituting the basis functions ϕ_i for v in (14), is

$$K(U_0)(U - U_0) = L(U_0),\qquad(20)$$

where $K(U_0)$ is called the stiffness matrix and $L(U_0)$ is called the load vector. The stiffness matrix is the negative Jacobian of L:

$$K(U_0) = -\frac{\partial L}{\partial U}(U_0).\qquad(21)$$

So (20) is now a linear equation for U given the previous approximate solution U_0. Thus, if U_0 were close enough to a solution, the linear equation (20) should find an improved approximate solution U, and this procedure can be iterated until a solution is found to acceptable accuracy. Clearly, the nonlinear solver by Newton's Method is central to COMSOL Multiphysics' PDE solver. Yet COMSOL Multiphysics automates all of the steps involved in generating the finite element analysis of a PDE. It symbolically forms the Jacobian of the nonlinear operator $L(U)$ if it can. If it cannot, it numerically assembles the Jacobian. If the PDE were itself linear, this is not too cumbersome. Yet assembling the stiffness matrix is a Herculean task — it was common that the finite element analysis, both meshing the elements and assembling the stiffness matrix was the central feature of many doctoral studies in the sciences and engineering not too long ago. For new combinations of PDEs, or even variations on the coefficients (quasi-linear rather than constant, for instance, as in §1.2), the bookkeeping to organize the assembly of the stiffness matrix is a daunting task. Furthermore, the sparse solver methods for (20) and time-integration required substantial programming effort to coordinate for a single problem. Yet COMSOL Multiphysics has done it as a set of subroutines (MATLAB functions) that coordinate multiple PDE systems (application modes) seamlessly.

Before explaining the implementation of boundary conditions we shall work through a simple ODE to highlight the concepts we have discussed so far. Following the worked example should guide you through the weak formulation and finding approximate numerical solutions to a given second order ODE.

Exercise 2.5.

A worked example of finite element calculations in detail. To elaborate the concepts described above a simple ODE is solved using the variational principles that forms the core to FEM. The problem is simple. Solve the second order ODE

$$\frac{d^2u}{dx^2} + 4u = 8x^2 \tag{22}$$

subjected to boundary conditions

$$u(0) = u\left(\frac{\pi}{4}\right) = 0$$

using the weak formulation.

This simple second order ODE has an analytic solution (Prove it!)

$$u(x) = \cos(2x) + \left(1 - \frac{\pi^2}{8}\right)\sin(2x) + 2x^2 - 1. \tag{23}$$

Since we know the analytic solution, a comparison would give the error of the approximate solution we find.

The first step in developing the weak formulation is to assume a weight function and a trial function. Take $U(x)$ as the trial function and $\phi(x)$ as the weight function. We discuss the exact forms of these functions later. The trial function $U(x)$ forms a solution to the ODE. Therefore, if we substitute $U(x)$ in (22), the resulting equation gives the residual:

$$R = \left|\frac{d^2U}{dx^2} + 4U - 8x^2\right|. \tag{24}$$

Subsequent steps really amount to the minimization of this residual. The minimization process starts by evaluating the *weighted residual*. To evaluate the weighted residual, multiply (24) by $\phi(x)$ and integrate over the domain (i.e. $0 \leq x \leq \pi/4$).

$$R(x) = \int_0^{\pi/4}\left(\phi\frac{d^2U}{dx^2} + 4\phi U - 8\phi x^2\right)dx. \tag{25}$$

Using integration by parts, one can simplify above equation to obtain

$$R(x) = \int_0^{\pi/4}\left(4\phi U - \frac{d\phi}{dx}\frac{dU}{dx}\right)dx - 8\int_0^{\pi/4}\phi x^2 dx. \tag{26}$$

To advance further we need to make some crucial assumptions. Since we are free to assign any function to $U(x)$ and $\phi(x)$ as far as they agree with the boundary conditions, we assume $U(x) = \phi(x)$. This is known as Galerkin's method. If $U(x) \neq \phi(x)$, then it gives the Rayleigh-Ritz formulation. We have to select an algebraic function of x to satisfy the boundary conditions $u(0) = u(\pi/4) = 0$.

$$\varphi(x) = U(x) = \phi(x) = u_1\varphi_1 + u_2\varphi_2 + \cdots + u_N\varphi_N = \sum_{i=0}^{N} u_i\varphi_i. \tag{27}$$

We assume N functions as follows.

$$\varphi_1 = x\left(\frac{\pi}{4} - x\right), \qquad \varphi_2 = x^2\left(\frac{\pi}{4} - x\right), \qquad \ldots, \qquad \varphi_N = x^N\left(\frac{\pi}{4} - x\right). \tag{28}$$

Therefore

$$\varphi(x) = \sum_{i=1}^{N} u_i x^i \left(\frac{\pi}{4} - x\right). \tag{29}$$

This selection satisfies the boundary conditions regardless the number of terms included in the series. Since $U(x) = \phi(x)$, the weighted residual becomes

$$R(x) = \int_0^{\pi/4} \frac{1}{2}\left(4\varphi^2 - \left(\frac{d\varphi}{dx}\right)^2\right) dx - 8\int_0^{\pi/4} \phi x^2 dx. \tag{30}$$

By substituting (29) in to (30) and evaluating the integral we obtain an expression for R independent of x. However, that expression contains N unknowns, $\mathbf{u}_i$. We have to evaluate values of the $\mathbf{u}_i$ so that the weighted residual is minimum.

From (29)

$$\frac{d\varphi}{dx} = \sum u_i \left(\frac{\pi}{4}ix^{i-1} - (i+1)x^i\right). \tag{31}$$

Therefore we have

$$R(u_i) = 2\sum u_i u_j \left(\frac{\pi}{4}\right)^{i+j+3}\left(\frac{1}{i+j+1} - \frac{2}{i+j+2} + \frac{1}{i+j+3}\right)$$

$$- \frac{1}{2}\sum u_j u_j \left(\frac{\pi}{4}\right)^{i+j+1}\left(\frac{ij}{i+j-1} - \frac{2ij+i+j}{i+j} + \frac{(i+1)(j+1)}{i+j+1}\right)$$

$$- 8\sum u_i \left(\frac{\pi}{4}\right)^{i+4}\left(\frac{1}{i+3} - \frac{1}{i+4}\right), \quad i,j = 1,2,3,\ldots,N. \tag{32}$$

Then we minimize the residual by taking derivatives of $\mathbf{R}$ w.r.t $\mathbf{u}_j$. For predetermined number N, this results in N algebraic equations that have to be solved simultaneously.

$$\frac{dR(u_i)}{du_j} = 0. \tag{33}$$

For $N = 2$ there are only two unknowns; u_1 and u_2. It produces two linear equations.

$$0.122u_1 + 0.048u_2 = -0.120 \,,$$
$$0.048u_1 + 0.033u_2 = -0.063 \,. \tag{34}$$

In matrix form

$$\begin{bmatrix} 0.122 & 0.048 \\ 0.048 & 0.033 \end{bmatrix} \begin{Bmatrix} u_1 \\ u_2 \end{Bmatrix} = \begin{Bmatrix} -0.120 \\ -0.063 \end{Bmatrix} \,. \tag{35}$$

It resembles the general form

$$[K]\{u\} = \{L\} \,, \tag{36}$$

where K is the Jacobian (stiffness matrix), and $\mathbf{u}$ is the vector of unknowns. L is the forcing vector (load vector). Solution to (35) gives us

$$u_1 = -0.554579 \,,$$
$$u_2 = -1.112560 \,.$$

Therefore the solution to (22) is

$$u(x) = -0.554579x \left(\frac{\pi}{4} - x \right) - 1.11256x^2 \left(\frac{\pi}{4} - x \right) \,. \tag{37}$$

If we go one step further by assuming $N = 3$, then we get three algebraic equations with three unknowns; u_1, u_2 and u_3. The resulting matrix equation is

$$\begin{bmatrix} 0.1216 & 0.0477 & 0.0228 \\ 0.0477 & 0.0328 & 0.0200 \\ 0.0228 & 0.0200 & 0.0139 \end{bmatrix} \begin{Bmatrix} u_1 \\ u_2 \\ u_3 \end{Bmatrix} = \begin{Bmatrix} -0.120 \\ -0.630 \\ -0.351 \end{Bmatrix} \,. \tag{38}$$

The solution for (38) is

$$u_1 = -0.588 \,, \qquad u_2 = -0.838 \,, \qquad u_3 = -0.349 \,.$$

Therefore the new solution for (22) becomes

$$u(x) = -0.588x \left(\frac{\pi}{4} - x \right) - 0.838x^2 \left(\frac{\pi}{4} - x \right) - 0.349x^3 \left(\frac{\pi}{4} - x \right) \,. \tag{39}$$

Figure 9 shows the plots of (37) and (39) together with the analytic solution (23). As we can clearly see, the approximate algebraic solutions can achieve good agreement with the exact solution if more terms of the series are included.

If you worked the example, by now you have a clear idea of the weak formulation of a solution. In next section we discuss the implementation of the boundary conditions in COMSOL Multiphysics.

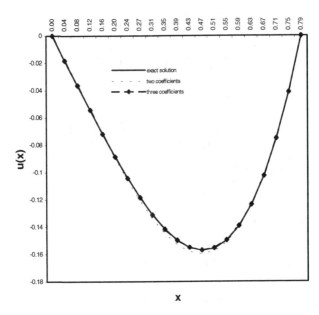

Figure 9. Plot of solutions to (39). Algebraic function with 3 components of the series is in good agreement with the analytic solution.

3.1. *Boundary conditions*

As described for the canonical case above, one should note that the stiffness matrix K is equivalent to Neumann boundary conditions. As we saw in Chapter One, pure Neumann conditions lead to a singular stiffness matrix, which COMSOL Multiphysics could not directly treat, since it resulted in the addition of an arbitrary and large constant to the solution found by projection methods on to the eigensystem of the stiffness matrix. One of the vagaries of FEM is the treatment of boundary conditions.

We could propose to treat boundary conditions much as is done with finite difference methods. The appropriate lines of the matrix equation are replaced by direct constraints on the unknowns u_i so that the order of the matrix is preserved. This has the unpleasant effect of breaking the sparsity of the stiffness matrix with regard to some boundary conditions, and thus artificially requiring full matrix solvers that are much less accurate and inefficient by comparison to sparse solvers for the same matrix equation. FEM has an elegant solution using Lagrange multipliers — a well known method for dealing with equality constraints in optimization problems. Suppose in addition to the PDE constraints, we have a series of boundary conditions that are to be satisfied in weak form for all $v \in V$. By applying the basis function expansion and writing the boundary integrals for each basis function, by analogy to the PDE constraints (19), we arrive at a vector equation for the boundary constraints:

$$M(U) = 0. \tag{40}$$

This constraint residual equation, as it is known, need not be N equations. Usually it is just a handful of equations in N unknowns, as not all basis vectors taken as test functions v contribute a boundary constraint. The linearized version of (40) reads similarly to (20):

$$N(U_0)(U - U_0) = M(U_0), \tag{41}$$

where N is the negative Jacobian of M:

$$N(U_0) = -\frac{\partial M}{\partial U}(U_0). \tag{42}$$

Now for the clever trick. The stiffness matrix equation is augmented by a vector of unknowns Λ, called the Lagrange multipliers, which multiply N^T, where the superscript T means transpose:

$$K(U_0)(U - U_0) + N(U_0)^T \Lambda = L(U_0). \tag{43}$$

Why is this clever? Well, if the constraint (40) is satisfied, then there is a unique set of Lagrange multipliers satisfying (43). (41) and (43) permit the simultaneous solution of more than just boundary conditions, however. Any constraint — internal pointwise, subdomain integral, edge or boundary that can be expressed in weak form can be treated by the Lagrange multiplier method.

Lagrange multipliers

So how do Lagrange multipliers ensure that $M(U) = 0$ is satisfied? By a variational principle. With $\Lambda = 0$ (43) is equivalent to the minimum principle for

$$\min_U \left\{ \frac{1}{2} U^T K U - U^T L \right\}. \tag{44}$$

If we wish to ensure the constraint (40) is satisfied simultaneously, then we add a weighted penalty to (44) for the extent to which $M(U) = 0$ is not satisfied. The weights are called Lagrange multiplers Λ.

$$\min_U \left\{ \frac{1}{2} U^T K U - U^T L + \Lambda \cdot M \right\}. \tag{45}$$

Now we use the linearization of $M(U) = M(U_0) + N(U_0)(U - U_0)$ to simplify (45). Note that constant terms do not contribute to the minimization.

$$\min_U \left\{ \frac{1}{2} U^T K U - U^T L + \Lambda \cdot N U \right\}. \tag{46}$$

The minimization (46) is then equivalent to (43), i.e. the solution to (43) minimizes (46). In the parlance of FEM, (46) is the "minimization in the energy," i.e. weighted by the stiffness matrix K. It should not be confused with the least squares minimization, which by analogy with (45) is

$$\min_U \left\{ \frac{1}{2} \|L(U)\|^2 + \Lambda \cdot M \right\}. \tag{47}$$

Linearization of L and M leads, after much re-arrangement and neglect of constant terms, to the condition

$$\min_U \left\{ \frac{1}{2} U^T K^T K (U - U_0) - U^T K^T L + U^T N^T \cdot \Lambda \right\}. \tag{48}$$

Thus, the solution U to (48) by a theorem in linear algebra, is the solution to the *normal equations* [6]:

$$K^T K (U - U_0) + K^T (K^T)^{-1} N^T \cdot \Lambda = K^T L, \tag{49}$$

which is the *least squares solution* to

$$K(U - U_0) + (K^T)^{-1} N^T \cdot \Lambda = L. \tag{50}$$

So the solution to (50) ensures that the constraint (40) $M(U) = 0$ is satisfied in the least squared error sense (47), whereas the constraint (43) satisfies (40) in the sense of lowest energy. COMSOL Multiphysics uses (43) rather than (50) for simplicity, rather than (50) for greatest accuracy. The distinction is important as the least squared error minimization (47) is defined for any general nonlinear operator $L(U)$, but the stiffness energy (45) is only sensible for K that is symmetric and positive definite. If this is not the case, then the Lagrange multipliers in (43) are merely a convenience, not a guarantee that the constraint (40) is satisfied in any approximate sense. (50) is a stronger condition, yet at the price of extra matrix manipulations. (47) is open to the criticism that $M(U)$ is not constrained to be a penalty, so a stronger condition is to explicitly consider each constraint as a penalty individually [7]

$$\min_U \left\{ \frac{1}{2} \|L(U)\|^2 + \sum_j \Lambda_j \|M_j(U)\|^2 \right\}. \tag{51}$$

This technique does not render its solution so succinctly as a matrix equation, as the constraint term involves N, M, K and Λ.

Weak terms

So if you were wondering how we treated the point source of vorticity in §2.1, it was by a weak term, which merely evaluated the integral

$$\int_{\Omega} v\delta(x)dx = v|_{x=0} \tag{52}$$

and made the appropriate contributions to the stiffness matrix and load vector in (44).

3.2. Basic elements

Fundamental to the FEM is the concept that any domain can be implemented as a collection of smaller subdomains of preferred shape. These subdomains are called *finite elements*. Corners of an element are called nodes at which the solutions to field variables are computed. There can be nodes in between corner points that are commonly called *edge nodes*. In COMSOL Multiphysics, when you generate the mesh, it subdivides the computational domain in to a selected form of elements and form of nodes accordingly. One can find more than a hundred types of elements in use. If you are a beginner, it is natural to be puzzled over the type of elements that should be used and the number of elements to be used.

The discretization process proscribes the type and the number of elements. The number of elements is directly connected with the accuracy of the solution. The higher the number of elements used, the lesser will be the error. However, having a large number of elements would be computationally expensive, demanding a large chunk of RAM and an extended runtime.

Defining an unnecessary number of elements is a very common practice. There is no formula that allows you to choose optimally exact number of elements. It is only by experience that you would be able to decide the right amount of elements to pack in a domain. Though the accuracy increases with the number of elements N, there will be a certain number N_c beyond which the sensitivity of accuracy becomes negligible.

Figure 10 shows the normalized error against the number of elements N. The number of elements doubles in each iteration. One can see that the last three points do not make any considerable improvement on the accuracy. However one can perform a few short runs to find out the appropriate

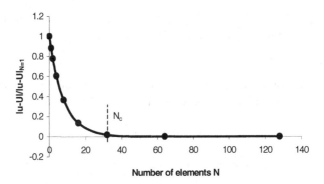

Figure 10.   Normalized error versus number of elements. The increase of N results in improving the accuracy but beyond certain number N_c, the effect become negligible.

number of elements to be used. There are instances where one is interested in a certain region of the domain rather than the whole domain. For instance, take the case of the flow past a cylinder. The boundary layer behavior around the cylinder is to be investigated. In such cases one can and should pack more elements around the cylinder having a lower density of elements in the far field. This way one can attain the accuracy required without increasing the number of elements. COMSOL Multiphysics allows such grid stretching. Another point worth mentioning at this juncture is the skewness of the elements in stretched grids. Skewed elements make the formation of the Jacobian impossible. Therefore great care should be taken in using stretched meshes. The type of the elements to be used depends on the problem that has to be solved. The dimensionality of the domain defines the dimensionality of the elements. The most simple is the 1-D element that represents a line segment between two nodes at each end. The most fundamental element in 2-D is a triangle where as in 3-D it is a tetrahedron. Table 6 shows some of the basic elements in use with respect to the dimensionality. Though we used straight lines, curvilinear segments between nodes would provide a more general form of the elements.

Though the geometry of the element is an important factor, the element types are categorized according to the interpolation polynomials used with them. According to this categorization there are three types of elements: Simplex, Complex and Multiplex. If the polynomials used have linear terms and constants with nodes at the corners, then the elements are called simplex. Complex elements use higher order polynomials (quadratic, cubic, quintic, etc.) with edge and internal nodes together with corner nodes.

Table 6. Basic elements.

Dimensionality	Shape
1-D	
2-D	 Triangle　　　　Rectangle　　　　Quadrilateral
3-D	 Tetrahedral　　Regular Hexahedral　Irregular Hexahedral

Multiplex elements have their sides in parallel with the coordinate axes and use higher order polynomials.

As mentioned above, complex elements use higher order polynomials. The combinations of polynomials and nodal configurations can be determined using the Pascal triangle, Pascal tetrahedron or Pascal hypercubes. Table 7 shows the Pascal triangle with polynomials up to fifth order. The polynomials that are selected should be complete: i.e. it should contain all terms up to the highest order. For example $c_1 + c_2x + c_3x^2$ is complete while $c_1 + c_2x^2$ is not since it does not contain the first order term. For any 2-D element, by taking all the terms above a selected horizontal line, one can easily obtain the complete polynomial up to the required order. In general there should be a node for each term of the polynomial. For example a cubic element should have four nodes along each side. But there can be more complex combinations. Table 8 shows the linear, quadratic and cubic elements for a 2-D triangular element.

COMSOL Multiphysics provides you with two major types of elements irrespective to the geometry, namely Lagrange and Hermite elements. These elements are so named because of the type of the interpolation

Table 7.   Pascal's Triangle for 2-D elements.

1	Constant
x y	Linear
x^2 xy y^2	Quadratic
x^3 x^2y xy^2 y^3	Cubic
x^4 xy^3 x^2y^2 x^3y y^4	Quartic
x^5 x^4y x^3y^2 x^2y^3 xy^4 y^5	Quintic

polynomials used in them. As the name suggests, the Lagrange polynomials are used as the basis functions in Lagrange elements. Suppose a field variable $\mathbf{u}(x)$ is expressed using Lagrange polynomials L_n over a 1-D element. Then,

$$\mathbf{u}(x) = L_1(x)u_1 + L_2(x)u_2 + \cdots + L_N(x)u_N, \qquad (53)$$

where u_n are the unknown coefficients. The $L_n(x)$ is given by

$$L_n = \prod_{M=1,M \neq N}^{n} \frac{x - x_M}{x_N - x_M}$$

$$= \frac{(x - x_1)(x - x_2) \cdots (x - x_{N-1})(x - x_{N+1}) \cdots (x - x_n)}{(x_N - x_1)(x_N - x_2) \cdots (x_N - x_{N-1})(x_N - x_{N+1}) \cdots (x_N - x_n)}. \qquad (54)$$

The expansion generates the polynomials of desired order. Lagrange elements are the most commonly used type in CFD. They provide the value of the variable at nodes.

Hermite elements use the Hermite polynomials to interpolate the values of the field variables. The main difference between Lagrange and Hermite elements is the degrees of freedom (DOF) available. In the case of Lagrange elements DOF are the values of the function at nodes (This consists values of all variables at the node). However in Hermite elements other than the function values at nodes, the first derivatives of the variables at corner points are available. Again, suppose a variable $u(x)$ to be determine over 1-D elements. Since the values at nth and $(n+1)$th nodes, $u(x_n)$ and $u(x_{n+1})$ and the derivatives $du/dx|_i$ and $du/dx|_{i+1}$ are known a polynomial of four unknowns should be used to approximate $u(x)$.

$$u(x) = \alpha_1 + \alpha_2 x + \alpha_3 x^2 + \alpha_3 x^3. \qquad (55)$$

Since x_i and x_{i+1} are known positions, one can easily write four equations:

Table 8. Types of complex elements with basis functions up to third order.

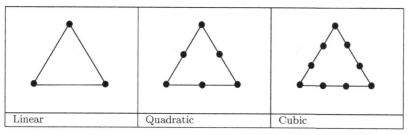

| Linear | Quadratic | Cubic |

two with function values at nodes and two with first derivatives at nodes.

$$
\left\{
\begin{array}{c}
u_i \\
du_i \\
u_{i+1} \\
du_{i+1}
\end{array}
\right\}
=
\left[
\begin{array}{cccc}
1 & x_i & x_i^2 & x_i^3 \\
0 & 1 & 2x_i & 3x_i^2 \\
1 & x_{i+1} & x_{i+1}^2 & x_{i+1}^3 \\
0 & 1 & 2x_{i+1} & 3x_{i+1}^2
\end{array}
\right]
\left\{
\begin{array}{c}
\alpha_1 \\
\alpha_2 \\
\alpha_3 \\
\alpha_4
\end{array}
\right\},
\tag{56}
$$

where du denotes the derivative of u w.r.t. x at the nodes. By matrix inversion α_n can be expressed in terms of nodal values and values of derivatives. The resulting equation is

$$
u(x) = u_i\varphi_1(x) + du_i\varphi_2(x) + u_{i+1}\varphi_3(x) + du_{i+1}\varphi_4(x), \tag{57}
$$

where $\varphi_i(x)$ are the Hermite interpolation functions (cubic functions in this case, also known as cubic splines).

If you observed closely, you could see that the cubic interpolation function in Lagrange elements needs four nodes whereas in Hermite elements only two nodes are required. Hermite elements are commonly used in solving load and stress distributions of trusses.

Further to these two types, COMSOL Multiphysics offers curved mesh elements and Argyris elements. The curved mesh elements are provided to facilitate the approximation of true boundaries with higher accuracy. The Argyris elements are fifth order Hermite elements using nodal values as well as derivatives up to second order. It also uses the normal components of ∇u at the midpoints of the sides. Argyris elements require determination of 21 constants of a quintic polynomial. In addition to predefined elements, COMSOL Multiphysics allows user defined elements. We advise interested readers to consult COMSOL Multiphysics manuals for detailed description on how to define a new class of elements.

Here we provided a sufficient description of basic elements to make a beginner comfortable with the jargon and using the elements with some understanding. With COMSOL Multiphysics, one does not need to become

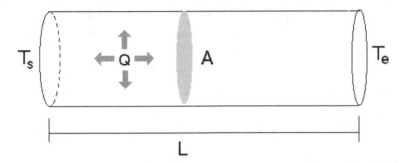

Figure 11.   Axial heat transfer along an insulated rod. Each end has temperatures T_s and T_e. The length of the rod is L and cross sectional area $A = 1$. Heat is generated within the rod at a constant rate of Q.

an expert in meshing techniques and development of elements. As you have already seen, once you generate the domain over which the differential equations are to be solved, meshing is just a click of a button away. However, if a reader is interested in understanding the basic concepts, development of elements and meshing techniques, refer to [8] and references therein.

Exercise 2.6.

Heat transfer in a uniform media. To explain the formulation and solution methods involved in FEM, a worked example is considered. In Chapter 1 §6.1, heat transfer in a nonuniform medium was treated. Here we consider similar problem with more simplifications. Instead of a variable heat transfer coefficient, we consider a constant coefficient and consider $f(x)$ to be uniform over the length of the domain. The length of the domain $L = 1$. Figure 11 shows the physical system. We assume the heat flow across any cross section (shown as A in Figure 11) normal to the centre axis to be uniform. Therefore the temperature varies only along the axis: hence reduced dimensionality.

The heat transfer in this problem is fully described by the equation Chapter 1, (31). With the uniform cross section $A = 1$, and constant source $Q = f(x)$ in (1.31). The equation becomes:

$$\frac{d}{dx}\left(k\frac{dT}{dx}\right) + Q = 0 ; \qquad 0 \le x \le 1, \tag{58}$$

the data for the problem are:

$$k = 3.3 \text{ J/°Cms},$$
$$Q = 10 \text{ J/sm},$$
$$T|_{x=0} = 1,$$
$$q|_{x=1} = 1.25 \text{ J/m}^2\text{s}.$$

Step 1. Variational formulation

This PDE is the strong form of the equation for heat conduction within a cylinder. The first step in FEM is to derive the weak form of the equations. To derive the weak form, equation (58) is multiplied by a weight function and integrated over the domain.

$$\int_0^1 w \left[\frac{d}{dx} \left(k \frac{dT}{dx} \right) + Q \right] dx = 0. \tag{59}$$

Integrating by parts (using the divergence theorem in 1-D) we obtain

$$\int_0^1 \left(\frac{dw}{dk} k \frac{dT}{dx} \right) dx = \left[wk \frac{dT}{dx} \right]_0^1 + \int_0^1 wQdx. \tag{60}$$

From heat transfer theory, Fourier's law gives the heat flux across a unit cross section is given by Fourier's law $q = -k\frac{dT}{dx}$. Therefore,

$$\int_0^1 \left(\frac{dw}{dk} k \frac{dT}{dx} \right) dx = -[wq]_0^1 + \int_0^1 wQdx. \tag{61}$$

From earlier sections, we know that the polynomial basis functions have to be used to approximate the unknowns w and T. Selection of these polynomials is the second step of the FEM procedure.

Step 2. Discretization and choice of polynomials

It is obvious that we are going to use 1-D elements. We can have simplex elements for simplicity, i.e. linear polynomials to approximate the unknowns. Figure 12 shows the descretization of the cylinder. It uses nodes to divide the length into equal segments. The higher the number of segments, the higher will be the accuracy. However in this example, to demonstrate the FEM formulation we use fewer segments with nodes at each end. The polynomial defined piecewise varies linearly between any two nodes. In general, suppose the temperature $T(x)$ is approximated as

$$T = a + bx. \tag{62}$$

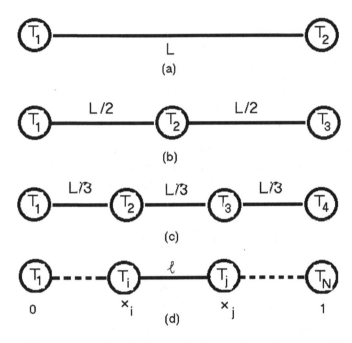

Figure 12.   Discretization of the cylinder. (a) is the schematic representation of the cylinder. (b) and (c) show two and three element discretization. In calculation we use (c). (d) shows the general case where the cylinder is divided to N elements. (c) is used to derive the general form of shape functions.

Now, suppose we have divided the length of the cylinder to $N - 1$ elements with N nodes as shown in Figure 12(d). A random element extending from x_i to x_j is considered. The temperatures at nodes are assumed to be T_i and T_j. From (62) we can write the temperatures at nodes i and j.

$$T_i = a + bx_i \,,$$
$$T_j = a + bx_j \,. \qquad (63)$$

Solving for a and b gives

$$a = \frac{1}{l}(T_i x_j - T_j x_i) \,,$$

$$b = \frac{1}{l}(T_j - T_i) \,, \qquad (64)$$

where $l = x_j - x_i =$ length of the element. Since we know a and b (62) can be rewritten as

$$T^e = \frac{1}{l}(x_j - x)T_i + \frac{1}{l}(x - x_i)T_j \,, \qquad (65)$$

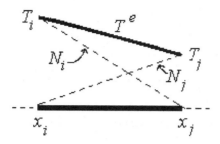

Figure 13.   Profiles of shape functions N_i and N_j and temperature profile T^e along the element constructed using shape functions.

the equation (65) is the linear approximation function for the element. It describes the temperature variation at any point within the element (hence the notation T^e). Instead of a and b, we now have temperature values at the nodes T_i and T_j as unknowns. Let $N_i = \frac{1}{l}(x_j - x)$ and $N_j = \frac{1}{l}(x - x_i)$. Then (65) can be rewritten as

$$T^e = N_i T_i + N_j T_j , \qquad (66)$$

N_i and N_j are known as the shape functions.
$N_i = 1$ at $x = x_i$ and $N_i = 0$ at $x = x_j$.
$N_j = 1$ at $x = x_j$ and $N_j = 0$ at $x = x_i$.
The temperature distribution along the element is determined by these two functions and end values. Figure 13 shows the profiles of N_i, N_j and the resulting temperature T^e. One can generate $T^{e'}$s for all elements. These element shape functions can be used to formulate the global shape functions. Figure 14 shows the definition of the global shape functions.

If we consider the first element there are two local shape functions: N_1^1 which is associated with node 1 and N_2^1 associated with node 2. For the second element again we have a local shape function associated with node 2 defined as N_2^2. Each global shape function is zero elsewhere except in the elements associated with the corresponding nodes. This enables us to define the global temperature variation.

$$T = \sum T^e = N_1 T_1 + \cdots + N_i T_i + \cdots N_N T_N = \sum_{j=1}^{N} N_j T_j . \qquad (67)$$

This completes the specification of the basis functions. Now we can return to the variational formulation again.

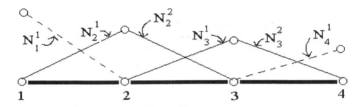

Figure 14.   Global shape functions.

Step 3. Assembling the element equations to form the global problem

In Step 2 we derived the approximation function for T. Galerkin's formulation assumes the weight function to be same as the approximation for the unknown variables. Therefore we have $w = T$. With this, we can substitute T and w in (61).

$$\underbrace{\int_0^1 \left[k\frac{d}{dx}\left(\sum N_i T_i\right)\frac{d}{dx}\left(\sum N_j T_j\right)\right] dx}_{\mathbf{I_k}}$$

$$= \underbrace{-\left[q\sum N_i T_i\right]_0^1}_{\mathbf{I_b}} + \underbrace{\int_0^1 q\sum N_i T_i dx}_{\mathbf{I_s}} . \tag{68}$$

For clarity we consider terms of (68) separately. The equation (68) actually gives the error of the approximate solution (refer to exercise 2.5). To minimise the error, equation (68) should be differentiated w.r.t T_i which are the coefficients of the polynomials. The minimization process converts $\mathbf{I_k}$ into the stiffness matrix $\mathbf{K}$. For four nodes (i.e. three elements as in Figure 12c) we can expand $\mathbf{I_k}$ as below. In what follows we indicate the limits as x_i and x_j. This is to reduce the complications that arising in evaluating the terms. The intervals in the integrals depend on global shape functions.

$$I_k(T)$$

$$= \int_{x_i}^{x_j} \left[\frac{d}{dx}N_1 T_1\left(\frac{d}{dx}N_1 T_1 + \frac{d}{dx}N_2 T_2 + \frac{d}{dx}N_3 T_3 + \frac{d}{dx}N_4 T_4\right)\right] dx$$

$$+ \int_{x_i}^{x_j} \left[\frac{d}{dx}N_2 T_2\left(\frac{d}{dx}N_1 T_1 + \frac{d}{dx}N_2 T_2 + \frac{d}{dx}N_3 T_3 + \frac{d}{dx}N_4 T_4\right)\right] dx$$

$$+ \int_{x_i}^{x_j} \left[\frac{d}{dx} N_3 T_3 \left(\frac{d}{dx} N_1 T_1 + \frac{d}{dx} N_2 T_2 + \frac{d}{dx} N_3 T_3 + \frac{d}{dx} N_4 T_4 \right) \right] dx$$

$$+ \int_{x_i}^{x_j} \left[\frac{d}{dx} N_4 T_4 \left(\frac{d}{dx} N_1 T_1 + \frac{d}{dx} N_2 T_2 + \frac{d}{dx} N_3 T_3 + \frac{d}{dx} N_4 T_4 \right) \right] dx .$$

$$(69)$$

In the minimization process we differentiate $I_k(T)$ w.r.t. each T_i and set each derivative to zero. This procedure generates a number of equations equal to number of nodes. For instance the differentiation of (69) gives

$$\int_{x_i}^{x_j} \frac{dN_1}{dx} \frac{dN_1}{dx} T_1 dx + \int_{x_i}^{x_j} \frac{dN_1}{dx} \frac{dN_2}{dx} T_2 dx$$

$$+ \int_{x_i}^{x_j} \frac{dN_1}{dx} \frac{dN_3}{dx} T_3 dx + \int_{x_i}^{x_j} \frac{dN_1}{dx} \frac{dN_4}{dx} T_4 dx .$$

$$(70)$$

Likewise, there will be three more equations. In matrix form it gives the stiffness matrix.

$$\frac{d}{dT} I_k(T)$$

$$= K$$

$$= \begin{bmatrix} \int_{x_i}^{x_j} k \frac{dN_1}{dx} \frac{dN_1}{dx} dx & \int_{x_i}^{x_j} k \frac{dN_1}{dx} \frac{dN_2}{dx} dx & \int_{x_i}^{x_j} k \frac{dN_1}{dx} \frac{dN_3}{dx} dx & \int_{x_i}^{x_j} k \frac{dN_1}{dx} \frac{dN_4}{dx} dx \\ \int_{x_i}^{x_j} k \frac{dN_2}{dx} \frac{dN_1}{dx} dx & \int_{x_i}^{x_j} k \frac{dN_2}{dx} \frac{dN_2}{dx} dx & \int_{x_i}^{x_j} k \frac{dN_2}{dx} \frac{dN_3}{dx} dx & \int_{x_i}^{x_j} k \frac{dN_2}{dx} \frac{dN_4}{dx} dx \\ \int_{x_i}^{x_j} k \frac{dN_3}{dx} \frac{dN_1}{dx} dx & \int_{x_i}^{x_j} k \frac{dN_3}{dx} \frac{dN_2}{dx} dx & \int_{x_i}^{x_j} k \frac{dN_3}{dx} \frac{dN_3}{dx} dx & \int_{x_i}^{x_j} k \frac{dN_3}{dx} \frac{dN_4}{dx} dx \\ \int_{x_i}^{x_j} k \frac{dN_4}{dx} \frac{dN_1}{dx} dx & \int_{x_i}^{x_j} k \frac{dN_4}{dx} \frac{dN_2}{dx} dx & \int_{x_i}^{x_j} k \frac{dN_4}{dx} \frac{dN_3}{dx} dx & \int_{x_i}^{x_j} k \frac{dN_4}{dx} \frac{dN_4}{dx} dx \end{bmatrix}.$$

$$(71)$$

K is a symmetric matrix and Galerkin's method forces this symmetry. $\mathbf{I}_b$ and $\mathbf{I}_s$ in (68) give rise to two $1 \in 4$ matrices:

$$f_b = - \begin{bmatrix} qN_1 |_0^1 \\ qN_2 |_0^1 \\ qN_3 |_0^1 \\ qN_4 |_0^1 \end{bmatrix} \quad \text{and } f_s = \begin{bmatrix} \int_0^1 QN_1 dx \\ \int_0^1 QN_2 dx \\ \int_0^1 QN_3 dx \\ \int_0^1 QN_4 dx \end{bmatrix} .$$

$$(72)$$

The compact equation is $[K]\{x\} = \{L\}$ where $\mathbf{F} = \mathbf{f}_b + \mathbf{f}_s$. The column matrix $\mathbf{f}_b$ contains the boundary terms and $\mathbf{f}_s$ contain the source terms. $\mathbf{x}$ is the vector of unknowns (nodal temperatures in our case). Components in $\mathbf{L}$, $\mathbf{f}_b$ and $\mathbf{f}_s$ have to be evaluated elementwise.

Step 4. Numerical manipulation

As we formulated the global problem in Step 3, the rest is down to matrix manipulation to evaluate the unknowns. As the first step we have to evaluate the components K_{mn} of stiffness matrix $\mathbf{K}$. K_{1n} corresponds to node 1. Therefore N_1^1 and N_2^1 are the only nonzero global shape functions.

$$K_{11} = \int_0^{0.33} k \frac{dN_1^1}{dx} \frac{dN_1^1}{dx} dx$$

$$= \int_0^{0.33} 3.3 \left(-\frac{1}{0.33}\right) \left(-\frac{1}{0.33}\right) dx = 10,$$

$$K_{12} = \int_0^{0.33} k \frac{dN_1^1}{dx} \frac{dN_2^1}{dx} dx$$

$$= \int_0^{0.33} 3.3 \left(-\frac{1}{0.33}\right) \left(\frac{1}{0.33}\right) dx = -10,$$

$$K_{13} = K_{14} = 0.$$

In evaluating terms in the second row we immediately make use of the symmetry of the matrix.

$$K_{21} = K_{12} = -10.$$

Upon evaluating the K_{22} we run into a problem — which shape function to use N_2^1 or N_2^2? The solution is simple. Since those two functions are defined over two elements, we have to integrate relevant function over the appropriate element considering the limits from 0 to 0.66 (or more generally $2l$).

$$K_{22} = \int_0^{0.66} k \frac{dN_2}{dx} \frac{dN_2}{dx} dx = \int_0^{0.33} k \frac{dN_2^1}{dx} \frac{dN_2^1}{dx} dx + \int_{0.33}^{0.66} k \frac{dN_2^2}{dx} \frac{dN_2^2}{dx} dx$$

$$= \int_0^{0.33} 3.3 \left(\frac{1}{0.33}\right) \left(\frac{1}{0.33}\right) dx + \int_{0.33}^{0.66} 3.3 \left(-\frac{1}{0.33}\right) \left(-\frac{1}{0.33}\right) dx$$

$$= 20.$$

K_{23} involves functions defined only over the second element. Hence N_2^2 and N_3^1 are to be considered.

$$K_{23} = \int_{0.33}^{0.66} k \frac{dN_2^2}{dx} \frac{dN_3^1}{dx} dx$$

$$= \int_{0.33}^{0.66} 3.3 \left(-\frac{1}{0.33}\right) \left(\frac{1}{0.33}\right) dx = -10.$$

$K_{24} = 0$ since N_4 does not share node 2. $K_{31} = 0$ according to the same line of reasoning. $K_{32} = K_{23} = -10$ and K_{34} is to be evaluated in the same manner we evaluated K_{23}. Again $K_{41} = K_{42} = 0$ as the shape functions do not share the node in question. $K_{43} = K_{34}$ by symmetry. K_{33} and K_{44} are to be evaluated in the same way we evaluated the K_{22}, considering the relevant shape function over the relevant domain. The completed stiffness matrix is given below.

$$K = \begin{bmatrix} 10 & -10 & 0 & 0 \\ -10 & 20 & -10 & 0 \\ 0 & -10 & 20 & -10 \\ 0 & 0 & -10 & 10 \end{bmatrix}. \tag{73}$$

This is the famous tridiagonal matrix in FEM. In this case it is only 4×4 since we have only four nodes. With full modeling, one would get a huge, sparse matrix of few thousands of components, yet still banded.

The next step is to evaluate the components in $\mathbf{f}_b$ and $\mathbf{f}_s$. In evaluating the terms in $\mathbf{f}_b$ it is important to identify only N_1 and N_4 remain nonzero at $x = 0$ and $x = 1$. In fact $N_1 = N_4 = 1$ at $x = 0$ and $x = 1$. All other shape functions are zero as far as start and end points are concerned. Therefore

$$\mathbf{f}_b = - \begin{bmatrix} qN_1|_0^1 \\ qN_2|_0^1 \\ qN_3|_0^1 \\ qN_4|_0^1 \end{bmatrix} = \begin{bmatrix} q_{x=0} \\ 0 \\ 0 \\ -1.25 \end{bmatrix}. \tag{74}$$

In computing the terms in $\mathbf{f}_s$, the integrals are to be evaluated taking into consideration where the global shape functions are defined.

$$f_s = \begin{bmatrix} \int_0^1 QN_1 dx \\ \int_0^1 QN_2 dx \\ \int_0^1 QN_3 dx \\ \int_0^1 QN_4 dx \end{bmatrix} = \begin{bmatrix} \int_0^{0.33} QN_1^1 dx \\ \int_0^{0.33} QN_2^1 dx + \int_{0.33}^{0.66} QN_2^2 dx \\ \int_{0.33}^{0.66} QN_2^1 dx + \int_{0.66}^{1.0} QN_2^2 dx \\ \int_{0.66}^{1.0} QN_4^1 dx \end{bmatrix} = \begin{bmatrix} 1.65 \\ 3.3 \\ 3.3 \\ 1.65 \end{bmatrix}.$$

$$(75)$$

By putting together (??), (74) and (75), the complete matrix equation is obtained.

$$\begin{bmatrix} 10 & -10 & 0 & 0 \\ -10 & 20 & -10 & 0 \\ 0 & -10 & 20 & -10 \\ 0 & 0 & -10 & 10 \end{bmatrix} \begin{bmatrix} T_1 \\ T_2 \\ T_3 \\ T_4 \end{bmatrix} = \begin{bmatrix} q_{x=0} \\ 0 \\ 0 \\ -1.25 \end{bmatrix} + \begin{bmatrix} 1.65 \\ 3.3 \\ 3.3 \\ 1.65 \end{bmatrix}. \qquad (76)$$

From here onward, matrix manipulation become the main focus. $q_{x=0}$ is to be evaluated once temperatures are estimated. This is possible since T_1 is known *a priori*. We leave solving (76) to the reader. However, after few manipulations we found

$$\begin{bmatrix} T_1 \\ T_2 \\ T_3 \\ T_4 \end{bmatrix} = \begin{bmatrix} 1 \\ 1.7 \\ 2.01 \\ 2.11 \end{bmatrix}.$$

There is an analytic solution for (58). The temperatures at nodes calculated using the analytic solution are $T_1 = 1$, $T_2 = 1.96$, $T_3 = 2.59$ and $T_4 = 2.89$.

Clearly, four element approximate solutions are not particularly accurate. The reader can readily implement this example in COMSOL Multiphysics for arbitrary accuracy. The purpose of this four element worked example is to make concrete all the steps that are automatically done by COMSOL Multiphysics upon specifying the problem (58) and using the default settings and options.

This example discussed the basics of FEM. However we left untouched many important issues. For an in depth study of FEM the reader is referred to [5] and [8]. The example is targeted to give an insight to what happens inside COMSOL Multiphysics when you set the problem and ask it to solve. The availability of software packages like COMSOL Multiphysics greatly reduces the need for understanding the fundamentals of the FEM. Instead of spending a considerable time on learning the method, one can concentrate on solving the problems and physics involved. However, it should be mentioned that an understanding of the core issues in FEM might help in describing the errors and interpreting solutions in some cases.

Exercise 2.7.

Steady state heat transfer in 3-D. In section §1.1 we considered the steady state heat transfer equation with a distributed source, the Poisson equation. Here, we demonstrate the 3-D solution without the source — Laplace's equation. There is nothing particularly new in this example except the demonstration of 3-D modeling. Since all of the models in this book are run on a relatively low performance PC, complicated 3-D modeling would tax its resources. Consequently, this is the only 3-D example in the book. In 3-D modeling it is especially important to conserve memory by taking full advantage of symmetries in your geometry. In this problem, we will model the steady heat transfer within a hexagonal prism with differentially heated (or cooled) basal and side planes. The basal planes are held at the hot temperature ($T = 1$) and the side faces are held at the cold temperature ($T = 0$). Since the steady state solution is sought, the thermal diffusivity is immaterial — as long as the medium is conductive, it achieves the same steady state. Figure 15 shows the 3-D geometry and mesh for our model.

Figure 15 does not resemble the hexagonal prism. But since the differential equation (2) (with $f(x) = 0$) and the geometry admit six-fold periodic symmetry, solutions to (2) on Figure 15 are periodically extendible to the full hexagonal prism. And all solutions to (2) on a hexagonal prism are periodically reducible to a solution on Figure 15. It is important when attempting to exploit geometrical symmetry to make sure that the equation and boundary conditions, in general the entire model, shares the symmetry property. Furthermore, one should be careful about solutions that break symmetries inherent in the model description and domain. Nonlinear problems can admit solutions that break the symmetry of the equations and the boundary and initial conditions — by bifurcations typically. All the different solutions to a nonlinear system must satisfy the symmetry conditions collectively, but may violate them individually. Here we are safe — Laplace's equation is linear and it is not an eigenvalue problem.

Start up COMSOL Multiphysics and enter the Model Navigator. Follow the instructions in Table 9. This application mode gives us one dependent

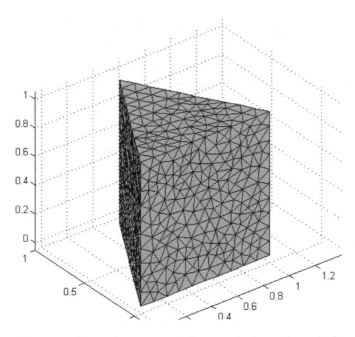

Figure 15.   One-sixth segment of a hexagonal prism and standard mesh.

variable u, in a 3-D space with coordinates x, y, z. Now we are in a position
to set up our domain. The drawing details using a work plan and then
extrusion to create give the equilateral triangular prism of Figure 15. This
is the second easiest drawing technique (extrusion) among those available;
the easiest is to select primitive 3-D objects. All of the interest in Laplace's
equation lies in the boundary conditions. Here, we set up a mixed set
of Neumann boundary conditions for the symmetry edges and Dirichlet
conditions for the fixed temperature faces.

Click on the Solve button ($=$) to arrive at a solution resembling Figure 15. This solution is not, however, the most general periodic solution
possible. To make the solution periodic, we need to alter our boundary
conditions. The conditions on domains 1, 2 (sides) must become periodic
boundary conditions to best approximate hexagonal symmetry. When this
example was first constructed in Femlab 2.2, there was considerable difficulty implementing periodic face conditions across boundary faces 1 and 2,
since the meshes had to be perfectly mapped on each face to machine preci-
sion. Although $\frac{\sqrt{3}}{2} \approx 0.866025$, this is not exact to machine precision. Now,
however, periodic boundary conditions are built-in with data entry on the

Table 9. 3-D hexagonal prism for steady-state heat conduction (Laplace's equation).

Model Navigator	Select 3-D. COMSOL Multiphysics|PDE Modes|Classical PDEs|Laplace's Equation Set Element: Lagrange-Quadratic OK
Draw Menu	Select work plane settings Accept the x–y plane and defaults Enter a triangle with vertices (0, 0), (1, 0), (0.5, 0.866025) by specifying line segments. Select all line segments Use the Draw menu to "coerce to solid" CO1 Select from the Draw menu: extrude. Accept the defaults, in particular the distance 1 in the z-direction
Physics Menu: Boundary settings	Select domain 1, 2 (sides) and choose Neumann Select domain 3, 4 (top, bottom) and choose Dirichlet and set $h = 1$, $r = 1$ Select domain 5 (back) and choose Dirichlet and set $h = 1$, $r = 0$. OK
Solve	Click on the Solve (=) button
Post-processing: Plot parameters	Default slice plot for temperature u is displayed

Table 10. Alteration to periodic boundary conditions for hexagonal symmetry.

Physics Menu: Periodic (boundary) conditions	Source tab: select boundary 1, enter u into the expression field, and click on the constraint name field to generate pconstr1 as the default name Destination tab. Select boundary 2, check box for "use selected boundaries as destination" and enter u into the expression field Source vertices tab: enter 1 2 4 sequentially Destination vertices tab: enter 1 2 6 sequentially OK
Solve Menu	Click on the solve (=) button
Post-processing: Boundary integration	Compute the expression $nx * ux + ny * uy + nz * uz$ across all five boundaries

Physics Menu: Periodic Conditions dialogue box. The periodic conditions are a special case of coupling variables, which are described in detail in chapters four and seven. Table 10 gives the instructions for implementing periodic boundary conditions along the faces 1 and 2. The recipe includes specifying the vertices in each face that are mapped to each other. This gives COMSOL sufficient information to specify the sense of the mapping.

An exercise for the reader. Compute the flux across all boundaries. nx, ny, and nz are the components (built-in) of the normal vector. $n \cdot \nabla u$ is the scalar formula for the normal derivative, which in heat conduction

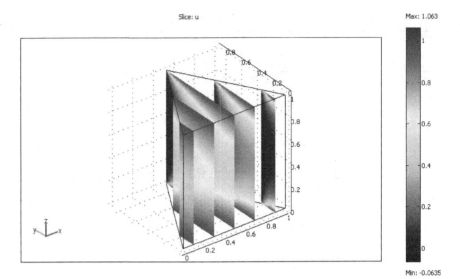

Figure 16.   Temperature profile within a segment of the hexagonal prism (slice plot).

problems is proportional to the normal flux. Typically, the total heat flux
is given by

$$Q = k \int_{\partial \Omega} n \cdot \nabla u \, ds \, .$$

What would you expect the sum of all the fluxes to be theoretically? Is
there an appreciable discrepancy? What do you believe is responsible if
there is? Does refining the mesh help?

4. Summary

The flexibility of COMSOL Multiphysics and FEM analysis in treating
higher dimensional problems and canonical PDEs was explored. The ease
with which point sources, quasi-linear terms, and periodic boundary conditions are treated was demonstrated. An overview of how the stiffness
matrix, load vector, and general (boundary) constraints are dealt with by
the FEM approach was presented. Worked examples of the FEM approach
illustrated the principles.

References

[1] A. Constantinides and N. Mostoufi, *Numerical Methods for Chemical Engi-
     neers with MATLAB Applications* (Prentice Hall, Upper Saddle River NJ,
     1999).

[2] R. B. Freiden, *Physics from Fisher Information: A Unification* (Cambridge University Press, 1998).

[3] W. B. Zimmerman and J. M. Rees, The wavelength of solitary internal waves in a stably stratified fluid layer, *Nonlinear Processes in Geophysics* **11**(2) (2004) 165–180.

[4] B. N. Hewakandamby and W. B. Zimmerman, Dynamics of surface spreading of insoluble surfactants on flowing thin films, *Physics of Fluids*, in press.

[5] J. N. Reddy, *An Introduction to the Finite Element Method* (McGraw-Hill Inc, New York, 1993).

[6] G. Strang, *Introduction to Applied Mathematics* (Wellesley-College Press, Massachusetts, 1986), p. 37.

[7] A. R. Mitchell and R. Wait, *The Finite Element Method in Partial Differential Equations* (Wiley-Interscience, New York, 1977).

[8] T. J. Chung, *Computational Fluid Dynamics* (Cambridge University Press, Cambridge, 2002).

Chapter Three

MULTIPHYSICS

W.B.J. ZIMMERMAN

Department of Chemical and Process Engineering,
University of Sheffield,
Mappin Street, Sheffield S1 3JD, United Kingdom
E-mail: w.zimmerman@shef.ac.uk

Multiphysics is a recent conceptualization to categorize modeling where different physicochemical mechanisms are prevalent in a given application, where these mechanisms are modelled by wholly different field equations. But to be multiphysics, the field equations must couple. In this chapter, we treat models of multiphysics for thermoconvection and nonisothermal chemical reactors as examples of the genre of multiphysics. Applications in later chapters show largely multiphysics modeling since "single physics" models are likely to be well studied in their core disciplines. We also take the opportunity to introduce the concept of parametric continuation, which is an essential mechanism for arriving at the solution to highly nonlinear problems by inching there by starting from nearby solutions in function space or even linear systems. Our thermoconvection model is then altered to treat the differential side wall heating of water between walls held at the freezing and boiling points (without boiling) and the full dependency of buoyant force on temperature. Simulations in large cavities show the beginnings of stratification in temperature. Next we treat a nonisothermal tubular reactor that couples mass and energy transport. Finally, we treat chemical reaction in the pores of a solid pellet with diffusion from a bulk flow.

1. Introduction

COMSOL Multiphysics makes a big selling point of *multiphysics* modelling as a key advantage of its software package. Not long ago I described one of the important features of the burgeoning research area of microfluidics as requiring skills in multiphysics modelling. A respected colleague asked pointedly, "What's that? Physics that happens on multiple scales?" So multiphysics is jargon that may not be uniformly recognized in the sciences and engineering. Not wanting to use the term wildly, we shall define multiphysics modelling here to mean any complete, coupled system of differential equations that has more than one independent variable of different physical dimensions (vector equations count as one equation). The COM-SOL Multiphysics definition is actually an operational definition — "Does COMSOL Multiphysics have a single application mode for it or can you only describe it by coupling more than one application mode?" In COM-SOL Multiphysics' Model Navigator, you can create a multiphysics model by coupling two or more application modes (under the multiphysics tab).

Of course either definition is a Byzantine notion, so let's make it concrete by examples. Are fluid dynamics multiphysics? Yes, but only on the technicality that pressure is an independent variable which has different units to velocity. Is it multiphysics in COMSOL Multiphysics? No, because there is a single Navier-Stokes application mode. Is heat transfer multiphysics? No, there is only one independent variable — temperature, and only one COMSOL Multiphysics application mode. What about thermofluids? Yes, as velocity, temperature, and combustion conversion are three independent variables with different units, and there are three transport equations coupled. Many typical research areas in chemical engineering are multiphysics: physicochemical hydrodynamics, magnetohydrodynamics, electrokinetic flow, multiphase flow, double diffusion, and separations.

COMSOL Multiphysics deals with specific common multiphysics applications by creating application modes that are a full description of single field models, but can be readily coupled to other application modes. The user provides the coupling by specifying PDE terms and boundary and initial conditions symbolically. COMSOL Multiphysics does the "bookkeeping" to make sure that application modes have different nomenclature for the field variables and derived quantities that are commonly computed. For two coupled application modes, COMSOL Multiphysics symbolically assembles the FEM description for each mode by forming the sparse matrices, including the user specified coupling terms. The bugbear of manually programming FEM is the assembly of the stiffness matrix — collecting all contributions. If the pre-made application modes do not cover the user's coupled system, then the user can adapt as many coefficient form, general form, or weak form systems as necessary to describe their dynamics.

A few examples will illustrate multiphysics modelling to much better effect. Chapter 8 of the book of Ramirez [1] has a wealth of multiphysics PDE models with chemical engineering applications. They are computed on simple domains with finite difference methods coded with full detail in MATLAB. We shall adapt several such examples to COMSOL Multiphysics models here. But first we will attempt some simple buoyant convection problems.

2. Buoyant convection

Coupling momentum transport and heat transport is a well studied area of transport phenomena. The governing equations are

$$\frac{\partial u}{\partial t} + u \cdot \nabla u = -\frac{1}{\rho}\nabla p + \nu\nabla^2 u + \frac{\alpha g}{\rho}T,$$

$$\nabla \cdot u = 0, \tag{1}$$

$$\frac{\partial T}{\partial t} + u \cdot \nabla T = \kappa\nabla^2 T.$$

Here, the dependent variables are described as follows: **u** is the velocity vector, p is the pressure, and T is the temperature. The independent variables are spatial coordinates (implied in the differential operators) and time t. Everything else is a parameter (ν, ρ, α, κ, g) with fixed value once the fluid and vessel geometry are selected. If there is no imposed moving boundary or pressure gradient, then the whole of the motion is created by temperature gradients and is termed buoyant (or free) convection. If there are imposed velocities or pressure gradients, then the application is termed forced convection. Either case can be studied by the same multiphysics mode created in COMSOL Multiphysics, but are historically considered different physical modes.

In buoyant convection, there are two dimensionless parameters that govern the dynamical similarity of the problem, the Prandtl number that is a function of the fluid, and the Rayleigh number that gives the relative importance of temperature driving forces to dissipative mechanisms:

$$Pr = \frac{\nu}{\kappa},$$

$$Ra = \frac{\alpha g(\delta T)h^3}{\rho\nu\kappa}, \tag{2}$$

where h is the depth of the fluid, δT is the applied temperature difference, α is the coefficient of thermal expansion, **g** is the gravitational acceleration vector (g is its magnitude), ρ is the density, ν the kinematic viscosity, and κ is the thermal diffusivity.

Batchelor [2] showed that differentially heating any sidewall automatically induces buoyant motion, so the canonical buoyant convection problem is the hot wall/cold wall cavity flow. This problem is always taken as a test case for the development of new numerical methods for transport phenomena. We will develop a COMSOL Multiphysics model for it in this section. This problem is treated in [3], but the variations on the theme treated here are original.

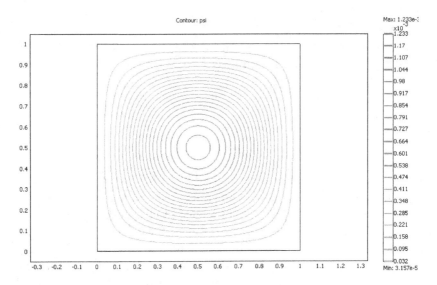

Figure 1.   Steady state streamlines of hot wall/cold wall buoyancy driven convection for $Ra = 1$.

Launch COMSOL Multiphysics and in the Model Navigator, select the Multiphysics tab. Follow the instructions in Table 1 to set up the buoyant convection model. It should be noted that this model starts with the hydrostatic case. Later we introduce the buoyant force. The rationale for starting simple and building up complexity comes from experience. Most complex systems have solutions that are far from trivial. Trying to jump to them from a standing start is asking a lot from the Newton solver. Moreover, the human element in coding complex systems must not be overlooked. If you build up you model piecemeal, adding each layer of modeling complexity on top of the next, not only will your intermediate solutions be closer to the whole system dynamics, but you will also be assured that each previous layer was correct, so likely causes of problems are the last thing you did. This little homily should serve as a warning — you only get "multiphysics" right if all the "single" physics is combined correctly.

The last mode, the coefficient form, will be used to solve directly for the streamfunction from the streamfunction vorticity Poisson equation:

$$\nabla^2 \psi = -\omega\,. \tag{3}$$

You may have noticed that the Incompressible Navier-Stokes application mode will print "flowlines." But to my eye, they are streaklines of randomly positioned particles, rather than the streamlines (contours of

Table 1. Buoyant convection of a fluid cavity with one hot wall and one cold wall. File name: buoyantconvection.mph.

Model Navigator	Select 2-D. COMSOL Multiphysics|Fluid Dynamics|Incompressible Navier-Stokes Click Multiphysics button. Add button Chemical Engineering module|Energy balance|Convection and Conduction. Add COMSOL Multiphysics|PDE Modes|PDE, Coefficient form Dependent variable: psi Application mode name: coef Add. OK
Options Menu: Constants	$T_0 = 0\ T_1 = 1$
Draw Menu	Specify objects: square. Accept defaults
Physics Menu: Boundary settings	Select Multiphysics menu. Select mode ns boundary 1–4 Use default boundary condition (No slip) OK Multiphysics menu Select mode chcc Physics Boundary settings Set boundary 1 with $T = T_0$ (Temperature boundary condition) Set boundary 4 with $T = T_1$ (Temperature boundary condition) Keep boundaries 2 and 3 as thermal insulation. OK Multiphysics menu Select mode coef Physics Boundary settings Accept defaults Diriclet $h = 1$, $r = 0$ settings OK
Physics Menu: Sub-domain settings	Select domain 1 (coef mode) Set $c = 1$; $d_a = 0$; $f = vx - uy$ OK Multiphysics Select mode ns Physics Subdomain settings Select subdomain 1 (ns mode) Set $\rho = 1$; $\eta = 1$; $F_x = 0$; $F_y = -1$ OK Multiphysics Select mode chcc Physics Subdomain settings Select subdomain 1 Set $\rho = 1$; κ (isotropic) $= 1$; $C_p = 1$; $u = u$; $v = v$ Init tab; set $T(t_0) = T_0 + (T_1 - T_0) * x$ OK
Mesh Menu	Select mesh parameters Point tab. Select points 1–4 Enter maximum element size 0.05 Remesh OK
Solve	Click on the Solve (=) button After the error, select Solve Menu Solver parameters Select advanced tab Set type of scaling: None OK

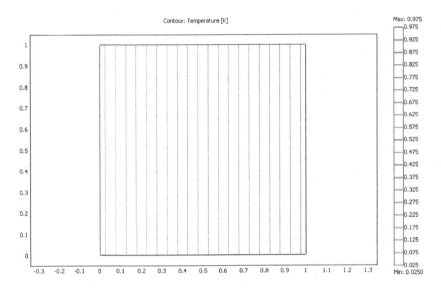

Figure 2.   Isotherms between 0 (left) and 1 (right) at steady state for $Ra = 1$.

streamfunction) that are traditionally interpreted in two-dimensional flow. Adding equation (3) is straightforward, and not particularly expensive to compute.

At this stage we will leave out the constants Ra and Pr. For simplicity throughout, we will keep $Pr = 1$, which is a good approximation for many gases. By enforcing the range of the temperature to lie between 0 and 1, i.e. a dimensionless temperature, all of the dynamics are controlled through the Rayleigh number.

There should be 960 elements. Click on the = button on the toolbar to Solve. In the first instance, you should get an error message:

```
Error:
    Failed to find a solution:
    No convergence, even when using the minimum damping factor.
    Returned solution has not converged.
```

Follow the instructions to change the scaling. Now plot the temperature profile. Is it what you expected? How does it compare with the initial condition.

Now plot the streamfunction. Surprised by the complexity? Now look at the scale. Why so small? Recall that we set $F_y = -1$ (gravity is in the negative y direction). This has the effect of adding hydrostatic pressure only. So there is no back action on the momentum equation from the imposed differential sidewall temperatures. So what we have here is a plot of velocity noise generated by round-off error. It is always important to look at the scale of contour/surface plots to assess whether we are interpreting noise!

You may have had some difficulty getting COMSOL Multiphysics to converge to a solution. When I originally wrote this example in FEMLAB 2.2, it converged fairly rapidly. Yet when done with FEMLAB 2.3, it took a long time. There are two contributions to the slow convergence — (1) the new scaling feature for the error estimate under the Solver Parameters, and (2) the lack of a pressure datum point. The first (scaling factor) was unexpected. Basically, COMSOL Multiphysics hopes to aid convergence by scaling each contribution to the error automatically. But since our velocity field has the true solution of a zero velocity field, numerically we find the approximate solution as noise around zero. The automatic scaling feature is trying to "resolve" the noise, which is not particularly sensible. Solution — turn the feature off! Select under Solver Parameters the Scaling box, and in the pop-up dialogue box, select "none" rather than the default automatic.

The second point about the pressure datum is discussed in detail in Chapter 5. However, COMSOL Multiphysics has made it very simple to implement a pressure datum. Under the Physics menu, select point settings and then domain (vertex point) 1. Check box for point constraint $p = 0$.

To implement a buoyant force that varies with temperature, edit the appropriate subdomain setting for the IC NS mode to be:

$$F_y = Ra\,T\,. \qquad (4)$$

Actually, the proper dimensionless term is as below, but given our scaling for temperature set in the constants, they are equivalent. If you wish to use temperatures with units, then edit T_0 and T_1 appropriately, and use this substitute for (4). Later, we will use temperatures with units, so I recommend its use:

$$F_y = Ra \times \left(\frac{T - T_0}{T_1 - T_0} \right)\,.$$

Add Ra to the constants list and set it to $Ra = 1$. Now use the Restart toolbar button, which uses the noise velocity field as an initial condition. This is actually a useful technique for introducing noise as an initial condition. Figures 1 and 2 show the streamlines and isotherms at steady state.

A quantity of central interest in thermal convection studies is the heat flux. The natural dimensionless measure of heat flux is the ratio of the total time averaged rate of heat transport to the conductive rate of heat transport, termed the Nusselt number:

$$Nu = \frac{\kappa(\delta T)/L + \langle \overline{uT} \rangle}{\kappa(\delta T)/L} = 1 + \frac{L\langle \overline{uT} \rangle}{\kappa(\delta T)} . \tag{5}$$

The overbar represents spatial average and the brackets time average. FEMLAB permits the computation of the separate terms in the numerator as subdomain integrations under the Post menu. Compute the dflux_T_chcc (conductive heat flux) and cflux_T_chcc (convective heat flux) integrated over the whole domain. What is the corresponding Nusselt number? Does this surprise you given the scale of the streamfunction in Figure 1? Now try simulating $Ra = 20000$. Do you get a converged solution?

2.1. *Buoyancy driven cavity flow: Parametric continuation*

Our solution strategy for the hot wall/cold wall problem to reach $Ra = 200000$ is to build up elements of the solution piecemeal. Were we to try to start at $Ra = 20000$ directly, we found that COMSOL Multiphysics cannot find a solution. Why not? For a nonlinear problem, the initial condition may not be in the "basin of attraction" for the desired solution, so Newton's Method could career far off. For it to work well, Newton's Method must start near a solution. For instance, the initial solution for hydrostatic pressure and velocity noise for $Ra = 0$ was an essential step. As a fully linear problem, it was readily solvable. It serves the important purpose of introducing an asymmetric velocity profile (due to the numerical noise of truncation error). This permits the solution for $Ra = 1$, which is qualitatively similar to the $Ra = 0$ in that it has a circulation, though a massive change in scale. Even then, though qualitatively similar to the $Ra = 1$ solution, $Ra = 20000$ was too far a leap from $Ra = 1$ to converge. The notion of traversing the solution space to introduce various topological features consistent with the target solution so as to be in its basin of attraction is similar to the established concept of parametric continuation. In parametric continuation, one restarts the simulation with a parametric value close to that of the saved, converged solution. Because the solution at the new parameter values is not expected to be much different than at the old parameter values, the Newton solver should converge rapidly. This methodology only fails if the new parameter is close to a bifurcation point — in which case multiple solutions are possible. The Jacobian used by Newton's Method is then very close to singular, so convergence may not be achieved. Or if it is, which of the multiple solutions that is selected may not be *a priori* predictable.

Parametric continuation is typically used for one of two purposes. One is to map the response of some feature of the solution over a range of parameters. The second is to reach a target solution for which jumping to the solution from any arbitrary initial condition is nonconvergent. So parametric continuation is metaphorically crawling along the limb of a tree, rather than expecting to jump and arrive safely. Parametric continuation can fail to converge as one ramps up a complexity parameter (like a Rayleigh or Reynolds number), and the complexity of the solution at smaller scales becomes unresolved. Thus, parametric continuation aids the identification of the parameter values where refining the mesh is important.

Using the COMSOL Multiphysics parametric solver

FEMLAB 2.2 did not have a built in parametric continuation feature, but FEMLAB 2.3 introduced it. In the predecessor to this book, I demonstrated building a parametric solver in Matlab calling the COMSOL Multiphysics subroutines to solve the PDE system at each stage. This is a powerful technique, since it gives the programmer total control over the parametric continuation process. But since the basic functionality exists with the parametric solver in the GUI, there is no need to re-invent the wheel. Here's how to do it in the GUI. Under the Solver Menu, select the Solver Parameters entry, and the Solver Parameters dialogue box pops up. Select the "Parametric Nonlinear" solver. On the General tab, enter the name of parameter: Ra. Then enter the list of parameter values: 2*logspace (1, 5, 20). The latter creates a list from $Ra = 20$ to $Ra = 200000$ of length 20 which is exponentially spaced. Click OK, and then launch the Solver with the = button from the toolbar. This will take some time to solve, so I recommend clicking the "log" tab so you can see the parametric solver step through the parameter space. The log is particularly important if a step crashes — either due to resolution problems or bifurcation difficulties. The last step of the log should look like this:

```
Parameter Ra = 200000
```

Iter	ErrEst	Damping	Stepsize	#Res	#Jac	#Sol
1	3.1	1.0000000	32	87	48	115
2	0.008	1.0000000	3.1	88	49	117
3	1.8e-008	1.0000000	0.008	89	50	119
4	1.5e-011	1.0000000	1.8e-008	90	51	121

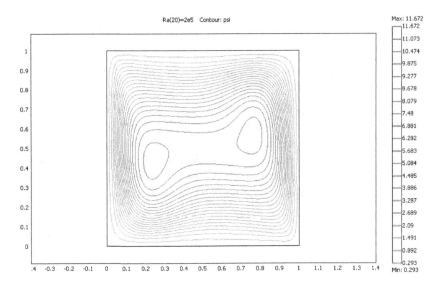

Figure 3.   Steady state streamlines of hot wall/cold wall buoyancy driven convection for
$Ra = 200000$.

The log shows that very few Newton iterations are necessary to achieve convergence. Ideally, the most efficient selection of parametric step gives two iterations — the minimum number of iterations the nonlinear solver of COMSOL Multiphysics performs on any solution to ensure convergence.

We are now in a position to assess why COMSOL Multiphysics did not converge on its original attempt to solve $Ra = 20000$ directly. Figures 3 and 4 show the streamlines and isotherms, respectively, for this highly convective flow. The streamlines are no longer topologically equivalent to the "concentric circles" flowfield seen in Figure 1 for the practically diffusive case of $Ra = 1$. The solution for the velocity field is a big jump in two senses — magnitude, where the volumetric flowrate at $Ra = 1$ was approximately 10^{-3} to $Ra = 200000$, where the volumetric flow has jumped to nearly 12 and structurally, since the topology of the flow has changed. The latter transformation is arguably more important in the difficulty of convergence directly. By using parametric continuation, each small parametric step did not result in such are large jump in solution space, so convergence was improved.

Post-processing parametric solutions

The whole suite of post-processing tools in the COMSOL Multiphysics GUI can be applied to the parametric solutions that we just found. My personal

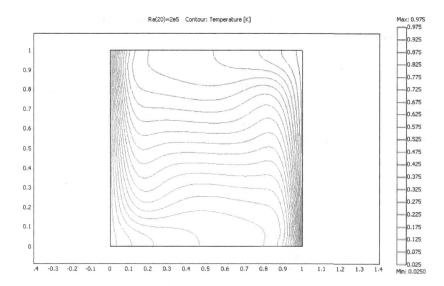

Figure 4. Isotherms between 0 (left) and 1 (right) at steady state for $Ra = 200000$.

favorite is to animate the parametric solutions so that the parameter value plays the role of time. For instance, pull down the Post Processing menu and select Plot Parameters. The contour tab should be set to show, with say, the temperature profile, and the velocity field (U_ns) selected on the surface tab (check boxes). Then, much as the solitary wave propagation example animation in chapter two, select the animate tab, and click on start animation (by default all parameter values are selected). The movie (rolling.avi) shows the (parametric) development of the rolling over of the temperature isotherms and the development of two wall circulation zones serving as the convective boundary layer.

Where the GUI has some difficulty is the computation of derived quantities, such as the Nusselt number, which is a global measure of the solution. You can compute the Nusselt number "by hand" using the subdomain integration dialogue box for dflux_T_chcc (conductive heat flux) and cflux_T_chcc (convective heat flux) integrated over the whole domain, as in the previous subsection. And this could be done, laboriously, for each parameter value, to produce the data for a plot of Nusselt number versus Rayleigh number. Do repetitious calculations, of course, was the reason for inventing the digital computer in the first place, so there must be a less laborious way of computing Nusselt numbers than repeating the same GUI steps. Both COMSOL Script and MATLAB serve as a "macro language" for COMSOL Multiphysics. Here we will show how to compute the Nusselt number by exporting the parametric solution to Matlab, and then automating the computation for all values of the Rayleigh number. In this case, integration coupling variables (see chapter four) could be used to do this post-processing in the GUI. More elaborate post-processing will require such Matlab or COMSOL Script programming nonetheless.

Under the file menu, select Export and then fem structure as 'fem.' In MATLAB, you can see the structure of the solution as follows:

```
>> fem.sol

ans =


       u: [8453x20 double]
   plist: [1x20 double]
   pname: 'Ra'
```

It is a simple matter to compute the subdomain integration of the convective flux, for instance, using the postint COMSOL Multiphysics command:

```
>> postint(fem,'cflux_T_chcc', 'dl',[1], 'solnum','end')

ans =

   16.3742
```

I am sure you must think I was very clever to sort through the User's Guide to learn the syntax for postint, but actually, COMSOL Multiphysics generated the above command. The last commands I executed in the COMSOL Multiphysics GUI was a subdomain integration of the convective and diffusive fluxes. COMSOL Multiphysics logs every command executed by the GUI, and you can access them by using the Save As feature on the File Menu and saving as a Model M-file. I then opened the m-file (parabuoyant.m) with a text-editor (MATLAB's m-file editor), and cut-and-pasted the command into the MATLAB GUI. That is actually more clever than sorting through the syntax of the COMSOL Multiphysics script language, since it generates commands that are known to work properly! (As an accomplished programmer, I am very good at generating commands that do not work properly until debugged!) The next important observation is that the postint command works with any of the parametric solutions by referring to them by solution number (solnum in COMSOL Multiphysics-ese).

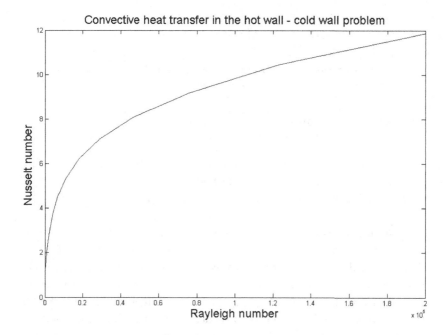

Figure 5.   Nusselt number versus Rayleigh number found by parametric continuation.

The postint command above used the special solution number 'end.' It works just as well with a number, say 10:

```
>> postint(fem,'cflux_T_chcc', 'dl',[1], 'solnum',10)

ans =

    1.5890
```

This solution corresponds to the 10th entry of the plist, i.e.

```
>> fem.sol.plist(10)

ans =

    1.5695e+003
```

Figure 5 was generated by the following "for loop" in MATLAB:

```
>> output=zeros(length(fem.sol.plist),3);
>> for j=1:length(fem.sol.plist)
output(j,1)=postint(fem,'cflux_T_chcc', 'dl',[1], 'solnum',j);
output(j,2)=postint(fem,'dflux_T_chcc', 'dl',[1], 'solnum',j);
output(j,3)=1+output(j,1)/output(j,2);
end
>> plot(fem.sol.plist,output(:,3))
```

The key feature is replacing the solution number 10 with the iteration variable j. A recurring theme in this book is to use a small number of COMSOL Multiphysics script commands to act on an GUI-created fem structure to achieve the desired computations and post-processing steps. Here, wrapping a for-loop around some COMSOL Multiphysics m-file generated commands was sufficient to save a great many repetitive GUI steps.

The user will probably want to invest some time in learning MATLAB programming tools and gaining a handle on the FEMLAB function library. Fortunately, FEMLAB's logging feature which records the FEMLAB commands issued by the GUI provides an excellent starting point for constructing your own FEMLAB programme. We have already given several applications of MATLAB programming with FEMLAB functions, but a full description of either MATLAB programming or FEMLAB functions is beyond the scope of this book. Chapter one (matrix operations) and the Appendix (vector calculus) provide only a rudimentary working capacity in MATLAB programming. We will continue to use this user defined programming style in the book, with sufficient explanation to guide the informed reader, and at least to inform the MATLAB novice of what power they are missing out on!

2.2. Variations on a theme: Nonmonotonic density

The governing equations for buoyant convection (1) assume the conventional Boussinesq approximation [4], i.e. that the velocity field is divergence free, that kinematic viscosity is constant, and that the only effects of density variation are felt by the body force in the Navier-Stokes equations, which was taken to depend proportionally to the coefficient of thermal expansion and temperature. The latter constraint, is too severe. The Boussinesq approximation only requires that density is a slowly varying function of position so that locally the velocity is divergence free. So a less restrictive

Table 2. Specific gravity of liquid water.

$T°C$	ρ	$T°C$	ρ	$T°C$	ρ	$T°C$	ρ
0	0.99987	20	0.99823	45	0.99025	75	0.97489
3.98	1.	25	0.99707	50	0.98807	80	0.97183
5	0.99999	30	0.99567	55	0.98573	85	0.96865
10	0.99973	35	0.99406	60	0.98324	90	0.96534
15	0.99913	38	0.99299	65	0.98059	95	0.96192
18	0.99862	40	0.99224	70	0.97781	100	0.95838

set of governing equations is

$$\frac{\partial u}{\partial t} + u \cdot \nabla u = -\frac{1}{\rho_0}\nabla p + \nu\nabla^2 u + \frac{\rho(T)g}{\rho_0},$$

$$\nabla \cdot u = 0, \tag{6}$$

$$\frac{\partial T}{\partial t} + u \cdot \nabla T = \kappa\nabla^2 T,$$

where $\frac{\rho(T)}{\rho_0}$ is a general function of temperature, and $\rho_0 = \rho(T_0)$. The Rayleigh number is no longer a constant, but depends on this function:

$$Ra(T) = \frac{gh^3}{\nu\kappa}\left(\frac{\rho(T) - \rho_0}{\rho_0}\right) = Gr'\left(\frac{\rho(T) - \rho_0}{\rho_0}\right), \tag{7}$$

where the gravity group Gr' now appears as a dimensionless parameter. The density function plays the role of a nonlinear expansivity (and possibly nonmonotonic).

So what does one of these expansivity functions look like? See Figure 6. Next, how do we organize this data so as to use it in the COMSOL Multiphysics GUI? First, re-load the saved MPH file for buoyant convection that you developed in Section 2 of this chapter (buoyantconvection.mph). Under the options menu, select functions, and then click the NEW button. In the pop-up New Function window, name the function watrdens and specify that the function is interpolation type by ticking the button, and pull down the selection of data from a file. In the file name entry, enter watrdens.txt. This text file is available from the author by request and has the form abbreviated below from the data in Table 2:

```
% Temperatures
273 276.98 278 283 288 291 293 298 303
% Densities
0.99987 1. 0.99999 0.99973 0.99913 0.99862 0.99823 0.99707
0.99567
```

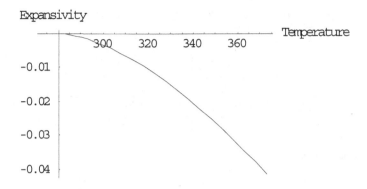

Figure 6.   Expansivity $\frac{\rho(T)-\rho_0}{\rho_0}$ versus temperature (K) for water.

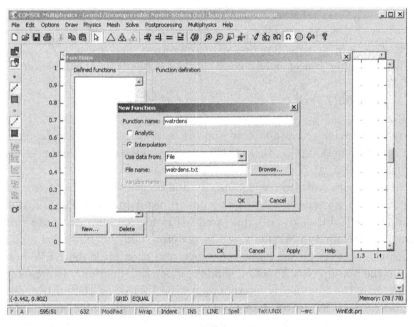

Figure 7.   Setting up the interpolated function watrdens(T) from text data file watr-dens.txt.

The percentage character (%) is a comment line. Personally, I would have preferred a column entry format, but large length double rows are possible with most text editors. Click the OK button and then the Function Window now has entries. My preferences are interpolation method by piecewise cubic and extrapolation method by extrapolation. Click OK to finish defining the interpolation function watrdens(T). Note that watrdens.txt has to be present in your path for it to be found. Typically, your COMSOL Multiphysics top directory is the safest spot. Since I am running MATLAB concurrently, COMSOL Multiphysics found watrdens.txt in the current MATLAB directory.

Now we are ready to use our new function. Pull down the options menu and select Add/Edit constants. Replace the Rayleigh number entry with the gravity group Gr, and set it initially to $Gr = 0$. Change the other constants to $T_0 = 273$ and $T_1 = 373$ to represent walls held at both extreme water temperatures at 1 atm. Now edit the ns mode (multiphysics menu) subdomain settings on the Physics Menu and set

$$Fx = -Gr * \text{watrdens}(T)$$

This solves the buoyancy problem. Reset $Gr = 1$ and click on the restart button to get a nontrivial solution. Previous versions of FEMLAB required the user to also specify the derivative of supplied functions. Here, we note that COMSOL Multiphysics does not need this information for interpolation functions it defines.

I used the parametric solver to migrate to higher Gr using automated parametric continuation. My hand made code for parametric continuation arrived to $Gr = 10^5$ after about a day on my fastest linux PC workstation as a background job without a pressure datum introduced in point mode to improve the convergence rate. With the GUI and the parametric solver, excellent convergence was found within 45 seconds of the entire parametric list $Gr = 2*\text{logspace }(1, 5, 20)$. Figure 8 shows the plot of Nusselt number versus $\log Gr^{1/3}$ which suggests that the 1/3 exponent is correct. This is a well known result experimentally for laminar free convection.

The temperature isotherms tell an interesting story in Figure 9 that the fluid is starting to stratify, with cold fluid under hot.

3. Unsteady response of a nonlinear tubular reactor

Ramirez [1] [6] reports a simulation of the adiabatic tubular reactor where heat generation effects are appreciable. Generally, tubular reactor design estimates follow from steady-state 1-D ODE simulations. In the model of Ramirez, the reactor starts up cold or is subjected to perturbations of its steady operation which convect through the system before returning to steady operation. In this regard, such transient effects are important considerations for the safe, stable and controlled operation of tubular reactors.

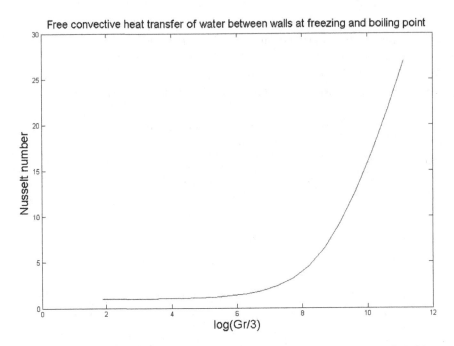

Figure 8. Convective heat transfer from a water filled cavity with vertical walls held at the freezing and boiling temperatures at 1 atm.

Ramirez treats first order chemical reaction with heat generation. Thus only the mass transport equation for one species and energy transport equation, coupled through the temperature dependence of the reaction flux and the heat generation by reactive flux, need be considered. Interestingly, Ramirez solved the highly coupled, nonlinear equations by a technique of quasilinearization with finite difference techniques. The solution at the current time and the linearization of the equations about that solution are used to predict the profiles of concentration and temperature at the next time step. The procedure is iterated until convergence at the new time step is achieved. The prediction and correction steps involve solution of sparse linear systems. This is, of course, the same procedure as used by FEMLAB, except it is the finite element approximation and the associated sparse linear system that is solved iteratively by Newton's method.

Here Θ is dimensionless temperature and Γ is dimensionless concentration. The governing equations are given here in dimensionless form:

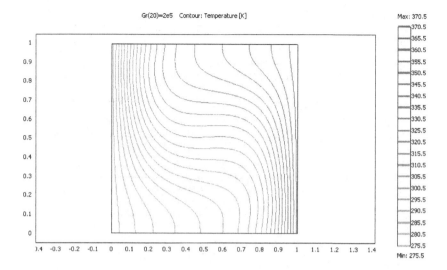

Figure 9.   Isotherms for water density model with $Gr = 200000$.

$$\frac{\partial \Theta}{\partial t} = \frac{\alpha}{D}\frac{\partial^2 \Theta}{\partial x^2} - r_2\frac{\partial \Theta}{\partial x} + B_1\Gamma \exp\left(-\frac{QQ}{\Theta}\right),$$

$$\frac{\partial \Gamma}{\partial t} = \frac{\partial^2 \Gamma}{\partial x^2} - r_2\frac{\partial \Gamma}{\partial x} - B_2\Gamma \exp\left(-\frac{QQ}{\Theta}\right),$$

(8)

subject to boundary conditions on the reactor inlet and outlet:

$$-\frac{\partial \Theta}{\partial x}\bigg|_{x=0} = r_1(1 - \Theta|_{x=0})\frac{\partial \Theta}{\partial x}\bigg|_{x=1} = 0,$$

$$-\frac{\partial \Gamma}{\partial x}\bigg|_{x=0} = r_2(1 - \Gamma|_{x=0})\frac{\partial \Gamma}{\partial x}\bigg|_{x=1} = 0.$$

(9)

The former are called Danckwerts boundary conditions [7]. The initial conditions for temperature are uniform everywhere at $\Theta = 1$. Ramirez [1] considers two different liquid phase reactions. The first is a reactor with an intermediate conversion at a single steady state. The second is a triple steady state. The Peclet numbers for heat and mass transfer are taken as identical for convenience, but this is not a realistic assumption which can be relaxed in COMSOL Multiphysics without any difficulty. This problem has a long history in the chemical engineering literature, and the equal Peclet numbers are a legacy of an analytic approximation by Amundson [8] which provides validation for the solution.

Table 3. Parameters for Case 1.

$r_1 = \dfrac{vL}{\alpha}$	30	$B_1 = \dfrac{(-\Delta H)kC_0L^2}{\rho c_p T_0 D}$	7.84×10^7
$r_2 = \dfrac{vL}{D} = Pe$	30	$B_2 = \dfrac{kL^2}{D}$	1.2×10^8
$\dfrac{\alpha}{D}$	1	$QQ = \dfrac{E}{RT_0}$	17.6

Table 4. Parameters for Case 2.

$r_1 = \dfrac{vL}{\alpha}$	30	$B_1 = \dfrac{(-\Delta H)kC_0L^2}{\rho c_p T_0 D}$	8.47×10^9
$r_2 = \dfrac{vL}{D} = Pe$	30	$B_2 = \dfrac{kL^2}{D}$	1.2×10^{10}
$\dfrac{\alpha}{D}$	1	$QQ = \dfrac{E}{RT_0}$	23

Amundson's technique combines the mass and energy equations by linear transforms:

$$n = \Theta,$$

$$\Gamma = \frac{B_2}{B_1}(n_{\lim} - n).$$

This, in turn, leads to equations (8) and (9), along with the BCs, being described identically, if and only if, the ratio of diffusivities is unity, $\frac{\alpha}{D} = 1$:

$$\frac{\partial^2 n}{\partial x^2} - Pe\frac{\partial n}{\partial x} + B_2(n_{\lim} - n)\exp\left(-\frac{QQ}{n}\right) = 0, \qquad (10)$$

subject to

$$-\frac{\partial n}{\partial x}\bigg|_{x=0} = Pe(1 - n|_{x=0})\frac{\partial n}{\partial x}\bigg|_{x=1} = 0. \qquad (11)$$

Here, $n_{\lim}$ is the limiting dimensionless temperature that can be achieved upon exhaustion of the reactant, $\Gamma = 0$. The Peclet number, Pe, is either the thermal or mass Peclet number (r_1 or r_2).

Amundson proposed a one dimensional search to the above boundary value problem, starting from a guess of $n(x = 1)$ and shooting back to $x = 0$. In both of the above cases, $n_{\lim} = 1.656$.

Let's first solve the single convection-diffusion-reaction equation (10) using COMSOL Multiphysics. Because it is a boundary value problem, FEM has a natural advantage here.

Start up COMSOL Multiphysics and enter the Model Navigator and follow the instructions outlined in Table 5. The mesh is extremely important here, as rapid variations near the boundaries due to the boundary conditions are expected.

Table 5. Nonlinear tubular reactor model.

Model Navigator	Select 1-D space dimension Select Chemical Engineering module|Mass balance|Convection and diffusion. OK
Draw Menu	Specify objects: Line $x : 0\ 1$ Name: interval OK
Options Menu: Constants	Enter the data from table Name Expression Pe 30 B_2 $7.84E + 7$ $n_{\lim}$ 1.656 QQ 17.6
Physics Menu: Boundary settings	Select boundary 1 Select boundary condition: flux; enter $N_0 = Pe * (1 - c)$ Select boundary 2 Set boundary condition: flux; keep default $N_0 = 0$ OK
Physics Menu: Sub- domain settings	Select domain 1 Set $u = Pe$; $R = B_2 * (n_{\lim} - c) * \exp(-QQ/c)$; keep default $D = 1$. Init tab; set $c(tw) = n_{\lim}$. OK
Mesh Menu	Select mesh parameters Boundary tab Select boundary 1 and 2, set maximum element size 0.0001 Select Global tab; set maximum element size scaling factor 0.01. Remesh OK
Solve	Click on the Solve (=) button View Log. 17 iterations are used for this highly nonlinear problem, but about a CPU second

This results in a 181 element meshing. Note that "Maximum element size" constrains the elliptic mesh generator to give appropriately large or small elements as directed by the Mesh tab page for the subdomain, boundary, or point.

The solution stage is interesting. It takes COMSOL Multiphysics 17 iterations to get there (this is a highly nonlinear problem), but it converges to 10^{-8} accuracy eventually. I played around with several meshes and

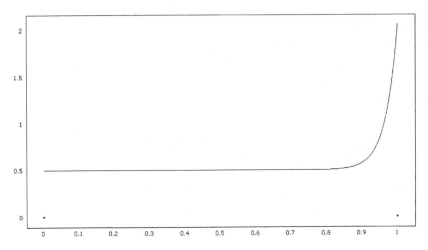

Figure 10.   Dimensionless temperature/concentration profile for Case 1 of a nonlinear tubular reactor.

parametric approaches before hitting on this one.   Figure 10 holds the steady state solution for this case.

For a transient stability study, it is necessary to have a steady state described by a function, and then to perturb that steady state. So we would like to create a function holding the data in Figure 10. One of the major criticisms of Femlab 2.x was the difficulty of extracting numerical data from the FEMLAB solutions. As we saw in chapter one, it is possible to extract numerical data using the fem structure directly from the fem.sol.u degrees of freedom, but only readily with Lagrange linear elements in 1-D. COMSOL Multiphysics includes a function, postinterp, which extracts numerical data by using the degrees of freedom and the Galerkin formulation to construct interpolated values of the variables solved for. With nodal points on the elements, these interpolations simplify to the "exact" numerical solution, but actually, the trial function expansion (see Eq. (15) of Chapter 2) is just as accurate within the finite element as it is on its nodes. So perhaps it is more accurate to refer to "postinterp" as a variable reconstruction function.

To extract data using postinterp in MATLAB, start by exporting the fem structure to MATLAB using the FILE menu.   An example is

```
>> postinterp(fem,'c',0.95)

ans =

    0.8483
```

Clearly, the first argument is the fem structure, the second is the variable to be reconstructed, and the third is the coordinate within the subdomain for which it is to be reconstructed. It is a simple matter to generate a vector of c-values at a vector of x-values:

```
>> postinterp(fem,'c',linspace(0,1,5))

ans =

    0.5000    0.5000    0.5000    0.5009    2.0721
```

Here the position coordinate is replaced by a vector of positions. The output is the vector of values at those positions.

COMSOL Multiphysics created a second way of exporting data from its graphics. Pull down the File Menu, select Export, and Current plot... This generates a dialogue box to save a text file which includes all the information plotted in the graph. I saved the data in Figure 10 as tubular.txt. It has the structure:

```
% Coordinates
0.0                    0.5
2.5000138E-5           0.5
5.0000275E-5           0.5
7.500041E-5            0.5

...

0.99997324             2.0704775
1.0                    2.072139
% Elements (lines)
1          2
2          3
3          4


...
```

```
% Coordinates
0.0                        0.0
1.0                        0.0
% Elements (lines)
1          1
2          2
```

The 182 nodal positions and values are the numerical data we are after, which is almost as good as an interpolation function. I chopped away all the comment lines (beginning with %) and all lines after the "numerical data," i.e. elements and coordinates, to achieve a text file with two columns of data. This is not difficult given modern text editors (such as MATLAB's m-file editor). Make sure to save it in the current MATLAB directory or change directories to where you save it. It can be read into MATLAB, for instance, with the command:

```
input=dlmread('tubular.txt'); input=input(:,1:2)
```

Annoyingly, dlmread finds an end column of zeros that is spurious, so the second command is to throw them out. We have already seen how to use the function creation tool built-in to the COMSOL Multiphysics GUI, so here, we will show how to create an m-file function. In the M-file editor (MATLAB file menu, New, M-file), create a new file initcond.m. Enter the following MATLAB commands and save it in the same directory as tubular.txt.

```
function a=initcond(x) input=dlmread('tubular.txt');
a=interp1(input(:,1),input(:,2), x,'spline')+0.05
   *sin(31.4159265*x);
```

The result is a function m-file initcond.m for the steady solution with superimposed period one-fifth sine perturbation, shown in Figure 11 generated from the MATLAB command:

```
plot(input(:,1),initcond(input(:,1)))
```

Now Pull down the Physics menu, then reselect Subdomain settings. Now enter initcond (x) as $c(t_0)$ on the init tab (replacing n_{lim}) and check the time-dependent solver in COMSOL Multiphysics under Solve Menu, Solver Paramters. Enter output times as 0:0.0001:0.003. Then under the Postprocessing menu, select Plot Parameters and Animate. Enjoy the experience of watching the perturbation dissipate and propagate out of the system.

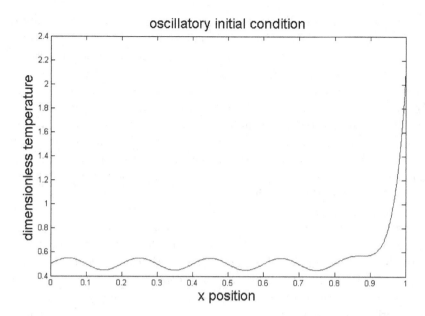

Figure 11. Steady solution with superimposed period one-fifth sine.

After creating this example, I learned of the marvellous new functionality entailed in the Solver Manager with COMSOL Multiphysics 3.2. There are essentially five different "solutions" that can be used as initial conditions. Previously, users were limited to either the initial condition stated on the Init tabs, or the last converged solution. Now you can have those two "solutions" as well as a stored solution, and two variations of expressions which are evaluated using solutions. One of these will avoid completely the export and import steps illustrated here. Enter $c + c\_perturbed$ on the Init tab for concentration, and then under the Solver Manager Initial Value tab, check the "Initial value expression using the current solution." Here $c\_perturbed$ would be the spatial oscillation. The effect is precisely the same as our storage of the steady state and creation of the perturbation in an m-file, but by far simpler.*

Now for Table 4. Change the Constants under the Options menu to reflect the higher B_2 and QQ coefficients. n_{lim} stays the same. Use the stationary nonlinear solver from the oscillatory initial condition. What does the converged solution look like?

---

*I am indebted to Anna Dzougoutov for this "trick of the trade."

Converging from a solution that is not so similar to the steady-state is not so straightforward. Try the initial condition $c(t_0) = 1$. Does it converge with the stationary nonlinear solver? Now try the time dependent solver and set output times to 0:0.0001:0.01, then animate the solution. As you can see, the time dependent solver is attracted to the steady solution, but the stationary nonlinear solver wasn't "close enough" in solution space to fall onto the solution. Now perhaps it is. Check the stationary nonlinear solver and click OK. Then click on the re-start button on the toolbar, which takes the initial solution as the last of the time dependent solutions. This should converge in about five iterations to the same solution found earlier.

Now try the initial condition $c(t_0) = n_{lim}$. With the time dependent solver, set output times to 0:0.001:0.1, then animate the solution. You should be able to watch the initial condition pass completely out of the reactor and converge to the steady state solution found by the two previous methods. This should be a clear signal that the time dependent solver may be an essential tool in attacking nonconvergence. Even in problems that have no inherent unsteady time scale, pseudo-time dependent solution may be essential to finding a converged stationary solution. If so, then we can be fairly certain that the steady state so found is stable, since it is attractive.

Exercise 3.1.

This chapter is entitled "Multiphysics." The problem statement is definitely for two physics modes (heat and mass transport with reaction), yet due to Amundson's technique, the problem could be simplified to "single physics" for $\frac{\alpha}{D} = 1$. Try implementing the calculation with the ChemEng Module modes convection and conduction (cc) and convection and diffusion (cd) with the same parameters, but as written in equations (8), (9). Take the initial condition to be uniform temperature $n = 1$. Solve for the steady state after a long time, or use the steady solver. Compare you results with the Amundson technique solution given here.

4. Heterogeneous reaction in a porous catalyst pellet

It would be an injustice not to draw on the COMSOL Multiphysics Model Library for an example of multiphysics. Although the chemical engineering curriculum does not contain many examples of multiphysics partial differential equations, the same cannot be said for the chemical engineering model library of COMSOL Multiphysics. In accord with this book's policy of complementary to the COMSOL Multiphysics manual set, however, we must treat any problem that we extract differently. This section is inspired by

the heterogeneous reaction modeling in a porous catalyst pellet, treated in
[9]. The model uses the incompressible Navier-Stokes application mode and
couples the results to the convection and diffusion mode through the multiphysics facility of COMSOL Multiphysics in two-dimensions (c.f. Eq. (7)),
adding the additional reaction term in the pellet subdomain, but without
convection (reaction-diffusion model). This clearly counts as multiphysics
by either definition given at the beginning of the chapter, since there are two
different PDEs with independent variables u, v, p, and c having different
units. The twist that we add to the model is to decouple the multiphysics.

Because the reaction is taken to be isothermal and constant density,
there is no back action coupling the concentration field into the Navier-
Stokes equations. The mass transport requires knowledge of the velocity
field to compute convection, but this does not change the momentum transport. So rather than computing both momentum transport and mass transport simultaneously, we can compute them sequentially. Why? Primarily
because of the computing efficiency. If one requires several solutions over a
range of mass transport/reaction parameters, but with the same flow field,
then computing the flow field only once and importing the velocity field
is the most computationally efficient method (or should be, if coded efficiently). Secondly, whatever platform you use to compute on is probably
memory limited if you want to refine the mesh. For instance, because we
computed the streamfunction explicitly in the buoyant convection example
earlier, it was not possible to refine the mesh further without running out
of system memory on a 1 Gb RAM linux PC workstation. The final reason
is that it illustrates further handles into the COMSOL Multiphysics GUI
through MATLAB programming, which is one of the reasons to read this
text.

We visited the Incompressible Navier-Stokes (2-D) mode in §2, and in
fact if we add a reaction source term to (1) and call concentration T rather
than c, then those equations describe the model perfectly.

Launch COMSOL and in the Model Navigator, select the Multiphysics
tab. Follow the instructions in Table 6 to set up the model. This example uses the Solver Manager to decouple the multiphysics of the problem.
Clearly, the incompressible Navier-Stokes mode does not involve any composition effects — the boundary conditions and physical properties are independent of concentration. However, the convective-diffusion transport and
reaction equation for concentration of the reagent is intimately dependent
on the flow field. The solver manager allows us to solve for the flow field
first, without solving for the concentration field. Then we can activate the

Table 6.   Porous catalyst pellet model. File name: pellet.mph.

Model Navigator	Select 2-D space dimension Select COMSOL Multiphysics|Fluid Dynamics|Incompressible Navier-Stokes Click Multiphysics button. Add Select Chemical Engineering module|Mass balance|Convection and diffusion|Transient Analysis. Add. OK			
Options Menu Axis/Grid Settings	**Axis tab**		**Grid tab**	
	X_{min}	-0.001	X spacing	0.001
	X_{max}	0.003	Extra X	0.0009
	Y_{min}	-0.001	Y spacing	0.001
	Y_{max}	0.007	Extra Y	0.0021 0.0039
	Uncheck Auto on the grid tab to manually specify the grid spacings			
Draw Menu	Deselect solid by double clicking on **SOLID** on the **StatusBar** Draw an arc B_1 (2nd degree Bezier curve) by clicking at the corners (0, 0.0039), (0.0009, 0.0039), (0.0009, 0.003), (0.0009, 0.0021) and (0, 0.0021) Select solid by double clicking on **SOLID** on the **StatusBar**			
Options Menu: Constants	$Ro = 0.66$; $mu = 2.6\mathrm{e}{-5}$; $vo = 0.1$; $D = 1\mathrm{e}{-5}$; $D_{eff} = 1\mathrm{e}{-6}$; $k = 100$; $clo = 1.3$			
Physics Menu: Boundary settings	Multiphysics: Select ns mode Boundary 2. Select inflow BCs with $u = 0$ and $v = vo$ Set boundaries 1, 4, 6 with Slip/Symmetry Set boundary 5 with outflow BC $p = 0$ Set boundary 7, 8 with No-Slip. OK Select chcd mode Select boundary 2; boundary condition; concentration $c_0 = clo$ Select boundaries 1, 3, 4, 6 with insulation symmetry Select boundaries 5; set convective flux boundary condition OK			
Physics Menu: Subdomain settings	Multiphysics: Select ns mode Select subdomain 1. Enter $\rho = ro$; $\eta = mu$ Select subdomain 2. Uncheck "Active in this domain" OK Select chcd mode Select subdomain 1 Enter D(isotropic) $= D$; $R = 0$; $u = u$, $v = v$ Select subdomain 2 Set D(isotropic) $= D_{eff}$; $R = -k * c\hat{}2$; $u = 0$; $v = 0$. OK			
Mesh Menu	Select mesh parameters Boundary tab Set max element size near vertices: Boundary 6: $1\mathrm{e}{-4}$ Boundaries 7 8: $5\mathrm{e}{-5}$ Remesh OK Refine the mesh once using the toolbar button			
Solve Menu Solver Manager	Solve For tab. De-select chcd mode. OK Solver parameters. Select stationary nonlinear solver. OK Click on the Solve (=) button			

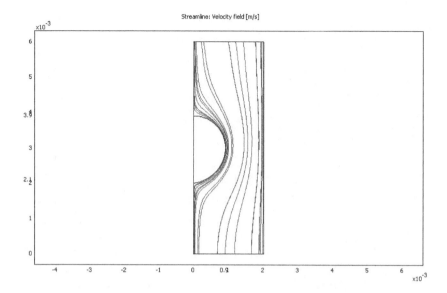

Figure 12.   Streamlines surrounding the pellet.

concentration model, but "freeze" the flow field by turning off its solution in the Solver Manager. This staging of solutions saves computational time and complexity. It would be most obvious in a transient analysis of the reaction — why solve for flow field at every step if it never changes?

The meshing in the pellet reactor example results in 7744 elements. There are 26143 degrees of freedom in the flow field solution. This results in streamlines as shown in Figure 12. After solving for the flow field, use the Solver Manager to de-select the Navier-Stokes mode and select the chcd mode as active. When using the Restart solver (Solve Menu entry or button on toolbar), the last converged solution is taken as the initial condition. Much greater flexibility comes from using the Solver Manager and Initial Value tab page. Here, the velocity field remains fixed at that found by the ns mode. The chcd mode has 15701 degrees of freedom. Figure 13 shows the concentration contours surrounding the pellet reactor.

Please make sure to save this model MPH file (say pellet.mph), as we will use it for a building block for an extended multiphysics model in chapter four.

5. Discussion

This chapter introduced the concept of multiphysics and then ran away from it. One of the key messages from §3 and §4 should be to simplify

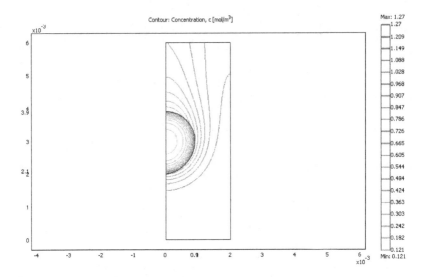

Figure 13. Contours of concentration (reduced by disappearance in pellet phase).

the analysis of your PDE system. For instance, in §3, multiphysics was not necessary due to a change of variable to eliminate one PDE equation. Similarly, in §4, multiphysics could be dealt with by sequential treatment of single physics. This has the virtue of separating the work. It should be clear that the major difficulty in solving the pellet reaction/diffusion/convection problem was resolving the fluid dynamics of the hemispherical obstruction. The concentration profile was resolved on the most coarse mesh chosen.

Without too much aggrandizement, this chapter also shows the ease of solution of highly nonlinear problems by parametric continuation which can be done in the GUI, and with somewhat more flexibility in Matlab or Comsol Script. The chapter also shows how to include variable physical properties and complicated field behaviors by interpolation functions programmed in MATLAB m-file functions.

Multiphysics is a recurrent theme in this text, largely because "single physics" is well studied. Thus, inherently, multiphysics descriptions are required for state-of-the-art research models. So several more examples will follow.

References

[1] W. F. Ramirez, *Computational Methods for Process Simulation*, 2nd ed. (Butterworth Heinemann, Oxford, 1997).

[2] G. K. Batchelor, Heat transfer by free convection across a closed cavity between vertical boundaries at different termperatures, *Quart. J. Appl. Math.* **12**(3) (1954) 209–233.

[3] M. G. Velarde and C. Normand, Convection, *Scientific American* **243**(1) (1980) 92–108.

[4] O. C. Zienkiewicz and R. L. Taylor, *The Finite Element Method. Vol. 3: Fluid Dynamics* (Heinemann-Butterworth, 2000).

[5] D. Hanselman and B. Littlefield, *Mastering MATLAB 7: A Comprehensive Tutorial and Reference* (Prentice Hall, Saddle River NJ, 2001).

[6] D. E. Clough and W. F. Ramirez, Stability of tubular reactors, *Simulation* **16** (1971).

[7] P. V. Danckwerts, Continuous flow systems. Distribution of residence times, *Chem. Eng. Sci.* **2** (1953) 1–18.

[8] N. R. Amundson, *Can. J. Ch. E* **43** (1965) 99.

[9] COMSOL Chemical Engineering Module, User's Guide, Version 2.2, pp. 2–74.

Chapter Four

EXTENDED MULTIPHYSICS

W.B.J. ZIMMERMAN[1], P.O. MCHEDLOV-PETROSSYAN[2*]
and G.A. KHOMENKO[2]

[1] *Department of Chemical and Process Engineering, University of Sheffield,
Newcastle Street, Sheffield S1 3JD United Kingdom*
[2] *Laboratoire d'Oceanografi,Cotiere du Littoral, ELICO, Universite du Littoral
Coted'Opale, MREN, 32, Avenue Foch, 62930, Wimereux, France*
[*] *Permanent address: NSS Kharkov Institute of Physics and Technology, Ukraine
E-mail: w.zimmerman@shef.ac.uk*

Extended multiphysics is a feature that is conceptually complicated and original
with COMSOL Multiphysics. The concept is the linkage of two or more logical
computational domains through coupling variables that can be used in either specifying the boundary conditions or subdomain PDE coefficients. The coupling variables can be found by subdomain or boundary integrations of internal or boundary
values. These naturally arise in the multiple scale modeling of physical phenomena
— the large scale model is coupled to subgrid cellular models, perhaps of a simpler parametric or lower spatial dimension. Extended multiphysics is ubiquitous
in process engineering, however, because unit operations are conceptually separate
domains, yet linked through at least inlet and exit conditions sequentially, but
frequently linked more subtly through process integration. So the whole field of
process simulation for optimization, design, retrofit, and control falls within the
remit of extended multiphysics. Integration with Simulink gives the possibility of
some unit operations being treated with distributed PDE models while others are
treated with lumped parameters, yet with nontrivial levels of coupling requiring
extended multiphysics modeling.

1. Introduction

If multiphysics, the subject of the last chapter, were a new concept to you,
extended multiphysics must be a more alien concept indeed. The chemical engineering model library of COMSOL Multiphysics now abounds with
extended multiphysics examples — the Maxwell-Stefan Diffusion in a Fuel
Cell model and the Monolith Reactor examples in the Chemical Engineering Model library [1] are especially recommended. This chapter introduces
extended multiphysics through simple linkages by scalar coupling variables.

Now that COMSOL Multiphysics has introduced auxiliary state variables through ODE Settings, it is not essential to introduce a second conceptual domain for simple extended multiphysics applications. Coupling variables can simply link the same domain or subdomains back on different parts of itself. Periodic boundary conditions shown in chapter two (solitary wave example) are an example of such a simple extended multiphysics application. Chapter seven gives a more thorough exploration of all three types of coupling variables, showing examples that illustrate how this unique feature of COMSOL Muiltiphysics makes convolution integrals, integral equations, integro-differential equations, and in general, inverse problems, treatable within finite element methods. Chapter nine shows how the concept can be applied to generic network modelling where the "nodes" are junctions in microfluidic processors and the links are "long" microchannels.

Initially I thought extended multiphysics was about coupling multiple scale models, as that is how it was done in [1]. This is a cutting edge area of research in multiphase flows/heterogeneous systems, because the dispersed phase can be treated as a point constraint in the domain of the bulk medium, but with information flowing in both directions. Usually the attempt is to treat such constraints parametrically, i.e. modeling the dependence of the small scale phenomena on bulk phase unknowns, and vice versa to complete the coupling of the scales. Usually, the small scale phenomena is too complex in its own right, for instance in the microhydrodynamics simulations of Grammatika and Zimmerman [2], to consider solving simultaneously with the bulk dynamics. So the coupling is through simple functional forms learned from simulations of the small scale dynamics, slaved to the large scale phenomena imposed on it. There are several drawbacks to the parametric slaving approach, but they are all summed up by "oversimplification." Fortunately, such models can be verified by experimentation that the physical systems can be well treated by the two scale approach. Traditional turbulence models are all heavily reliant on multiple scale modeling by parametrization. Since the multiple scale modeling techniques are specialized, perhaps extended multiphysics is not such a useful feature after all. To take advantage of it for complex modeling may require high performance computing.

Only belatedly did it occur to me that chemical engineering is awash with applications for extended multiphysics. First, let's give an operational definition for extended multiphysics in the COMSOL sense: a model is categorized as extended multiphysics if it requires description of field variables in two or more *logically* disjoint domains. They are not likely to be

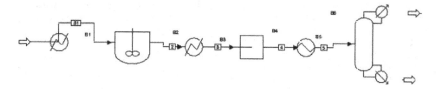

Figure 1. Flowsheet for a linear array of unit operations.

physically disjoint domains since the physics must be coupled in some respect to warrant solving the problems in each domain jointly. COMSOL allows the user to use several different geometries/application mode pairs in building up an extended multiphysics model.

So why is it that chemical engineering is awash with extended multiphysics? Look no further than your nearest flowsheet, say Figure 1.

For the process in Figure 1,

> "Cyclopropane at 5 bar and 30°C is fed at a rate of 1 kmol/h. It is heated to a reaction temperature of 500°C by a heat exchanger before entering as CSTR (continuously stirred tank reactor). The reactor has a volume of 2 m^3 and maintains the reaction temperature of 500°C. The isomerization reaction:
>
> $$C_3H_6 \rightarrow CH_3CH = CH_2 \tag{1}$$
>
> is first order with a rate constant of $k = 6.7 \times 10^{-4}\text{s}^{-1}$ at 500°C. The products of the reaction are then cooled to the dew point by a second heat exchanger before entering a compressor. The compressor increases the pressure to 10 bar, the pressure at which separation of reacted propylene and unreacted cyclopropane will occur. The compressed gas is then condensed to bubble-point liquid and feed to a distillation column. The column has 31 ideal stages with the feed onto the stage 16. It operates with a total condenser and a molar reflux ratio of 8.4 producing a distillate flow rate of 0.292 kg-mol/h."

Sound familiar? Such scenarios populate modules on "Process Engineering Fundamentals." Why is it extended multiphysics? Each unit operation constitutes its own logical domain, connected to the others by entry and exit points. In the conceptual design stage of such a plant, the unit operations are treated by simplified models to permit facile exploration of the configuration space. Process integration by means of recycle and heat exchanger networks adds greater complexity to the flowsheet, and greater scope for economies in operating and capital costs. Eventually, however, the process engineer has to give detailed designs for such plant. These days that includes process simulation, typically including optimisation, parametric sensitivity studies and transient analysis. And even if the plant were designed a generation ago, process studies of this nature are common for retrofit and optimisation. In many cases, plants were over designed by 30–50% (since such flexibility is a common safeguard in design), so now that the plant is operational, efficiency savings of 30–50% should be achievable. Thus has grown the burgeoning field of process systems optimisation. And this is a regime for extended multiphysics. If any of the unit operations in Figure 1 are to be modelled in detail, that usually involves a spatial-temporal PDE where the simplified model used in design might have been a lumped parameter model. For instance, suppose the reactor in Figure 1 is CSTR reactor which is jacketed by a bath of its product liquid (at 500°C) before entering the heat exchanger proper. Temporal fluctuations in the reactor temperature propagate through the bath to the heat exchanger, requiring control action, which in turn leads to transients in the compressor operation. These feed into the distillation column. Presuming the separated unreacted cyclopropane is recycled back to the feed to the reactor, the temperature fluctuations into the distillation column will have translated into composition fluctuations in the recycle stream, which will then affect the reactor conversion, starting the whole cycle again. The plant should be designed to dampen fluctuations back to the set point, rather than reinforce them. Extended multiphysics is in play at every level of process coupling. In the linear flowsheet of Figure 1, it is possible to isolate the modelling of each unit operation, since the entry and exit points are the only overlaps. It is still extended multiphysics if you want to link them up in COMSOL, but the linkages are simple. But if process integration enters in, then the linkages may be more thorough. For instance, in distillation columns, differential heating and cooling of stages can be done to influence separation efficiency (with multiple entry and exit points for various "fractions"). These streams can be crossed for heat integration and recycled for reactor integration. Thus, "lumped variables" of unit operations become distributed constraints for others. That COMSOL can be called by Simulink for greater detailed modelling of some unit operations is a feature that allows better plant simulation. The commercial plant simulation packages, such as AspenPlus and HYSYS, have implemented links to computational fluid dynamics packages to improve detailed simulation of selected unit operations. This trend will be come a flood, as it is less expensive and safer to simulate "what if" scenarios than to implement them on real plant. Examples of extended multiphysics will make the concept

Table 1. Lumped parameter catalyst pellet reactor model.

Model Navigator	Open pellet.mph (saved from chapter three)
Physics Menu: Subdomain settings	Select chcd mode and for subdomain 2 Uncheck active in this subdomain
Options Menu: Constants	Add constant $mtc = 1$ to the list
Physics Menu: Boundary settings	Multiphysics: Select chcd mode Boundaries 7 and 8. Choose flux BCs Specify $N_0 = mtc * (C\text{-}c)$
Options| Integration coupling variables| Boundary variables	Select boundaries 7 and 8. Enter Name: flux Expression: $(cx * nx + cy * ny)/0.002881$ Keep default for global destination
Physics| ODE Settings	Enter name: C Expression: $Ct + \text{flux} + k * C\char`^2$
Solve Menu Solver Manager	Solve For tab. De-select ns mode. Select ODE and chcd modes. OK Solver parameters. Select time dependent solver. Specify output times, 0:0.1:2. Click on the Solve (=) button
Postprocessing Menu| Cross-section plot parameters	General tab. Select all times Point tab. Expression: C OK

clearer. We will start with a 1-D convection-diffusion-heterogeneous reaction model for a fixed bed supported catalyst system.

2. Phenomenological modelling of a phase

Chemical engineers developed the useful concept of phenomenological modelling or lumped parameters for heat and mass transfer — describing fluxes across phase boundaries by heat and mass transfer coefficients. Frequently these are referred to as "boundary conditions of the third kind" or Biot-type boundary conditions. The driving force for the flux N is the difference in concentrations c and C across the phase boundary of surface area A:

$$N = KA(C - c),\tag{2}$$

Concentration in dispersed phase

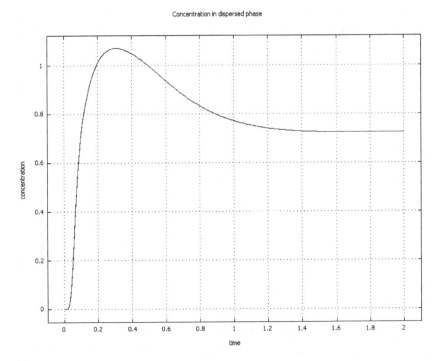

Figure 2.　Surface concentration C for phenomenological catalyst pellet reactor model.

K is termed the mass transfer coefficient, and it rolls up all the phenomena that are important in the overall mass transfer. This could include diffusion, phase equilibrium, and convection in either phase. The sign of N is determined by the modeller depending on the orientation defined as "inward" for each phase. Phenomenological modelling should be distinguished from distributed or differential modelling, which describes the detailed variable variation locally on a domain. Phenomenological modelling is also called "bulk" modelling, since it usually decribes the average value over the phase. Typically, if the phase is well-mixed, the detailed variation of the field variable is small, and phenomenological modelling is sensible. If the phase is not "homogenenized" by mixing, the detailed variation of the field variable can be large, and thus phenomenological modelling might not be accurate enough — an estimate. The key to phenomenological modelling are the transfer coefficients such as K. Sometimes, bulk models are reasonable due to the time scales of different processes/phenomena being rolled up into the transfer coefficient relative to those being included in the detailed model. The best description of the distinction between bulk and distributed mass transfer models is the textbook of Cussler [3].

In this section, we will modify the catalyst pellet reactor model of the last section to treat the case that the pellet is a well-mixed phase with a single concentration C, but the liquid is a distributed phase. It is probably stretching the bounds of imagination as to why the concentration should be uniform in on the catalyst surface, but it is possible if the surface is subject to rapid surface diffusion of the species by comparison to the time it takes for mass transfer to the surface. For this model to be reasonable, surface reaction should be relatively slow, i.e. k is small. In such circumstances, it is reasonable that (2) holds with C an average surface concentration, but c is the distributed concentration along the boundary. It is now straightforward to write a conservation equation for C in the pellet phase:

$$\frac{dC}{dt} = N - kC^2 . \tag{3}$$

The LHS of equation (3) represents the rate of accumulation of the reactive species on the pellet surface. The RHS represents the influx of the reactive species less its disappearance by binary reaction. The reaction is irreversible. The key things to note about (3) are that it is an ordinary differential equation and that it is on a conceptually different domain (0-D) than the liquid transporting the reactive species into the reactor. Thus, coupling the solution of (3) to the Navier-Stokes and convective-diffusion equation in the liquid phase with a Biot-type boundary condition (2) is conceptually extended multiphysics.

The advent of ODE Settings in Comsol Multiphysics for describing auxiliary state variable equations makes setting up this model especially easy. We don't actually have to set up a second domain or 0-D geometry as in the examples shown in Chapter one and two. Start up Comsol Multiphysics and follow the instructions in Table 1 to run the bulk model for the catalyst pellet reactor. This model modifies the MPH file developed at the end of Chapter three. Please save the model, as, say pellet_biot.mph, as we will change the kinetics in the next subsection.

2.1. *Bioreactor kinetics*

In this section, we will try a different variation. In this section, a similar approach will be used to model reaction of a passive scalar occurring in a single cell. The reaction kinetics will be taken as typical of bioreactors —

Langmuir-Hinshelwood:

$$r = \frac{kc_s}{(1 + \sigma c_s)^2},$$ (4)

where r is the rate of disappearance by reaction, which only occurs within the cell. σ represents the finite capacity of the cell to hold the substrate concentration, which saturates at a value controlled by this parameter. The usual rate controlling step, however, is the transfer of the nutrient from the medium across the cell membrane. The overall mass transfer process is usually modelled with a first order resistance, with the flux j given by

$$j = K_{\text{eff}}(c_i - c_s).$$ (5)

At steady state, the rate of disappearance by reaction is equal to the flux of nutrient across the cell membrance, i.e.

$$\int_{\partial\Omega} K_{\text{eff}}(c_i - c_s) = r.$$ (6)

Thus, the boundary condition on mass transport on the cell wall involves the concentration c_i on the boundary and the concentration within the cell itself, which is taken to be uniform. So the extended multiphysics here is to treat c_s in an additional 0-D space with reaction occurring only there, and coupling between the two spaces through the flux into the cell and through the boundary condition (5). Equation (6) can be seen as modeling the cell as a continuously stirred tank reactor (CSTR) with effective influx given by the integral, and irreversible reaction. The boundary condition (5) is ubiquitous in the chemical engineering literature, nevertheless, to the authors' knowledge, these examples are the first higher dimensional model that incorporates it as a boundary condition in a nontrivial way. If c_i is constant, (5) represents a simple mass transfer coefficient boundary condition of the Biot type that is easily included in any pde engine. But with c_i integrably coupled to the dynamics in the bounding domain, the implementation here, made possible by extended multiphysics in COMSOL Multiphysics, is unique. Attempts by one of us to implement this boundary condition in other finite element solvers were once abandoned due to the complexity of the coding.

Exercise 4.1.

With $K_{\text{eff}} = 1$ and $\sigma = 40$, alter pellet_biot.mph to implement Langmuir-Hinshelwood kinetics.

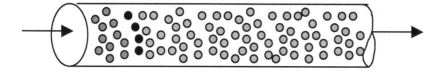

Figure 3. Schematic of a fixed bed with reaction largely localized.

3. Heterogeneous reaction in a fixed bed with premixed feed

Recently Mchedlov *et al.* [4] proposed a general lumped parameter model for heterogeneous reaction in a dispersed phase. The model focuses on situations where mass transfer is asymmetric, i.e. some species have greater mass transfer coefficients with the dispersed phase than others. Any number of physicochemical interactions could lead to this situation, but invariably it is in only slow flows, as through porous media, where kinetic asymmetry can survive. Turbulence usually leads to equal mass transfer coefficients for each species. Consider the reaction

$$u + v \rightleftarrows w \tag{7}$$

which only occurs in the dispersed phase. The lumped parameter model gives three convection-diffusion-mass transfer equations in the bulk phase, which for steady operation read as:

$$U\frac{\partial u}{\partial z} = D_u \frac{\partial^2 u}{\partial z^2} - j_u(u, \tilde{u}),$$

$$U\frac{\partial v}{\partial z} = D_v \frac{\partial^2 v}{\partial z^2} - j_v(v, \tilde{v}), \tag{8}$$

$$U\frac{\partial w}{\partial z} = D_w \frac{\partial^2 w}{\partial z^2} - j_w(w, \tilde{w}).$$

The fluxes j take the traditional mass transfer coefficient form

$$j_u = \kappa_u(u - \tilde{u}),$$
$$j_v = \kappa_v(v - \tilde{v}), \tag{9}$$
$$j_w = \kappa_w(w - \tilde{w}).$$

At steady state, these fluxes $j_u = j_v = -j_w$ are all equal due to stoichiometry and thus give two constraints on the bulk variables u, v, w and on the disperse phase concentrations $\tilde{u}$, $\tilde{v}$, $\tilde{w}$. The sixth constraint is on the surface reaction, which is presumed to be in equilibrium (fast reaction kinetics and

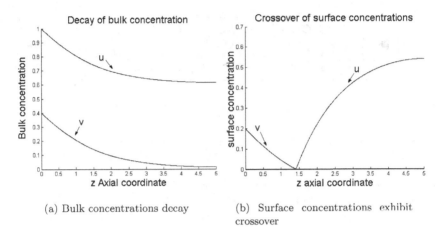

(a) Bulk concentrations decay

(b) Surface concentrations exhibit crossover

Figure 4.   Crossover in a tubular, heterogeneous catalytic reactor.

nearly irreversible):

$$\tilde{u}\tilde{v} - K\tilde{w} = 0 .\tag{10}$$

The boundary conditions will be taken as fixed concentrations of u and v at the inlet, no w, and outlet conditions with convection much greater than diffusion. For simplicity, since there are so many parameters, we will test just kinetic asymmetry of the mass transfer parameters and fix unit diffusivities $D_u = D_v = 1$, mobile product $k_w = 100$ and $D_w = 0.001$, one of the reactants to have unit mass transfer coefficient $k_v = 1$, and this leaves free parameters as the velocity u and mass transfer coefficient of the most resistive reactant, k_u, reactor length L, and equilibrium constant K. Since industrial interest lies in reactions that favor the products, we shall take $K = 10^{-5}$ as a nearly irreversible reaction. Initially, let's consider a reactor of length $L = 5$, velocity $u = 0.5$, and mass transfer asymmetry with $k_u = 0.2$. The inlet conditions will be $u_0 = 1$ and $v_0 = 0.4$.

Now to set up the COMSOL Multiphysics model. Table 2 gives the instructions to set up the premixed heterogeneous tubular reactor model. The differential variables U, V, W have Dirichlet boundary conditions at the reactor entry (boundary 1) and Neumann conditions at the outlet. The Neumann BCs for U, V, W require the normal component of Γ to vanish at the outflow boundary, presuming that the reactor is long enough that reaction has effectively finished, and diffusion has had long enough time to smooth out gradients. For the surface variables, however, $\Gamma = 0$ was specified, so entering zero Neumann conditions is a non-constraint ($0 = 0$).

Table 2. Premixed heterogeneous tubular catalytic reactor model. File name: premix.mph.

Model Navigator	Select 1-D space dimension Select COMSOL Multiphysics|PDE modes|PDE general Set dependant variables: U V W US VS WS. OK
Draw Menu	Specify objects|Line. Set x: 0 5 Set name: reactor. OK
Options Menu: Constants	<table><tr><td>u_0</td><td>v_0</td><td>D_u</td><td>D_v</td><td>D_w</td></tr><tr><td>1</td><td>0.4</td><td>1</td><td>1</td><td>0.001</td></tr><tr><td>u</td><td>K</td><td>k_u</td><td>k_v</td><td>k_w</td></tr><tr><td>0.5</td><td>1e-5</td><td>0.2</td><td>1</td><td>100</td></tr></table>
Physics Menu: Boundary settings	Select boundary 1. Select Dirchlet type boundary condition R tab Set $r_1 = u_0 - U$, $r_2 = v_0 - V$, $r_3 = -W$, $r_4 = r_5 = r_6 = 0$ Apply Select boundary 2. Select Neumann boundary type boundary condition. OK
Physics Menu: Subdomain settings	Select subdomain 1 Γ Tab: set $\Gamma_x = -D_u * U_x$; $\Gamma_2 = -D_v * V_x$; $\Gamma_3 = -D_w * W_x$; $\Gamma_4 = \Gamma_5 = \Gamma_6 = 0$ F tab: Set $f_1 = -u * U_x - k_u * (U - US)$ $f_2 = -u * V_x - k_v * (V - VS)$ $f_3 = -u * W_x - k_w * (W - WS)$ $f_4 = k_u * (U - US) - k_v * (V - VS)$ $f_5 = k_u * (U - US) + k_w * (W - WS)$ $f_6 = US * VS - K * WS$ d_a tab: $da_{11} = da_{22} = da_{33} = 1$; $da_{44} = da_{55} = da_{66} = 0$ Init tab $u(t_0) = u_0$; $v(t_0) = v_0$ OK
Mesh Menu	Select mesh parameters Boundary tab. Select boundaries 1 and 2 Enter maximum element size: 0.0001 Global tab. Set maximum element size scaling factor: 0.01 Remesh. OK (gives 2372 elements)
Solve Menu Solver Parameters	Select stationary Nonlinear solver Nonlinear tab: Check "highly nonlinear problem" OK

Note that the Dirichlet type boundary conditions in COMSOL Multiphysics are flexible enough to enter Dirichlet BCs for U, V, W, but Neumann conditions for the algebraic variables US, VS, WS at the reactor entrance. Please make sure to save the MPH-file (say as premix.mph) as we will modify it later in the chapter.

This model turns out to be highly nonlinear. That's why it takes 18 iterations to converge. Another reason for difficulty in convergence is that this model mixes differential equations for the bulk variables with algebraic constraints for the surface variables. Algebraic constraints are "stiff," meaning that they introduce effectively an infinitely fast time-scale or infinitely short length scale over which the system evolves. It took us (Deshpande and Zimmerman [5]) many years to realize that the neglected length scale here is that for accumulation. If there is a finite accumulation length scale, then the algebraic constraints have to be relaxed — they become differential equation constraints. This leads to a small "accumulation boundary layer" near the entrance by the end of which the algebraic constraints hold. In practise, resolving such boundary layers could require a finer grid, so there is a computational trade-off between stiffness and resolution.

Figures 4(a) and 4(b) show the behaviour of the reactant concentrations. Figure 4(b) in particular requires interpretation. Because of kinetic asymmetry, $\tilde{u}$ and $\tilde{v}$ vanish in different sections of the reactor. Because v has a greater mass transfer coefficient it populates the surface initially, $\tilde{u}$ reacts instantaneously as it arrives on the surface. As u is in bulk excess, however, eventually $\tilde{v}$ reacts away as well, until we reach the crossover point, where both surface reactants vanish. This is actually the point of greatest molecular efficiency, since any molecule of u or v that arrives on the surface reacts here. The theory of Mchedlov *et al.* [4] predicts the existence in parametric space of a crossover point, and gives a good approximation of its position X based on nearly irreversible reaction. Clearly, the actual profile requires solution of a two point boundary value problem with three conditions at either end. The system of equations (8)–(10) is a combined differential-algebraic system, which is inherently "stiff" due to the difficulty in satisfying the three nonlinear algebraic constraints simultaneously. Mchedlov *et al.* achieved it by shooting methods with stiff ODE integrators. The COMSOL solution naturally permits the satisfaction of two point BVPs and analytically determines the Jacobian of the nonlinear system, automatically with its symbolic tools. Mchedlov *et al.* determined the general Jacobian for their system, but due to the simple stoichiometry, used elimination to reduce the problem to a third order reaction-convection-diffusion system with highly nonlinear constraints. In terms of programmer effort, the COMSOL solution took an evening, the shooting method took several months.

3.1. *Reactor-separator-recycle extended multiphysics*

You would be forgiven for asking where in the above heterogeneous reactor model is the extended multiphysics. Although we saw rather clever use of COMSOL to solve a differential-algebraic system, there is not yet any

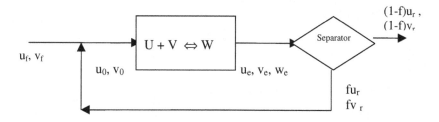

Figure 5. Simple flowsheet with heterogeneous reactor, separator and recycle.

extended multiphysics coupling. So now let's consider our reactor as part
of a very simple flowsheet with a separator and recycle. See Figure 5 for
the flowsheet.

The feed rates are taken as u_f, v_f. The reactor inlet rates are u_0,
v_0. The reactor exit rates are u_e, v_e, w_e. The separator is taken as an
ideal separator, but with a temporal response. This response is due to
the "stored" concentrations u_c and v_c that create the buffering effect. For
instance, a buffer tank where product w phase separates. The recycle rates
are u_r, v_r. With steady operation, the separator outlet rates must equal the
inlet rates. However, we are interested in the temporal response potentially,
so we will model the separator as a buffer tank with an effective capacitance.

Zimmerman [6] derived a model for imperfect mixing in buffer tanks due
to stratification effects. A model flow configuration in a buffer tank with
a two layer flow stratification was considered. The lower, denser stream
is presumed to short-circuit to the outlet, driving a recirculating cavity
flow in the upper layer. As the upper layer can be argued, due to strong
convective dispersion, to be well mixed, mass transfer to the upper layer
from the dense stream is the limiting step. In analogy with a plug flow
reactor, a shell balance on the material fluxes in a slug of the lower stream
leads to a lumped parameter mixing model with two limiting conditions:
(1) no mixing at infinite superficial velocity of throughput; and (2) perfect
mixing with infinite mass transfer coefficient. The time dependence of the
model is readily described as

$$u_r = Eu_e + (1 - E)u_c \,,$$
$$\frac{du_c}{dt} = \frac{F}{V}(u_e - u_c)(1 - E) \,, \tag{11}$$

F and V are the volumetric throughput and the volume of the buffer tank,
respectively. A similar set of equations holds for v. E is the lumped

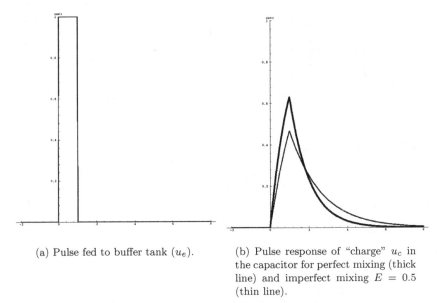

(a) Pulse fed to buffer tank (u_e).

(b) Pulse response of "charge" u_c in the capacitor for perfect mixing (thick line) and imperfect mixing $E = 0.5$ (thin line).

Figure 6.   Pulse response in an RC-circuit analogue model for mass transfer in a buffer tank.

parameter that describes the capacitance of the buffer tank. The latter of equations (11) is the the equation for the voltage response of the capacitor u_r in a driven RC-circuit with loading $(1-E)u_e$ and RC time constant $1-E$ [7]. Perfect mixing, analogous to a stirred tank model, occurs when $E = 0$, which then has the fastest possible response time constant. Figure 6(a) shows clearly that the concentration u_c in the upper layer is "charged" as the pulse passes and "discharges" after the pulse in the lower stream has passed. The outlet concentration (Figure 6(b)), however, for the imperfect mixing cases $E > 0$ shows large changes in concentration u_r due to the combination of the inlet stream short circuiting and the mass transfer to or from the upper reservoir, consistent with the first equation of (11).

Simple mass balance can be used to compute the reactor inlet concentrations:

$$\begin{aligned}
u_0 &= u_f + f u_r\,, \\
v_0 &= v_f + f v_r\,.
\end{aligned} \tag{12}$$

The nontrivial impact comes on reactor throughput. Taking the reactor to have unit area, the superficial velocity is given by:

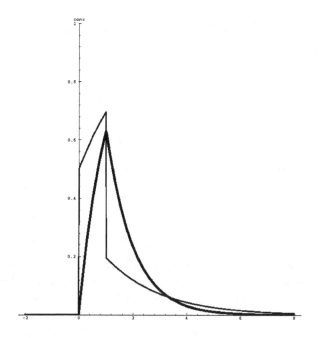

Figure 7. Outlet response of buffer tank to pulse in inlet to the pulse for perfect mixing (thick line) and imperfect mixing $E = 0.5$ (thin line).

$$U = U_{\text{no recycle}} \frac{u_0 + v_0}{u_f + v_f}. \tag{13}$$

By the way, the fraction retained in the recycle, f, might not be unity in case the system needs a purge. It may be the case that the product take-off has the effect of stabilizing the system since the throughput U is greater than unity (the no recycle throughput), but a purge may be necessary to avoid an infinite recycle ratio due to build up of reactants or trace impurities. It is always good chemical engineering design practice to include a purge, and then minimize it in operation.

Implementing the changes for recycle

This exercise is a *tour de force* in coupling variables. One strategy is simply to create a 0-D domain (as we have done many times before) to implement the ODE in time (11) and appropriate couplings to import the reactor outlet concentrations and merge the recycle with the feed stream into the reactor inlet. However, the advent of ODE Settings for state variables makes it unnecessary to introduce the second domain. Also, many intermediate

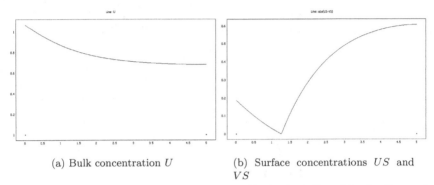

(a) Bulk concentration U (b) Surface concentrations US and VS

Figure 8.   Steady state reactor concentration profiles in the bulk and in the dispersed catalyst phase for 10% recycle.

coupling variables that were introduced in the Femlab 2.x version of this example were solely for the purpose of moving information availability from boundaries to/from subdomains. The global destination option for variables removes the need for such intermediates for information passing.

Pull down the Multiphysics menu and open premix.mph, which we created earlier in this chapter. Follow the instructions in Table 3 to implement the feedback loop of the recycle stream. Note that the initial conditions for U, V, US, VS are nontrivial profiles. One of the difficulties with differential-algebraic time-dependent systems is finding consistent initial conditions, i.e. initial conditions that do not result in very rapid initial changes. Since differential-algebraic systems are "stiff" (having an infinitesimal scale), it is important to choose initial conditions that only cause slow variation. By selecting $U = US$ and $V = VS$, initially, we have mitigated the "stiffness" in the algebraic constraints — they are initially satisfied. Ideally, we should have defined two general modes here: one for the three bulk variables and one for the three surface variables. The initial profile of the surface variables could be solved for while holding the bulk variables constant. The resulting surface profiles would be "consistent" with the initial bulk profiles.

In the first instance, we replace the boundary conditions (u_0, v_0) at the inlet (boundary 1) with values given by scalar expressions $(u_{\text{inlet}}, v_{\text{inlet}})$ and make the outlet concentrations (boundary 2) by defining coupling variables $(u_{\text{exit}}, v_{\text{exit}})$. Then we must define scalar expressions $u_{\text{inlet}} = U_{\text{feed}} + f * u_c * (1 - lump) + f * lump * u_{\text{exit}}$ and $v_{\text{inlet}} = V_{\text{feed}} + f * v_c * (1 - lump) + f * lump * v_{\text{exit}}$. Note that (u_r, v_r) are substituted for here. These scalar expressions

Table 3. Heterogeneous catalytic reactor, separator, and recycle flowsheet model.

Model Navigator	Load premix.mph
Draw Menu	Specify objects|Line. Set x: 0 5 Set name: reactor. OK
Options Menu: Constants	Replace u_0 by U_{feed}, v_0 by V_{feed}, u by lump Add $f = 0$ as a named constant. OK
Physics Menu: Subdomain settings	Init tab. Set $U(t_0) = U_{\text{feed}} * (5 - x)/5$; $V(t_0) = V_{\text{feed}} * (5 - x)/5$; $US(t_0) = U_{\text{feed}} * (5 - x)/5$; $VS(t_0) = V_{\text{feed}} * (5 - x)/5$;
Physics Menu: Boundary settings	Select boundary 1. Select Dirchlet type boundary condition R tab Set $r_1 = u_{\text{inlet}} - U$, $r_2 = v_{\text{inlet}} - V$, $r_3 = -W$, $r_4 = r_5 = r_6 = 0$ OK
Options Menu: Integration coupling variables|Boundary variables	Select boundary 2 Enter name: u_{exit}. Expression: U. Keep global destination Enter name: v_{exit}. Expression: V. Keep global destination OK
Options Menu: Scalar expressions	Name: u_{inlet} Expression: $U_{\text{feed}} + f * u_c * (1 - lump) + f * lump * u_{\text{exit}}$ Name: v_{inlet} Expression: $V_{\text{feed}} + f * v_c * (1 - lump) + f * lump * v_{\text{exit}}$ Name: u Expression: $0.5 * (u_{\text{inlet}} + v_{\text{inlet}})/(U_{\text{feed}} + V_{\text{feed}})$ OK
Physics Menu: ODE settings	Name: u_c Expression: $(u_{\text{exit}} - u_c) * (1 - lump)$ Init: u_{exit} Name: v_c Expression: $(v_{\text{exit}} - v_c) * (1 - lump)$ Init: v_{exit}
Solve Menu	Solver Parameters|Advanced Tab: Use weak solver Solver Manager|Initial value tab. Check initial value expression. Solve

are globally defined, but are not closed as the depend on (u_c, v_c), which are the "stored capacity" state variables in the buffer tank, according to (11). These state variables need to be defined in ODE Settings.

The instructions in Table 3 give the steady state equations (algebraic) equations, i.e. $\frac{du_c}{dt} = \frac{dv_c}{dt} = 0$. In order to create the time-dependent version, as in (11), the ODE Settings expressions must have the additional contributions $-uct$ and $-vct$, the COMSOL-ese for these terms. Note, however, that these are only defined with respect to the time dependent solver.

Because this model uses coupling variables, it is advisable to use the Weak solution form in the Solver Parameter settings (this is now the automatic setting). As we learned in chapter two, it is only because of COMSOL's unique implementation of constraints in the weak form that extended multiphysics is possible. The general or coefficient solvers may not bring out the full coupling of the model.

With no recycle and steady state, COMSOL finds the same solution for the reactor as before. You can ramp up the recycle ratio gradually using the Re-Start button on the Toolbar. By $f = 10\%$, a modest variation in the position of the crossover point is noted. The effect of the recycle is to load even more U into the reactor, as well as to speed up the throughput somewhat.

Figure 8 shows the steady state result for the bulk concentration U and the surface concentrations US and VS. Because US effectively vanishes where $VS >> 0$, and vice versa, these two profiles are efficiently illustrated by $abs(US - VS)$. One should note that the 10% recycle results in larger U values overall, i.e. $U_{\text{inlet}} = 1.068152$ and $U_{\text{exit}} = 0.87$. One would expect that a 10% recycle should not cause variations in observed quantities much different than 10%. A case in point is the crossover length, which was about 1.4 in the no recycle case, and is about 1.3 in Figure 8(b).

Exercise 4.2.

The transient solution is a hard problem, but potentially more interesting as the "capacitor" takes a long time to charge in the buffer tank. Simply, the mass transfer flux and surface reaction constraints are not time evolution equations. COMSOL Multiphysics naturally gives these equations in the weak form a d/dt accumulation term in the time dependent solver. Without the time evolution terms, these algebraic constraints make the system rather stiff. The additional coupling variables also subtract from the "sparsity" of the system, thereby making the sparse matrix solvers strain harder to converge. So don't be surprised if the solver steps in extended multiphysics problems take longer. In this exercise, follow the advice about how to turn the steady state recycle model into a transient model. Try to estimate how long it takes for the steady state to set up.

Exercise 4.3.

In Chapter 3§3, we added an oscillatory disturbance to the steady state solution of a tubular reactor and observed the transient response. Store the

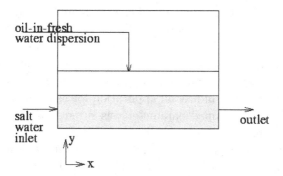

Figure 9.   Schematic of a stratified buffer tank with potentially two types of inlets and one outlet.

steady state solution to this heterogeneous reactor with buffer tank process, then create m-file functions for the initial conditions, adding an oscillatory disturbance to U only. How many m-file functions do you need to specify the steady state condition with oscillation? Comment about the transient response of the system. Alternatively, use the export function to get the data for the steady-state concentrations and build COMSOL interpolation functions.

Exercise 4.4.

The "sluggishness of the buffer" tank model depends to a large extent on the ratio F/V in (11), which is an inverse time scale. In the COMSOL model, implicitly, F/V was taken as unity. Explicitly add the quantity named *FoverV* as a parameter, and explore the transient response when varying *FoverV*.

4. Primacy of the buffer tank

In the previous section, the "main" physics were in the 1-D heterogeneous reactor, and the buffer tank, due to being modelled by a lumped parameter, was treatable by a 0-D capacitor model. Where lumped parameter models work, it is always a boon, since the dimensionality of the model is smaller and the equations generally simpler in form than the distributed system model that treats the physics more exactly. It begs the question, however, of where do you get a lumped parameter model from, and how do you get the lumped parameter dependencies. Generally, the lumped parameter model comes from analysis and simplification of a higher dimensional, distributed model. For instance, mass transfer coefficients come from solving film theories of convection and diffusion in a boundary layer flow. The lumped parameter, the mass transfer coefficient, can be predicted from the shape of the particle and the strength of the laminar flow. In turbulent flows, the functional form of the mass transfer coefficient is found from empirical correlations. The buffer tank lumped parameter model of §2 was developed for a specified industrial application for assessing concentration fluctuations, and the lumped parameter was fitted from samples of inlet and exit conditions.

Certainly to treat a specific industrial unit operation, semi-empiricism is a reliable approach. In the case of the buffer tank that inspired the work in [6], fluid density varied significantly with solute concentration (salinity), and thus the "capacitance" effect of the buffer tank was expected to be influenced by the rate of forced convection (throughput) F/V, viscous and mass diffusivity, and by the strength of free convection causing stratification, characterized by Reynolds, Prandtl and solutal Rayleigh numbers, respectively. Since the buffer tank lumped parameter model of §2 only includes the throughput effects explicitly, the dependence of the lumped parameter E on Reynolds, Prandtl and solutal Rayleigh numbers is unknown. It is just taken as a constant found from representative conditions. Whether or not a lumped parameter model is sufficient depends on the type and accuracy of the predictions required from the process model.

If the buffer tank is small, or shocks in solute concentration fluctuations are prevalent upstream, the lumped parameter model may be insufficient in predictive powers. Greater detail in the modeling would then be warranted.

In this subsection, we develop a two-component mass and momentum transport model for dense solute in a 2-D buffer tank. It is set up so that it can be augmented with a 1-D heterogeneous reactor with recycle model from the previous section. No lumped parameters are ever used in this model, as the detailed distributed effects of convection on mass transfer by coupling with diffusion and back action of density variations on convection are computed directly.

We will build up the model piecemeal, starting with the Incompressible Navier-Stokes model, then successively adding one mass transport mode, solutal Rayleigh effects, and then a second mass transport mode.

Component 1: Navier-Stokes flow field for cavity flow driven by free stream.

Table 4 gives the instructions to set up a flow field in the buffer tank cavity which is driven by the flow in the lower layer. After setting up this straightforward model, we wish to use this solution as the initial condition

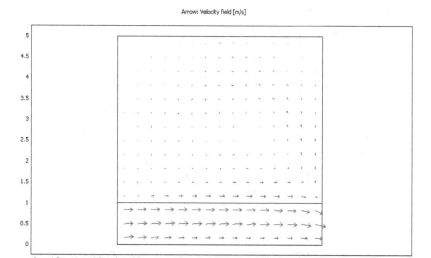

Figure 10. Cavity flow driven by a free current passing below.

for a more complex mass transport and buoyant convection model. This will ensure that the buffer tank is already flowing naturally when the density change in the fluid is introduced. Previously, with Femlab 2.x, it was necessary to store the solution as Matlab m-file interpolation functions and read them in through the Init tab. With the added functionality in the Solver Manager, it is now possible to avoid this data export and function import step. The "flow only" solution can be taken as the initial condition in the GUI. Save the model as buffer.mph for modification.

Component 2: Passive scalar convection and diffusionequation. Purists will note that the Chemical Engineering Module comes with a convection and diffusion mode (chcd). It is very good for the implementation of convective flux boundaries. Here, we implement the convective-diffusion equation in a general mode as the model does not require specialisation of the chcd mode. Follow the instructions in Table 5 to modify buffer.mph for mass transport effects.

Our instructions in Table 5 have two Solve Menu stages. The first sets up the solution to the flow only problem, which it holds in memory as the current solution. The second stage "turns on" all the variables and solves the transient problem by restarting with the initial condition that is the cavity flow only. Please save your model as, say, buffer_solutal.mph. We'll build more on it in the next section.

Table 4. Cavity flow driven by lower free stream. File name: buffer.mph.

Model Navigator	Select 2-D dimension Select COMSOL Multiphysics|Fluid Dynamics|Incompressible Navier-Stokes. OK
Draw Menu	Select specify objects|rectangle. Define $R1$ with width 5, height 1 and corner at (0,0). Now specify $R2$ with with width 5, height 4, and corner at (0,1)
Physics Menu: Boundary settings	Set up boundary 1 inflow/outflow BC; $u = 1$; $v = 0$ Set up boundary 2, 3, 5, 7 no slip Set up boundary 6 outflow/pressure $p = 0$. OK
Physics Menu: Subdomain settings	Select domains 1 and 2. Set $\rho = 1$; $\eta = 0.01$
Mesh	Click on the toolbar triangle for the default mesh
Solve Menu	Click on the solve (=) tool on the toolbar

Figure 11 shows the time history of with snapshots through time $t = 20$ (diffusive time scale) of the free convection velocity and concentration profiles. Although the density stratification is weak, it is apparent that denser fluid stays below lighter fluid. Times $0 - 1$, when animated, show the evolution of the gravity current as it spreads out along the bottom of the tank. The density front drives motion above and in front of it. Since $c_1 = 1$ fluid entering is denser than the $c = 0$ fluid next to it, it literally falls over. Rottman and Simpson [8] have conducted laboratory experiments that beautifully illustrate the formation of gravity currents. Although at some time after $t = 1$, the gravity current finds its way over to the constant pressure exit (whereupon it falls out), the gravity current continues to be the mechanism for driving the pseudo-steady flow. The fluid to the right is denser than the fluid to the left, so it just keeps on falling over. The initial push of fluid up and around that started the upper recirculation layer cycling does not maintain it. Rather, instead, it is the viscous drag from the gravity current layer that maintains the circulation above, much as how the free stream drives cavity flow.

The case of purely gravity current driven motion in a tank has not been studied before, so the two clear observations resulting from this model must be made. Firstly, the time to uniform concentration is extremely slow. The density variation with concentration not withstanding, one would expect nearly uniform concentration profiles after a few diffusion times, but in fact

Table 5. Modifications for buoyant convection and mass transport.

Model Navigator	Open buffer.mph Multiphysics tab|Model Navigator Select COMSOL Multiphysics|PDE Modes|General Mode. Dependent variables: c. Add. OK
Physics Menu: Boundary settings	Multiphysics: select g mode Set up boundary 1 Dirichlet BC; $R = 1 - c$ Set up boundary 2, 3, 5, 6, 7 Neumann BC. $G = 0$. OK
Physics Menu: Sub-domain settings	Multiphysics: select g mode Select domains 1 and 2. Set $F = -u * cx - v * cy$ Multiphysics: select ns mode Set $F_y = -0.25 * c$. OK
Mesh	Click on the toolbar triangle for the default mesh
Solve Menu (flow only IC)	Solver Manager|Initial value Tab|Check Initial value expression Solver Manager|Solve For Tab|Select only ns Solver Parameters|Select Stationary Nolinear Solver Click on the solve (=) tool on the toolbar
Solve Menu (full transient)	Solver Manager|Initial value Tab|Check Current solution Solver Manager|Solve For Tab|Select both ns and g Solver Parameters|Select time dependent solver Set times to 0:0.1:20 Click on the RESTART tool on the toolbar

there were still substantial gradients after $t = 50$. This is clearly due to the buoyant force opposing diffusive mixing, even in the presence of free convection which should, supposedly, enhance the mixing by dispersion. It is actually well known in the wave tank community that the ideal solution of fresh water/salt water can be used to set up any stable stratification density profile desired, simply because diffusion is such a weak mechanism that the profile is persistent. Turbulent mixing is another matter entirely. So the self-similar profile observed in Figure 11 for both velocity vectors and concentration is indicative of the long-lived nature of the transient intermediate approach to uniform mixing. It makes a mockery of "steady-state" analysis, since it is not clear that steady state is ever achieved in finite time nor is it clear that the uniformly mixed state will result at all.

Fick's law, which models the nonequilibrium transport of species, would have us believe that the equilibrium endgame has concentration uniformly diffused everywhere from a steady source. In fact, there are two greater complications that preclude this. The first is that it is not concentration that is diffusing at all, but rather chemical potential, and in an external gravitational field. At equilibrium, these two potentials must be balanced.

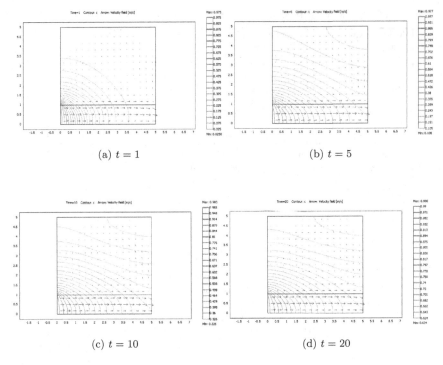

(a) $t = 1$ (b) $t = 5$

(c) $t = 10$ (d) $t = 20$

Figure 11.   History of forced convection in the buffer tank. Early times do not show the establishment of an upper recirculation zone, which does not occur until about time $t = 5$. Forced convection drags the lighter fluid with it at early times. It is only once stratification develops that recirculation follows. The later times are apparently self-similar in profiles.

So a permanent concentration gradient is maintained against a gravitational field. This fact is responsible for the difference in composition between air at sea level and at Mile High Stadium. In a buffer tank, it is probably meaningless, as the gradient in concentration is minute. The second complication that is probably more important in most chemical plants is that few solutions are exactly ideal, and many show significant volume change on mixing. Zimmerman [9] has shown that non-ideal solutions can have the structure of their stratification selected on chemical equilibrium grounds, and that only ideal solutions can ever be expected to form uniform mixtures at equilibrium.

The second observation is of the form of the velocity profile established — recirculation layer over a current. This is exactly the form postulated by

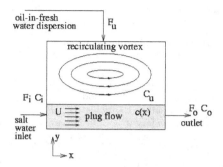

Figure 12. Plug flow across the tank bottom driving an upper recirculation layer.

Zimmerman [6] for which the lumped parameter model of imperfect mixing in the buffer tank was derived, equations (11). Figure 12 shows the idealized flow configuration for a denser current driving an upper recirculating layer. The lumped parameter model presumes that the recirculation is strong enough that the upper layer becomes well mixed, according to a theory of Batchelor [10], and thus a single concentration characterizes it. In fact, it seems that the upper recirculation is weak, yet the concentration gradients are small in the upper layer.

As the assumptions are all qualitatively met by the finite element model, it would seem likely that the lumped parameter model would be an acceptable approximation in the case of purely free convection through the buffer tank. Were the point of this chapter to verify the applicability of the lumped parameter buffer tank model, then we could run a parametric study fitting $E(Pr, Ra)$ for the free convection regime. The easiest route to fit E would be to compute outlet concentration by boundary integration over boundary 6 and fit the value of E which best fits the predictions of (11) to the simulated outlet times series. Since it is unlikely that a buffer tank would be operated under a purely free convection regime, it would not necessarily be useful information.

It is actually the case that the recirculation layer above is much stronger in this model, since forced convection imparts more momentum to the upper layer than free convection. It should not come as striking, however, that the flow configuration for gravity driven and pressure driven flows are broadly similar. Only at early times, while the transient flow field is still establishing, does the forced convection flow differ from the gravity current driven flow qualitatively. Before diffusion has had much time to act, fluid in the upper layer is just dragged along by viscous forces, yet it is heavy

Table 6.   An additional scalar transport mode (for species V).

Model Navigator	Open buffer_solutal.mph Multiphysics tab|Model Navigator Select COMSOL Multiphysics|PDE Modes| General Mode Dependent variables: c_2. Add. OK
Physics Menu: Boundary settings	Multiphysics: select g_2 mode Set up boundary 1 Dirichlet BC; $R = 1 - c$ Set up boundary 2, 3, 5, 6, 7 Neumann BC. $G = 0$ Set up boundary 6 outflow/pressure $p = 0$. OK
Physics Menu: Subdomain settings	Multiphysics: select g mode Select domains 1 and 2. Set $F = -u * cx - v * cy$

enough to fall back into the lower layer and fall out the constant pressure outlet. Once the upper layer becomes significantly stratified, however, the fluid dragged by the current has enough momentum to "turn the corner" and establish the upper recirculation layer. Thereafter, the profiles look self similar for both concentration and velocity vectors, and in qualitative agreement with the basis of the lumped parameter model for imperfect mixing in buffer tanks. So one would expect (11) to hold on average across the outlet, with $E(Re, Pr, Ra)$ as best fit "capacitance" constant found from time series analysis. Such analysis is beyond the scope of this chapter, but would be fruitful for modeling systems response in a complex flowsheet. In the next subsection, we link the 2-D model for the buffer tank to a 1-D model of the heterogeneous reactor, thereby justifying the description of the buffer tank modeling here in a chapter on "extended multiphysics."

Exercise 4.5.

Compute the average outlet concentration at a number of times for a pulsed inlet concentration, i.e. $U = 1$ for $t \in [0, 1]$ and $U = 0$ thereafter. Compare qualitatively the collected data for outlet concentration to Figure 7. Is the behaviour closer to perfect mixing or imperfect mixing with $E = 0.5$?

5. Linking the 2-D buffer tank to the 1-D heterogeneous reactor

The 2-D buffer tank model of §3 has unit flow rate inlet and unit concentration inlet conditions. However, neither is the outflow from the heterogeneous reactor a unit flow rate nor a unit concentration. Nominally the flow rate was set at $u = 0.5$ in §2, but the throughput varies with the recycle rate. So u, v and p will need to be scaled as initial conditions if the steady tank/heterogeneous reactor solution is to be used as an initial condition for the linked unit operations. If u_2 is the actually reactor outflow rate, then as initial conditions $u' = u_2 * u$, $v' = u_2 * v$, $p = \rho * u_2^2 * p'$. The last is the inertial scaling for pressure, appropriate at $Re = 100$.

It should be noted that we have two scalars to model in the 2-D buffer tank, since the concentration of species U and V can both vary independently. So we need to add another scalar transport mode. Table 6 is essentially a reproduction of the first scalar mode. Adding a solutal buoyant force that is linear in c_2 is simple (and we'll just change F_y in the ns mode in Table 7).

At this stage in my first (Femlab 2.2) development of this model, it was important to set up storage for the initial reactor condition as the steady-state found in §3 in Table 2. Here, due to the flexibility of the Solver Manager, we will merely reproduce the whole of the heterogeneous reactor model in a second, 1-D geometry. Table 7 reproduces such a sub-model. One might note that COMSOL Multiphysics does not have an obvious facility for merging models or copy modes. For instance, "copying" mode g to create mode g_2 would have saved much time. "Merging" buffer_solutal.mph with premix.mph would save the labor of duplicating the same GUI command steps as in Table 7. It could probably be done with the model m-file and a suitable cut-and-paste and global replacement editor utilities "by hand." But that sounds just as much trouble as running through the same steps again.

There are several things to note about the model. First is that, in comparison with the earlier algebraic model in Table 3, this distributed buffer tank model has no constant *lump* nor variables u_c and v_c. The coupling variables have the same form, except $lupm = 0$, effectively, since its role is played by the pde system of equations in the buffer tank (ns, g, and g_2 modes). Second is that the initial model is run with $f = 0$, so that the recycle is turned off. This has the effect of integrating the reactor and the buffer tank gently — finding the steady state operation of both, but effectively decoupled as the initial conditions are compatible. Adding only a slight amount of recycle $f = 0.001$ maintains convergence. Possibly, this is due to the dispersed phase model which might not always conserve mass due to an accumulation boundary layer in the dispersed phase being neglected. Higher recycle rates cause the model to career off from physically realistic operation. So this example should just be seen as indicative of how

Table 7.   2-D buffer tank model linked with heterogeneous reactor.

Multiphysics Menu	Model Navigator. Add geometry button Select 1-D space dimension Select COMSOL Multiphysics|PDE modes|PDE general Set dependant variables: $U \ V \ W \ US \ VS \ WS$ Set mode name: reactor. Add. OK
Draw Menu	Specify objects|Line. Set x: 0 5 Set name: reactor. OK
Options Menu: Constants	(table below)

U_{feed}	V_{feed}	D_u	D_v	D_w
1	0.4	1	1	0.001
f	K	k_u	k_v	k_w
0.0	1e-5	0.2	1	100
Init$U$	Init$V$			
0.6147	0.0147			

Physics Menu: Boundary settings	Reactor mode. Select boundary 1. Select Dirchlet type boundary condition R tab Set $r_1 = U_{\text{inlet}} - U$, $r_2 = V_{\text{inlet}} - V$, $r_3 = -W$, $r_4 = r_5 = r_6 = 0$ Apply Select boundary 2. Select Neumann boundary type boundary condition. OK Multiphysics: select mode g Set boundary 1 BC: $R = U_{\text{out}} - c$ Multiphysics: select mode g2 Set boundary 1 BC: $R = V_{\text{out}} - c_2$ Multiphysics: select mode ns Set boundary 1 BC (inflow/outflow): $u = vel$

Physics Menu: Subdomain settings	Reactor mode. Select subdomain 1

	U	V	W
F	$-vel * Ux - ku*$ $(U - US)$	$-vel * Vx - kv*$ $(V - VS)$	$-vel * Wx - kw*$ $(W - WS)$

Copy other entries from Tables 2 and 3
Multiphysics: select mode g
Select subdomains 1 and 2; Init tab: $c(t) = $ Init$U$
Multiphysics: select mode g2
Select subdomains 1 and 2; Init tab: $c_2(t) = $ Init$V$
Multiphysics: select mode ns
Select subdomains 1 and 2; Set $F_y = -0.25 * (c + c_2)$
OK

Options Menu: Integration coupling variables| Boundary variables	Multiphysics: reactor. Select boundary 2 Enter name: U_{out}. Expression: U. Keep global destination Enter name: V_{out}. Expression: V. Keep global destination OK Multiphysics: mode g. Select boundary 6 Enter name: u_{exit} Expression: c. Keep global destination Enter name: v_{exit} Expression: c_2. Keep global destination

Table 7 (*Continued*)

Options Menu: Scalar expressions	Select mode reactor Name: U_{inlet} Expression: $U_{\text{feed}} + f * u_{\text{exit}}$ Name: V_{inlet} Expression: $V_{\text{feed}} + f * v_{\text{exit}}$ Name: vel Expression: $0.5 * (u_{\text{inlet}} + v_{\text{inlet}})/(U_{\text{feed}} + V_{\text{feed}})$ OK. Repeat definitions of scalar expression for mode g
Mesh parameters	Choose mode reactor Set maximum element size 0.005. Remesh
Solve Menu Solver Parameters	Advanced tab|Scalings: None Advanced tab|Solution form: Weak OK. Solve
Options Menu: Constants	Set $f = 0.001$
Solve	Advanced tab|Scalings: Automatic Restart

to link two unit operations with complex spatial-temporal behaviour. The linear linkage works fine at steady state, but the recycle model with $f \neq 0$ appears to be numerically unstable.

Exercise 4.6.

In the no recycle model ($f = 0$), alter the initial condition so that $U(t_0) = U(f = 0) + 0.02 * \sin(31.4159265 * x)$. Does this oscillation grow or decay? What effect does the buffer tank have on the oscillation?

References

[1] COMSOL Chemical Engineering Model Library, User's Guide, Version 3.2, pp. 67, 294.
[2] M. Grammatika and W. B. Zimmerman, Microhydrodynamics of flotation processes in the sea surface layer, *Dynamics of Oceans and Atmospheres* **34** (2001) 327–348.
[3] E. L. Cussler, *Diffusion: Mass Transfer in Fluid Systems* (Cambridge, London, 1997).
[4] P. O. Mchedlov-Petrossyan, G. A. Khomenko and W. B. Zimmerman, Nearly irreversible, fast heterogeneous reactions in premixed flow, *Chemical Engineering Science* **58** (2003) 3005–3023.
[5] K. B. Deshpande and W. B. Zimmerman, *CES 2005* (2005).
[6] W. B. Zimmerman, The modelling of imperfect mixing in buffer tanks, *Mixing VI, IChemE Symposium Series*, Vol. 146, ed. H. Benkreira (1999), pp. 127–134.
[7] M. E. Valkenburg and B. K. Kinariwala, Linear circuits, *Prentice-Hall Computer Applications Series*, ed. F. F. Kuo (1982), p. 162.

[8] J. W. Rottman and J. E. Simpson, The formation of internal bores in the atmosphere: A laboratory model, *Q.J.R. Met. Soc.* **115** (1989) 941–963.

[9] W. B. Zimmerman, The effect of chemical equilibrium on the formation of stable stratification, *Appl. Sci. Res.* **59** (1998) 298.

[10] G. K. Batchelor, On steady laminar flow with closed streamlines at large Reynolds number, *J. Fluid Mech.* **3** (1956) 177–190.

Chapter Five

NONLINEAR DYNAMICS AND LINEAR
SYSTEM ANALYSIS

W.B.J. ZIMMERMAN

*Department of Chemical and Process Engineering, University of Sheffield,
Newcastle Street, Sheffield S1 3JD United Kingdom
E-mail: w.zimmerman@shef.ac.uk*

Eigensystem analysis of the linearized operator derived by FEM analysis (the stiffness matrix) is a powerful tool for characterizing the local stability of transient
evolution of nonlinear dynamical systems governed by pdes and for parametric
stability of stationary, nonlinear problems. Here we discuss how to perform such
an analysis in the context of two complex systems — Benard convection and viscous
fingering instabilities. The later are simulated from "white noise" initial conditions
added to a base flow. The linear stability theory in both cases assumes that the
noisy initial conditions include all frequencies, and thus whichever eigenvalue has
the largest real part corresponds to the eigenmode that grows most rapidly. FEM
eigenanalysis is shown to reproduce the predictions of linear stability theory with
good agreement, but is more general in regimes of applicability.

1. Introduction

Modelling versus simulation

So far, we have been concerned with the use of FEM for computational
modeling. The model could be expressed as a well posed mathematical
system, typically PDEs with boundary and initial conditions, possibly algebraic constraints. Such systems are theoretically deterministic, i.e. the
state of the system can be known up to any arbitrary accuracy at any
given time. By simulation, something different is usually understood —
the physics of the system includes some element of randomness in its temporal development. So we don't expect a simulation to be perfectly accurate
in all details. Simulations are expected to mimic the microscopic behaviour
of complex systems, typically by posing interaction rules for subsystems
from which the global, coordinated behaviour of the whole system emerges.
Where the low level interaction rules of the system are poorly physically

based, the simulation predictions about global emergent properties must be validated by experiment, perhaps even semi-empirically fitted.

Equivalence?

With the above classification scheme, computational modeling and simulations would appear to be wholly distinct — models are deterministic and physically based; simulations are stochastic and semi-empirically based. This dichotomy blurs, however, with modern understanding of complex systems. Billings and coworkers [1], for instance, have developed a data analysis technique for patterns in spatio-temporal systems that can identify the best PDE system within a candidate class that captures the nonlinear dynamics in experimental systems. The technique finds a rule based description for cellular automata that is consistent with the complex system pattern development. By limiting the types of PDE terms available to the model, an inverse mapping from interaction rules to PDE description can be elucidated. So, the common usage of muddling the terms "modelling" and "simulation" is justified by this functional equivalence. No doubt this is the "new kind of science" that Wolfram [2] is espousing; dynamics can be equated to simulation schemes (new science) which are equivalent to (nonlinear) *pde* systems derivable from physical laws (old science). Where the new science wins is that the applicable physical laws may be two complicated to describe in full *a priori*, but those that are being expressed in the complex system may be easily identifiable by finding the interaction rules that are consistent with the global emergent behaviour. Koza and coworkers at Stanford [3] have long been proponents of the view that the trick is to find the computer program that meets the physical requirements. Genetic programming is an approach to letting the program consistent with the observations to assemble itself.

Bridging the gap: Nonlinear dynamics

The gap between modeling deterministic systems and simulating stochastic ones is bridged by the nature of nonlinear dynamics and complex systems. The principle feature of a class of nonlinear dynamics — chaotic systems — is that of extreme sensitivity to initial conditions. States that are not particularly far initially in some sense become very far apart eventually. Before chaos theory became better understood in the late 1970s, it was conventional wisdom that in dissipative systems, equilibrium states or periodic oscillations would be the time asymptotic attractors for all initial states of the system. When a dynamical system is extremely sensitive to initial noise or uncertainty, such a system is termed complex. The paradigm in fluid dynamics is turbulence. Shear instability of flows "at high Reynolds number" lead to complexity of the motion at millimeter scales all the way up to thousands of kilometers in the atmosphere, for instance. Even though the temporal state of the system is theoretically deterministic, our uncertainty in the initial state is such that the system is indistinguishable from a stochastic one for practical purposes.

So is there any point in using PDE based models to describe complex systems for which the complexity is practically indescribable? Of course, we can derive or pose PDEs for the dynamics of the statistics (traditional turbulence modeling) or to collate the statistics of the dynamics. In meteorology, the latter is termed ensemble forecasting, and it is an attempt to quantify likely behavior of emergent properties, rather than to average out uncertainty.

Stability and eigenanalysis: Time asymptotic behavior

Key to the evolution of nonlinear systems is the notion of stability. A state of a system, $u_0(t)$, is said to be stable if small perturbations, δ, do not displace the new state of the system, $u(t)$, very far from the original state. The concepts of a "state," how you measure "small" and "how far" two states are separated need to be precisely defined for stability (and therefore instability) to be a useful concept.

The operational definition of a state $u(t)$ is simply to list all of the degrees of freedom necessary to uniquely define a recurring pattern in the system. For a FEM model, this means giving the time dependence of a solution which is typically either stationary or periodic. The exception is that a chaotic attractor is also a "state" of a dynamical system, deterministically known as a solution trajectory $u(t)$ from an initial state, but not uniquely defined as the attractor is an "asymptotic state" — many initial conditions are attracted after a long time to this state. In fact, the states of FEM models are easier to describe than for the underlying PDE system, which is inherently infinite dimensional. Once the trial functions and finite elements are chosen, a FEM model is finite dimensional and the degrees of freedom necessary to define a state is just the space of all possible solution vectors.

In terms of FEM models, it is also straightforward to describe the stability of a solution trajectory $u(t)$. Consider the FEM operator that maps the solution at time t to the solution at time $t + \Delta t$:

$$N\{u(t)\} = u(t + \Delta t) \, . \tag{1}$$

Conventionally, for small time steps, this operator can be linearized, so that when applied to the perturbed system, we can compute

$$N\{\mathbf{u}(t) + \delta\} = \mathbf{u}(t + \Delta t) + \mathbf{L}\delta \,, \tag{2}$$

where L is the Jacobian of N

$$\mathbf{L}_{ij} = \frac{\partial N_i}{\partial u_j}(\mathbf{u}) \,. \tag{3}$$

If L is a Hermitian matrix (if real, then symmetric), then the principal axis theorem says that the evolution of the perturbations can be exactly described in terms of the eigenvalues λ_i and the normalized eigenvectors ϕ_i of L as follows:

$$\mathbf{u}'(0) = \delta \,,$$

$$\mathbf{u}'(\Delta t) = \sum_{i=1}^{N} \delta \cdot \phi_i \exp(\lambda_i \Delta t)\phi_i \,, \tag{4}$$

where $\mathbf{u}'$ is the change in the system trajectory due to the occurrence of the initial disturbance δ. Due to the exponential growth rate, one would expect that from any initial condition, the mode associated with the eigenvalue λ with largest real part would eventually dominate the long term evolution of the disturbance to the state $\mathbf{u}$. It simply grows the fastest or decays the least. If the state $\mathbf{u}$ were either stationary or periodic, then if there is any eigenvalue with positive real part, then (4) will grow without bound. So the system is unstable. In fact, as we have defined the state $\mathbf{u}(t)$, even a chaotic attractor is unstable according to this criteria. The difficulty with a chaotic attractor is defining unequivocally what the asymptotic state $\mathbf{u}(t)$ is. Consequently, an instantaneous point in phase space $\mathbf{u}$ that is part of a chaotic state $\mathbf{u}(t)$ is found to always have at least one unstable direction ϕ_i, but since it is difficult to distinguish between the time evolution of the state $\mathbf{u}(t)$ and the perturbation, δ, the global stability of the attractor cannot be found by local, linear analysis. The eigenvalues λ_i from the local analysis of a chaotic attractor are called Lyapunov exponents. Since negative real parts for λ_i imply that a trajectory $\mathbf{u}(t)$ is decaying, at least one Lyapunov exponent must have a positive real part at each point of a chaotic attractor.

As an aside, equation (4) helps us understand what it means for a perturbation to remain small and the degree to which two trajectories are close. A straightforward measure of closeness of two trajectories, $\mathbf{u}_1(t)$ and

$u_2(t)$, is the distance formula (or error):

$$E\{\mathbf{u}_1, \mathbf{u}_2\} = \|\mathbf{u}_1 - \mathbf{u}_2\| = \left(\frac{1}{N} \sum_{i=1}^{N} [u_{1i} - u_{2i}]^2 \right)^{1/2}, \qquad (5)$$

where the sum is over the N degrees of freedom that defines a solution vector. For instance, the Newton solver attempts to converge successive solution approximations by sending E to zero. Equation (4) implies that to be small $E\{\mathbf{u}', 0\} < \varepsilon$ must hold for some tolerance ε for all time t. If all $Re\{\lambda_i\} < 0$, this is achieved for $\varepsilon \gg \delta$. If any $Re\{\lambda_i\} > 0$, this can never be achieved.

The error norm of (5) is just one of many weighted errors that can be defined, e.g.

$$E_W\{\mathbf{u}_1, \mathbf{u}_2\} = \|\mathbf{u}_1 - \mathbf{u}_2\| = \left(\frac{1}{N} \sum_{i=1}^{N} w_i [u_{1i} - u_{2i}]^2 \right)^{1/2} \qquad (6)$$

also defines a measure for any set of weights $w_i > 0$. The choice of all weights equal in (5) only makes sense for a convergence criteria if all degrees of freedom are expected to range over the same scale. One of the rationales for dimensional analysis of physical models is to condition all degrees of freedom to range over a unit scale. In any unconditioned model, the range of scales expected *a priori* for different degrees of freedom would not be expected to be identical. COMSOL Multiphysics introduced the "automatic scaling of variables" feature, that estimates the appropriate weights w_i automatically or permits user pre-defined scales. The release notes point out that in a structural mechanics application, displacements might be submillimeter, yet stresses could be megapascals. Without scaling of variables, numerically small quantities would have degrees of freedom contributing little to convergence criteria, and numerically large quantities would be unduly restricted by convergence criteria using formula (5).

In summary, excepting the case of chaotic states, linear theory can identify whether a stationary or periodic state is unstable. Regardless, it also identifies the mode(s) that are asymptotically attractive for the perturbation. For instance, if an eigenvalue is complex, then the frequency of oscillation of the perturbation can be predicted, along with the decay or growth rate. Furthermore, the eigenvector associated with the eigenvalue with greatest real part should be the pattern of degrees of freedom that a disturbance evolves into. Using FEM models, representing these eigenmodes is straightforward. They are elements of the space of all possible solutions, so any postprocessing that can be done on a solution can be done to an eigenmode as well. COMSOL Multiphysics, for instance, can be "tricked" into displaying and analyzing eigenmodes as though they were solutions.

Chapter organization

This chapter can only be a survey of the range of models that can be used in simulations. The theme of the chapter is to illustrate how features of the MATLAB/COMSOL Multiphysics computational engines can be used for simulation. A strong undercurrent, however, is awareness of how nonlinear dynamics is important in computational modeling. Our first case study, Rayleigh-Benard convection, is simply a stationary nonlinear system for which convergence is difficult to achieve because of the inseparability of parasitic time dependent solutions excited due to numerical noise. Undoubtedly, users of COMSOL Multiphysics have already found a straightforward application of nonlinear dynamics theory — conditioning the computational model on the basis of dimensionless parameters in the system. Our second case study illustrates the importance of resolving all scales of the complex system which naively range from the large scale of the geometric boundaries (dimensionlessly this is termed $O(1)$ or order unity since the lengths are usually scaled by a geometric length) down to some small scale set by nonlinear processes coupled with dissipation. If the parameter that characterizes nonlinearity is called R, and complex behaviour increases with increasing R, then one expects creation of complexity down to lengths $O(R^{-1})$. Thus, in regions generating complexity, the mesh should be gridded with resolution $O(R^{-1})$. Novice modellers routinely fail to recognize that no satisfactory solution may emerge if all the physics generating complexity are not resolved. Some physical processes routinely generate large complexity parameters. Buoyant convection usually has large Rayleigh number Ra. Pipeline flows are almost always at large Reynolds number Re. Heat transfer almost always has large Peclet number Pe. Simply, given the small values naturally occurring for transport coefficients, human scale flows lead to large complexity parameters. Convergence to a solution does not guarantee that the dynamics of the model are resolved. Careful modeling requires mesh refinement studies until a claim that "refining the mesh does not change the result appreciably" is fully justified. Even experienced modelers can fall into the trap of unresolved computational models due to the large complexity parameter problem. For instance, if there are still unresolved motions, but little "sub-grid" energy transfer, it is convenient to think that laminar solutions to the buoyant convection problems in double diffusion are, for example, able to ignore small scale dynamics. Chashechkin *et al.* [4] argue cogently that there is never a stationary solution to the double diffusion problem with vertically heated walls. Internal boundary currents are automatically excited, leading to sharp fine structure layering the flow. This feature is not captured by high solutal/thermal Rayleigh number convection since it is not possible even with typical high performance computing resources. So the "large eddy" simulations with low subgrid fluxes may still be unresolved, even if there is little change on mesh refinement — fine structure may influence global dynamics.

2. Rayleigh-Benard convection

Rayleigh-Benard convection is certainly the canonical problem for nonlinear dynamics and flow stability. You can visualize it by heating vegetable oil in your kitchen, sprinkling cocoa powder on the surface of a thin layer of oil heated from below in a frying pan. The hexagonal patterns are clearly visible unless your cocoa powder has congealed. Still an excellent reference on the history of the problem can be found in Drazin and Reid [5]. The gist of the problem is that a vertically decreasing temperature profile and no flow is identically a solution to the boundary value problem stated as

$$\frac{\partial u}{\partial t} + u \cdot \nabla u = -\frac{1}{\rho}\nabla p + \nu\nabla^2 u + \frac{\alpha g}{\rho}T \,,$$

$$\nabla \cdot u = 0 \,, \tag{7}$$

$$\frac{\partial T}{\partial t} + u \cdot \nabla T = \kappa\nabla^2 T \,,$$

(c.f. equations (1), Chapter 3) with boundary conditions of

$$T|_{z=0} = T_0 \,,$$
$$T|_{z=1} = T_1 \,, \tag{8}$$

where the bottom temperature is usually greater than the top. The dimensionless groups that matter are still the Prandtl number and the Rayleigh number:

$$Pr = \frac{\nu}{\kappa} \,,$$

$$Ra = \frac{\alpha g(\delta T)h^3}{\rho\nu\kappa} \,, \tag{9}$$

where h is the depth of the fluid, δT is the applied temperature difference, α is the coefficient of thermal expansion, $\mathbf{g}$ is the gravitational acceleration vector, ρ is the density, ν the kinematic viscosity, and κ is the thermal diffusivity.

You can be forgiven for thinking that we have just turned the hot wall-cold wall problem on its side. In the case of *vertical* heated walls, motion is automatically induced even with an infinitesimal temperature difference. Fluid along a hot vertical wall must rise. For *horizontal* heated walls, however, a steady, linear temperature profile with no motion is an exact solution to the system. If the heating is from above, that makes perfect sense as hot light fluid will lie over colder dense fluid — gravity supplies a buoyant restoring force to any fluid element that might be displaced vertically. For heating from below, however, the stratification has dense fluid over light. Buoyant forces should overturn this top heavy profile. Yet, if viscosity and thermal conductivity are strong enough, they resist the motion. Stability theory identifies a critical Rayleigh number above which convection cells form. Below that critical Rayleigh number for the onset of instability, Ra_c, dissipation still damps out motion.

So let's explore these two situations by finite element analysis.

2.1. *Heating from above*

We will save time by just altering our buoyant convection example for the hot wall-cold wall problem by changing the boundary conditions. However, in order to visualize convective rolls in two dimensions (hexagonal cells are a 3-D phenomenon), we need to have an aspect ratio of about 3:1 for width:height of the layer or greater. Deleting an element of the geometry, however, has the annoying difficulty of erasing the model settings in the COMSOL Multiphysics GUI for that element. So here, we use a little MATLAB manipulation to "trick" COMSOL Multiphysics into accepting a changed geometry without losing the model settings. Table 1 contains the instructions to set up the COMSOL Multiphysics model.

The new mesh generates 4928 elements. As you can see in Figure 1, the new geometry does not change the old solution, but the old solution is effectively lost upon remeshing, since the two are widely disparate. Nonetheless, the exporting and importing of fem structures, with a swapping of geometries in MATLAB, make it possible to change the geometry without losing the model settings. Care should be taken that the new geometry has the same boundary and subdomain numbers as the original geometry,

Table 1. Buoyant convection: heating from below. Aspect ratio 3:1. Name this model benard.mph.

Model Navigator	Open buoyantconvection.mph from Chapter 3
File Menu	Export fem structure as ... fem1
Draw Menu| Draw mode	Select R_1 and delete it
	Select specify objects|rectangle. Define R_1 with width 3, height 1 and corner at $(-1.5, 0)$
File Menu	Export fem structure as ... fem
MATLAB command line	>>fem1.geom=fem.geom
	>>fem1.draw.s=fem.draw.s
File Menu	Import fem structure ... fem1
Options Menu| Constants	Add $Pr = 1$ and change $Ra = -100$
Physics Menu: Boundary settings	ns mode: set boundary 1 and 4 slip/symmetry BC;
	chcc mode: set boundary 1 and 4 heat flux BC; $q_0 = 0$
	set up boundary 2 temperature BC $T = T_1$
	set up boundary 3 temperature BC $T = T_0$
Physics Menu: Sub-domain settings	ns mode. Select domains 1
	Set $\eta = 1$
	Set $F_y = Ra * Pr * (T - T_0)/(T_1 - T_0)$
	Init tab: set $u(t_0) = 100 * \sin(x)$;
	$v(t_0) = 100 * \cos(x) * y * (1 - y)$
	chcc mode: Init tab: set $T_1 - (T_1 - T_0) * y$
Mesh	Click on the toolbar triangle for the default mesh
	Refine the mesh twice
Solve Menu	Set the stationary nonlinear solver
	Click on the solve (=) tool on the toolbar
	Solver parameters: Set time dependent solver
	General tab. Set output times 0:0.1:5
	Solve

else the new model will not be "similar" to the old one. Here, replacing a square domain with a rectangular one does not change the mappings of the vertex, boundary, or subdomain indices.

The my first attempt at this model was poorly convergent, but the addition of a point datum removes the pressure degeneracy. This problem has no imposed pressure on the boundary, so the solution is unique up to an arbitrary additive pressure constant. The pressure equation

$$\nabla^2 p = f\left(\frac{\partial u_i}{\partial x_j}\right) \tag{10}$$

is therefore singular. The Point Settings under the Physics menu set the pressure datum as $p = 0$ at the origin. This removes the singularity and gives the pressure field uniquely.

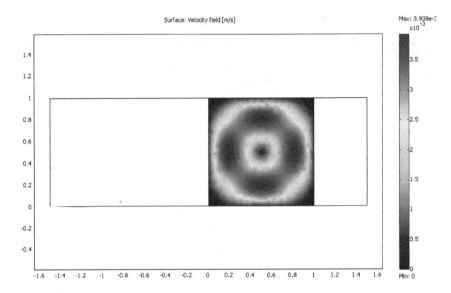

Figure 1.   Manipulating the geometry without losing the modelling settings.

The second reason for the slow convergence is that the velocity field should be identically zero as the solution. However, noise around zero interacts strangely with the new feature of the solver that permits the scaling of the estimated error using the nonlinear and time-dependent solvers. So this feature must be disabled.

Pull down the Solver menu and select Solver Parameters. Click on the Settings button under "Scaling of variables." Check the None option. Now select the Stationary Nonlinear solver, and solve. The convergence should be faster since the solver is not trying to distinguish noise around the zero (no flow) solution of the velocity field. Fluid dynamicists routinely nondimensionalize their PDE systems, i.e. scaling the variables which results in new variables that have no units — pure numbers about the size of unity. By manually scaling the PDE system, the dimensionless parameters, like Reynolds numbers, Prandtl numbers, Rayleigh numbers, etc. tell the typical size of the terms in the equations. If these numbers are large or small compared to unity, then the solution typically requires fine resolution in some parts of the domain to converge. A common requirement is that near boundaries, the mesh needs to be finer than far away from the boundaries, as these are the regions that are most likely to be rapidly changing with extreme dimensionless parameter values. The automatic scaling of variables

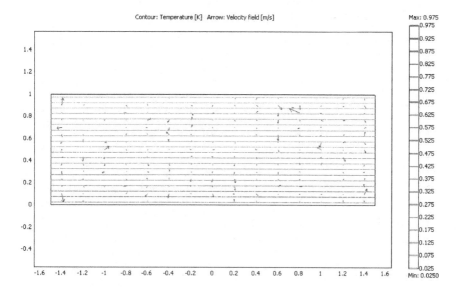

Figure 2. No flow solution to buoyant convection problem with heating from above, $Ra < 0$. The velocity vectors are randomly oriented and at the noise level. The isotherms are horizontal and parallel, appropriate for purely thermal conduction.

feature of COMSOL Multiphysics attempts to achieve the desirable result that the matrix solutions are well conditioned. In general, this is a good thing. However, in cases where the solution for a variable is naturally zero it might fail.

Another aspect in which scaling of variables contributes is in the estimation of errors which is used to measure convergence. Without scaling, it is conceivable that the units used for expressing quantities result in a huge disparity in the numeric values. Reporting energy in ergs might lead to large values but reporting lengths in kilometers might lead to small numbers. Thus, the "sum of squared errors" approach to error estimation might completely ignore length errors as small and focus on bringing down the error in energy for best convergence. Properly scaled variables give roughly equal weighting to the convergence of all the degrees of freedom.

There are several features of the model of Table 1 which are of interest. The initial conditions for the temperature field are set to give parallel horizontal isotherms consistent with purely thermal conduction of heat, as seen in the solution shown in Figure 2. The velocity field is intentionally set to "spinning" with only the boundary conditions satisfied by the initial profile. These artificial rolls are completely dissipated away by friction in

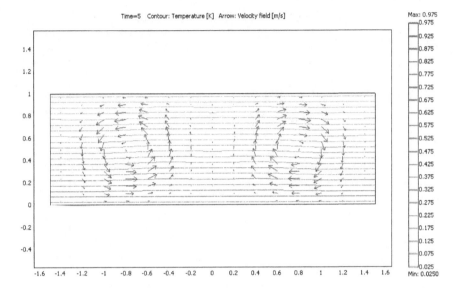

Figure 3.   Isotherms and velocity vectors for the decay of the initial rolls in the buoyant convection model with heating from above. The rolls are long lived (here at time $t = 5$) they are persistent, and the structure has taken an asymptotically self-similar form.

the steady state solution, which is ideally the zero flow case. Figure 2 just shows residual, randomly oriented velocity vectors on the scale of noise.

The transient solution found in the last solver step in the model of Table 1 is an attempt to understand how the initial condition breaks down and approaches the steady state. If the post-processing is animated, the movie shows that the temperature profile propagates waves by convection, and the structure of the decaying rolls changes in character. By time $t = 5$, the velocity field has a structure consistent with Figure 3.

2.2. *Internal gravity waves*

The automatic scaling setting fails for a subtle reason. With $Ra < 0$, any perturbation or numerical error excites a small amplitude internal gravity wave — an inherently time dependent phenomenon. So the stationary nonlinear solver cannot converge to the "internal waves" that are inherent in the Newton iterations. The automatic scaling setting senses that the proper velocity scale is that of the noise, and therefore tries to resolve and converge the internal gravity waves. Since these are small if the numerical error is small, they can be ignored, which is what happens if you disable the automatic scaling for the velocity field. That there are wave like solutions

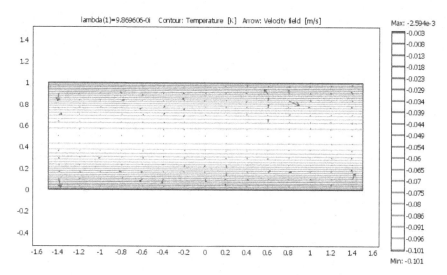

Figure 4. Isotherms and velocity vectors for the slowest decaying mode $\lambda(1) = 9.8696$ of the linear stability study of the rest state (no flow) with the static thermal conduction temperature profile with $Ra = -100$.

can be discerned from an eigenfunction analysis of the solution. The initial value problem of the previous subsection clearly shows that before the flow decays to the zero flow solution, long-lived structures are prevalent — the slowest decaying part of the buoyant convection dynamics.

Comsol Multiphysics allows us to explore these solutions in two ways by eigenfunction (linear system) analysis. The GUI has built-in an eigenvalue solver. Earlier versions of FEMLAB only used the eigenvalue solver for linear problems specified in the eigenfunction mode. The COMSOL Multiphysics eigenvalue solver is more robust — it will linearize the PDE system about the solution selected in the Solver Manager, assembling the stiffness and constraint matrices K, N and D (see chapter two) required in the block matrix for the time-dependent Newton solver:

$$-\lambda \begin{bmatrix} D & 0 \\ 0 & 0 \end{bmatrix} \begin{bmatrix} U \\ \Lambda \end{bmatrix} + \begin{bmatrix} K & N^\dagger \\ N & 0 \end{bmatrix} \begin{bmatrix} U \\ \Lambda \end{bmatrix} = 0. \tag{11}$$

It is important to note that the full transient description of the problem with da time-dependency coefficients must be specified for the eigenvalues λ to be computed correctly. These coefficients contribute to the D matrix. Clearly, if D is not defined (identically zero), then λ cannot be found.

It is very simple to use the built-in eigenvalue solver. Follow these simple steps:

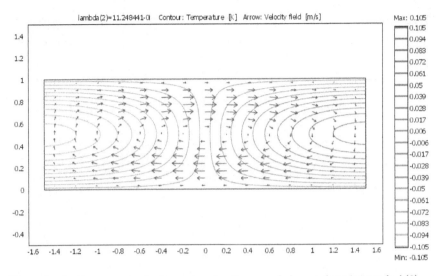

Figure 5.   Isotherms and velocity vectors for the second slowest decaying mode $\lambda(2) =$ 11.2484 of the linear stability study of the rest state (no flow) with the static thermal conduction temperature profile.

(1) Make sure you have the desired solution about which to compute the eigen system stored in any of the initial conditions, stored solution, or current solution.

(2) Use the Solver Manager Initial Value Tab to select the appropriate solution to use.

(3) Use Solver Parameters to specify the eigenvalue solver.

(4) Solve with the Solve (=) button on the toolbar.

This sequence make certain that you know which solution you are studying the linear stability of. If you looked carefully at the Solver Manager Initial Value Tab, you spotted the possibility of using any of five different variations of the problem as the initial solution about which the linear systems analysis can be conducted. The classical texts in linear stability typically study the stability of the steady state solution. But you can study the stability of any state in the transient solution. The meaning is perfectly clear — these are the timescales at which the dynamics of that solution are currently evolving.

This ability to select the solution about which the linear stability is conducted is crucial to the proper interpretation of the eigenvalues and eigenvectors. Figure 4 shows the eigenvectors found by the built-in eigenvalue

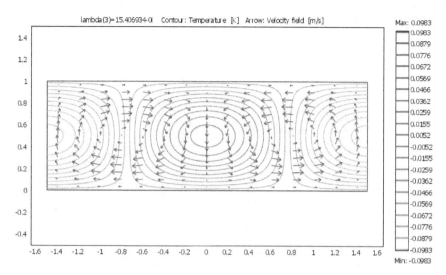

Figure 6. Isotherms and velocity vectors for the slowly decaying mode $\lambda(3) = 15.4069$ of the linear stability study of the rest state (no flow) with the static thermal conduction temperature profile.

solver. The postprocessing commands interpret the degrees of freedom of the eigenvector as linked velocity and temperature fields. These fields, when added to the base solution, must satisfy the linearized form of the boundary conditions. If the boundary conditions are linear to start with, then of course they will be satisfied exactly. It is clear from the form of the solutions themselves what these linearized boundary conditions are. The disturbance temperature field has no normal derivative at each boundary. The disturbance velocity field vanishes at each boundary. Close inspection of the temperature field shows that at horizontal bounding surfaces (top and bottom), the temperature value vanishes. Figure 4 is clearly a "temperature mode." The velocity field is just noise with no organized structure. But the temperature field is clearly only vertically dependent and, with inspection, very close to a sine. Not surprisingly, the eigenvalue is numerically close to π^2, which is analytically the eigenvalue found by normal mode analysis for $Ra = 0$.

Figure 5 makes clear that the second slowest decaying mode, which is only slightly faster in decay rate, is a "one roll mode." Traditionally, linear stability is done by projection onto normal modes, typically sinusoids, with known length scale(s). By constructing the eigenvectors using the finite element method as we have done here, there are no predetermined shapes

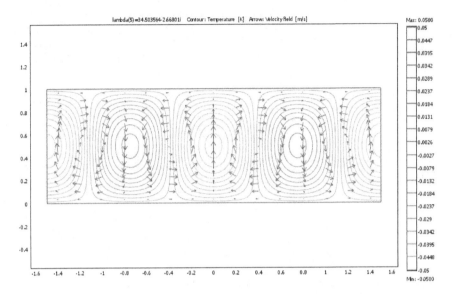

Figure 7. Isotherms and velocity vectors for the slowest decaying *oscillatory mode* $\lambda(5) = 34.5036 - 2.6680i$ of the linear stability study of the rest state (no flow) with the static thermal conduction temperature profile.

imposed. Consequently, the eigenvectors reflect the current dynamics of the problem. The characteristic length scales of the motion are not imposed, but solved for. Figure 6 is a "two roll mode," and therefore has characteristic horizontal length 1.5 and characteristic vertical length 1. Selecting between these modes as to which is likely to be seen in a simulation of decaying flow is problematic on decay rate alone. Our transient study, however, showed that the mode that was selected was indeed the "two roll mode," but that may have been due to its compatibility with the initial profile rather than winning the competition of decay rates.

The eigenvalue solver of COMSOL Multiphysics computes six eigenvalues as its default using the Arnoldi [18] method for sparse systems. Table 2 shows the first fourteen such eigenvalues, computed by the COMSOL Multiphysics/MATLAB command calls using the exported FEM structure:

$$>> \text{sol2=femeig(fem, 'U', fem.sol.u, 'Eigpar', 12)};$$

The arguments are described in the Reference Manual, however, it should be clear that fem is a fem structure, "U" specifies that the next argument is a solution vector; fem.sol.u is that solution found for $Ra = -100$; "Eigpar" is a flag that says the next argument describes the requested

Table 2. First twelve eigenvalues of the steady state buoyant convection problem when heated from above. $Ra = -100$.

9.8696	11.2484	15.4069	22.5541	$34.5036 \pm 2.6680i$	35.0367
36.0728	38.2728	39.4785	40.6278	$42.9201 \pm 5.9984i$	44.0741

eigenvalue solver parameters (in this case the smallest 12 eigenvalues in magnitude).

This generates a structure sol2 with substructures sol2.u and sol2.lambda. You probably find that sol2.lambda has twenty elements, and that sol2.u is a matrix with twenty columns and a huge number of rows. Each column is an eigenvector associated with the same numbered eigenvalue. Femeig uses iterative sparse methods for generating eigenvalue/eigenvector pairs. By default, the smallest magnitude eigenvalues were selected.

One should note that all the eigenvalues found have positive real part, $Re\{\lambda_i\}$. Since it is clear from the transient solution already conducted that the perturbations decay, this sign corresponds to decay, not growth. One should bear in mind that the eigenvalues found by COMSOL Multiphysics are decaying with positive real part and growing with negative real part. Since the complex eigenvalues come in conjugates, the sign of the imaginary part is not material. However, the existence of imaginary components is equivalent to identifying oscillatory solutions. The interpretation of the eigenvalues here should be that the eigenvector would be expected to grow with a growth factor $\exp(-\lambda t)$ for small amplitudes of the eigenvector. So imaginary components are complex exponentials, equivalent to sines and cosines — oscillations. Nonlinear effects will dominate for large amplitude contributions of the eigenvector, c.f. (4). One should note that all the eigenvalues of the $Ra = 0$ problem are real since the linear operator is symmetric (self-adjoint). The coupling between velocity and temperature fields due to buoyancy makes the operator nonsymmetric, introducing the possibility of oscillatory eigenvectors.

Figure 7 shows the eigenvector (temperature and velocity fields) associated with the slowest decaying oscillatory mode. The salient features of Figure 7 are that there are four hot and cold regions with fluid falling in the cold region and rising in the hot region. Each region occupies a unit width approximately, with a halfwidth of transition zone. Thus to see the whole structure, the aspect ratio must be about 3:1. The parametric solver can be used to explore regions of $Ra < 0$, but there are no situations where growing modes are excited. The best that happens is that for

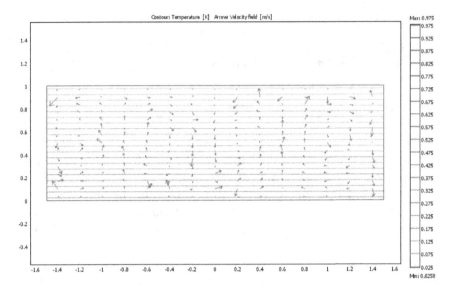

Figure 8. Aspect ratio 1:3 simulation with $Ra = 1970$ for full solution of temperature and velocity vectors. Note that the average velocity field $U_{-}ns$ is still $O(10^{-4})$.

large negative Ra, the decay rate diminishes. In a perfect (inviscid) fluid, it would be identically zero. As Ra increases in magnitude, dissipative effects become relatively weaker than the buoyancy, so the internal gravity waves become long-lived structures. Haarlemmer and Zimmerman [6] used wave tank studies to characterize the mixing properties of large amplitude internal gravity waves that are initially seeded in a concentration stratified fluid. They review the geophysical importance of this transport mechanism.

2.3. Heating from below

Heating from below changes the nature of the dynamical problem. As we found when heating from above, complex eigenvalues, equivalent to damped propagating waves were found. This is because vertical convection is opposed by the stable stratification of light fluid over heavy fluid, but gravity waves can propagate horizontally. If a patch of fluid is displaced vertically, it oscillates around its equilibrium position and can propagate right or left without loss of the original energy in an inviscid fluid. Lord Rayleigh [7] showed that when heating from above, the state of the fluid at rest is unconditionally stable. The same argument works in reverse to show that when heating from below, an inviscid fluid cannot remain at rest. But the

Table 3. Periodic boundary conditions with the Rayleigh-Benard problem. File name: rayleigh.mph.

Model Navigator	Open benard.mph
Options Menu|Constants	Set $Ra = 1$. Set $Pr = 1$
Physics Menu: Periodic conditions| Periodic boundary conditions	Select boundary 1 Enter expression:u. Click in the constraint name box to generate default name:pconstr1 Destination tab: select boundary 4 Expression u Source vertices:1 2. Destination vertices 3 4 Select boundary 1 Enter expression:v Constraint name:pconstr2 Destination tab: select boundary 4 Expression v Source vertices:1 2. Destination vertices 3 4 Select boundary 1 Enter expression:T Constraint name:pconstr3 Destination tab: select boundary 4 Expression T Source vertices:1 2. Destination vertices 3 4 OK
Physics Menu: Subdomain settings	ns mode. Init tab. Set $u(t_0) = v(t_0) = 0$
Physics Menu: Boundary settings	Set boundaries 1 and 4 to outflow/pressure
Mesh	Click on the toolbar triangle for the default mesh Refine the mesh
Solve Menu	Set the stationary nonlinear solver Click on the solve (=) tool on the toolbar Set the eigenvalue solver. Click on the restart solver

state of rest can persist to high Rayleigh number in a viscous fluid with heat conduction — the dissipative mechanisms oppose the overturning motion until the heating differential is strong enough. The theory of Reid and Harris [8] describes the critical Rayleigh number for cells with upper and lower rigid boundaries occurs at $Ra_c = 1708$ with a wavenumber of 3.117. The motion that is most unstable above Ra_c is supposed to be the onset of stationary cells in 3-D, and convection rolls in 2-D. Since the linear operator, and thus its FEM approximation as in (3), is self-adjoint, then all the eigenvalues are real. It follows that the unstable mode is not propagating, but stationary and growing in strength until it saturates.

To permit the possibility that waves might propagate, however, we need to change the horizontal boundary conditions from the earlier simulation which had no flux and slip boundary conditions on the horizontal bounding planes. Gravity waves cannot propagate through such planes, since they are transverse and require up-and-down motion. Furthermore, the model was stationary, so although complex eigenvalues are possible, propagation was prohibited. To implement periodic boundary conditions, a minor change is necessary.

Recipe for Periodic Boundary Conditions

Table 3 holds the recipe for periodic boundary conditions in a flow problem. The periodic constraints are a special case of extrusion coupling variables and Dirichlet constraints. The extrusion coupling variable makes the value of the quantity on the periodic boundary available at its counterpart, and then simultaneously imposes that the value (or some function of it in a generalized periodic constraint) matches the value on its counterpart. Extrusion coupling variables can map to the same dimension size. Furthermore, the boundary condition in Physics: Boundary settings must be consistent with the additional periodic constraint. In general mode, Neumann or natural conditions are specified on each boundary. Since these are the natural boundary condition, they can be thought of as "nonconstraints," which are then replaced with the periodic constraint. As we saw with the solitary wave example in chapter two, periodic boundary conditions do let information pass as expected when using this recipe.

The recipe is slightly altered for flow problems. There are no "nonconditions" for velocity variables. The natural boundary conditions are pressure outflow conditions. The jury is actually still out on the question of whether the pressure itself needs to be specified as a periodic quantity with such a boundary constraint. I have run the model with and without the pressure periodicity condition and noticed no difference in the results. I follow the philosophy — "when in doubt, leave it out." So I recommend leaving out the pressure periodicity condition. Pressure is a difficult concept in the Navier-Stokes equations. I often describe it as a "Lagrange multiplier" for the continuity equation. Pressure is whatever value it needs to be so that the continuity equation is satisfied in the incompressible hydrodynamics model. There are no thermodynamics in the hydrodynamics equations — incompressibility means density is constant. But density is an essential thermodynamic variable. So the pressure in hydrodynamics is a purely mechanical response to the velocity field — it adjusts so that mass is conserved. The upshot for modelling is that if velocity fields are constrained to be periodic, the pressure field is highly constrained as well. But is it naturally periodic if the velocity field is periodic? Hassell and Zimmerman [9] show that repetitive inserts in microfluidics can be modelled by the periodic constraint that the pressure field upstream is δp higher than that at the periodic face downstream, but in this case, the velocity fields are still exactly periodic. However, those authors ensure that the pressure drop δp is computed to be consistent with the imposed steady volumetric flow rate. This argument suggests that the pressure periodicity condition is only required if it is nontrivial.

Because the domain is about as long as the most dangerous mode (wavenumber 3.117 implies that we have nearly a period of $2\pi/3.117 \approx 2$, i.e. one wave per two units length), a domain of length 3 is sufficiently long to encompass one period of the unstable mode at supercriticality.

Our solution strategy is to compute the $Ra = 1$ solution first using the linear solver, and then use the Parametric Solver to continue to high Rayleigh number, finding the unstable mode visually from plots of the velocity field. At first I thought that this does not yield a visually unstable flow, even up to $Ra = 10000$ (see Figure 8). Why not? $u = v = p = 0$ and $T = y$ is a perfectly acceptable numerical solution, and the model finds solutions with small dimensionless convective flows, with velocity magnitudes of average $O(10^{-4})$, for all values of the Rayleigh number attempted. Professor Bruce Finlayson and chemical engineering student Michael Johnson (private communication) pointed out that since the Nusselt number scales with the Rayleigh number, these are actually giving rise to appreciable convective heat flux. However, there is no specific threshold of Ra which is apparently an abrupt change in Nusselt number. To find Ra_c something else must be tried. The obvious strategy is to use transient integration to determine if, after a sufficiently long time, random small magnitude initial conditions have grown expontially large as in (4). The problem with this is that COMSOL Multiphysics' Parametric Solver only applies to stationary models. The other solution is to compute the eigenanalysis for the system at each value of Ra in a parametric continuation of Ra to high Rayleigh numbers. We will do this two ways: one in the GUI, exporting solutions to the MATLAB workspace; the other in a MATLAB m-file with continuation implemented in a MATLAB loop structure. The results are edifying about the nature of the femeig command in the FEMLAB programming library.

2.4. GUI methodology

Figure 8 was generated from solving the Benard problem using parametric continuation in the GUI. The linear solver for the $Ra = 1$ problem was used, which is well conditioned. Parametric Solver was used to continue to high Rayleigh number. For eigenanalysis, we export our solution to MATLAB using the export FEM structure feature under the file menu.

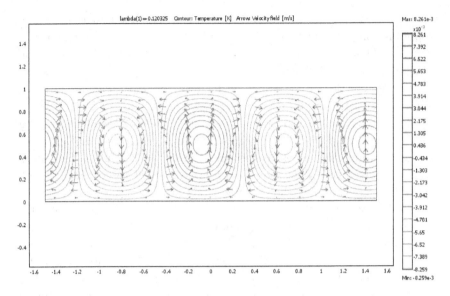

Figure 9. Aspect ratio 1:3 simulation with $Ra = 1970$ for the eigenvector of temperature and velocity vectors associated with the fastest growing eigenvalue, $\lambda = -0.120325$. Clearly the field variables have spatial periodicity 4. The scale of either temperature or velocity is arbitrary, but the ratios are fixed.

The data structure for a parametric solution is different than for a single, stationary solution. For instance, for the case of a parametric solution [1801:100:10001], fem.sol is an array with three elements: u (the solution), plist (parameter list), and pname (the continuation parameter). Execute the following on the MATLAB command line:

```
>> fem.sol
        u: [6966x83 double]
    plist: [1x83 double]
    pname: 'Ra'
```

A transient solution has a similar structure, but with a tlist of output times. To access any of the solutions, the appropriate column is requested. For instance,

>>sol2=femeig(fem, 'U', fem.sol.u(:,83), 'Eigpar', 12);

yields the 12 smallest magnitude eigen pairs of the FEM operator for the 83rd solution, appropriate for the parameter $Ra = 10001$. This feature would work very nicely if the FEM structure were robust in substituting

$Ra = 10001$ in the stiffness matrix computed by femeig. Unfortunately, femeig takes the last specified value of the parameter Ra in the FEM structure as a constant, which may have no relation to the final value in the parametric solver.

Table 4. Decay rates.

1951	0.044447
1952	0.035951
1953	0.027477
1954	0.019062
1955	0.010725
1956	0.0040364
1957	−0.0055668
1958	−0.014916
1959	−0.023515
1969	−0.032065

In our case, $Ra = 1$ was specified as a constant, so the eigenfunction computed is for $Ra = 1$ about the 83rd solution vector, which is still substantially close to zero everywhere. So, although the parametric solver is a good way to find solutions at high complexity parameter, it is not particularly good at interrogating them with eigenanalysis. Figures 8 and 9 were generated by using parametric continuation to solve up to $Ra = 1970$, and then changing solver to the stationary nonlinear solver then performing eigenanalysis with the restart solver (using the current solution). There is only one qualitative difference between Figures 5 and 7 — the same number of rolls. This computation was done entirely in the GUI. Previous versions of FEMLAB required back-and-forth shuffling of the FEM structure which is no longer necessary with options of initial value selection among stored and current solutions in the Solution Manager. There is now an unstable eigenvector with negative eigenvalue $\lambda = -0.l20325$. The critical condition must have been passed, i.e. $Ra_c < 1970$.

2.5. Matlab m-file methodology

But what do you do if you want to vary a parameter over a range of values and compile results for each individual parameter value? You still have to write your own looping structure in a MATLAB m-file. For instance, suppose we wish to find the critical Rayleigh number for a neutrally stable largest eigenvalue. We would need to compute femeig on each successive value of Rayleigh number, then substitute the old solution as the first guess for the new solution at higher Rayleigh number.

Table 4 gives the decay rates of the most dangerous eigenvector identified by FEM eigenvalue analysis of Rayleigh convection in a 3:1 periodic cell. These were computed by the m-file stability.m was generated by exporting the FEM structure of the Benard problem (rayleigh.mat from rayleigh.mph), and then modifying the end to put a looping structure around the stationary nonlinear solver and eigensolver:

```
load rayleigh.mat fem
%%%%%%%%%%%%%%%%%%%%%%%%stability.m%%%%%%%%%%%%%%%%%%%%%%%%%%%
Rayleigh=[100:100:4000]; output=zeros(length(Rayleigh),8);
%%%%%%%%%%%%%%%%%%%%%%%%%%%%Looping Structure%%%%%%%%%%%%%%%%%%
for j=1:length(Rayleigh)

% Define constants
fem.const = {'T0','0', ...
  'T1','1', ...
  'Ra', Rayleigh(j),...
  'Pr','1', ...
  'Re','1'};

% Multiphysics
% fem=multiphysics(fem);

% Extend mesh
% fem.xmesh=meshextend(fem);

% Evaluate initial value using current solution
init = asseminit(fem,'u',fem.sol);

% Solve problem
fem.sol=femnlin(fem, ...
                'init',init, ...
                'solcomp',{'psi','u','T','p','v'}, ...
                'outcomp',{'psi','u','T','p','v'}, ...
                'report','off');

% Save current fem structure for restart purposes
fem0=fem;

% Integrate on subdomains
```

```
I1=postint(fem,'cflux_T_chcc',...
        'cont',    'internal',...
        'edim',    2,...
        'solnum', 1,...
        'phase',  0,...
        'geomnum',1,...
        'dl',      1,...
        'intorder',4,...
        'context','local');

% Integrate on subdomains
I2=postint(fem,'dflux_T_chcc',...
        'cont',    'internal',...
        'edim',    2,...
        'solnum', 1,...
        'phase',  0,...
        'geomnum',1,...
        'dl',      1,...
        'intorder',4,...
        'context','local');

output(j,1)=Rayleigh(j); output(j,2)=I1; output(j,3)=I2;
% Solve eigenvalue problem
sol2=femeig(fem, ...
                'init',fem0.sol, ...
                'solcomp',{'psi','u','T','p','v'}, ...
                'outcomp',{'psi','u','T','p','v'}, ...
                'Eigpar',10, ...
                'report','off');
output(j,4)=sol2.lambda(1); output(j,5)=sol2.lambda(3);
output(j,6)=sol2.lambda(5); output(j,7)=sol2.lambda(7);
output(j,8)=sol2.lambda(9); fem.sol=fem0.sol;

end

save bifurc.mat fem sol2;

dlmwrite('bifurc.dat',output,',');
quit
```

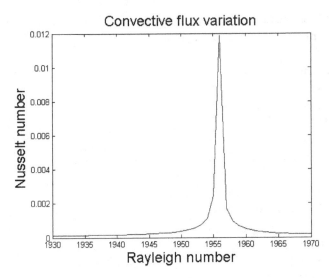

Figure 10. Dimensionless convective flux (Nusselt number) variation for aspect ratio 3:1 over a range of Rayleigh numbers.

The m-file script computes fluxes and the first ten eigenvalues for Ra in [100:100:4000], which shows a crossover between Ra values 1900 and 2000. Table 4 shows the eigenvalues homing in on the critical value of $Ra = 1956$ for aspect ratio 3:1. Saving the solution and fem structure, as well as the eigenvalues in a mat-file permits the re-loading of the final solution in the COMSOL Multiphysics GUI by importing from MATLAB into COMSOL Multiphysics. stability.m was computed as a UNIX background job from the command line:

```
comsol matlab path -ml nosplash -ml nodesktop -mlr
stability.m >err 2>err &
```

since it takes a good fraction of an hour to execute. The save command permitted subsequent perusal of the solution. The m-file script computes the total convective and conductive fluxes for each Rayleigh number solution. The critical Rayleigh number (circa 1956) corresponds to both the zero eigenvalue, but also an abrupt increase in convective heat transfer.

2.6. *Agreement with thin layer theory*

Recall, the theory of Reid and Harris [8] analyzes flow cells with upper and lower rigid boundaries. They predict that describes the critical

Table 5. Decay rates.

eigenvalue 1705	0.024
1706	0.016
1707	0.009
1708	0.002
1709	−0.006
1710	−0.014
1711	−0.022
1712	−0.029
1713	−0.037

Rayleigh number occurs at $Ra_c = 1708$ with a wavenumber of 3.117 for layer extending to infinity horizontally. Since our layer has aspect ratio 1:3, we would not particularly expect agreement. Davis [10] computes the 3-D solution for finite aspect ratio boxes, and finds substantially higher critical Rayleigh numbers, approaching the theoretical predictions only at high aspect ratio. For this reason, we have reproduced our simulations here for aspect ratio 1:10. The computed eigenvector shows a periodicity ten in a ten unit periodic layer for the critical mode, which is in agreement for the theoretical estimate of the wavenumber. Table 5 for the eigenvalues leaves no doubt that the critical mode occurs near Rayleigh number 1708. The triumph of eigenanalysis coupled with FEM is to find numerically the critical Rayleigh number and approximate length scale associated with the critical mode. Although the linear stability theory for this problem is not cumbersome, for many situations with nontrivial and three-dimensional base states, that cannot always be claimed. COMSOL Multiphysics, through eigenanalysis, provides a consistency check on linear stability analysis of stationary states. The numerical technique is far more robust, however. Eigenanalysis can be conducted on any solution, even of transient problems. Recall equation (4) shows that for self-adjoint operators, the eigenanalysis predicts the time asymptotic dynamics of the linearized system. For nonself-adjoint operators, it has been demonstrated that pseudo-modes that are not eigenmodes can grow rapidly before the time asymptotic eigenmodes dominate. Trefethen *et al.* [11] identified spiral pseudo-modes as leading to transitions to turbulence in Couette and Poiseuille flows at much lower Reynolds Numbers than anticipated by linear theory. This was confirmed experimentally. The extent to which eigenanalysis of transient flow problems identifies the fastest growing pseudomode in transient models for instantaneous states is a largely unexplored area, for nonself-adjoint operators. Here we have shown that for self-adjoint

Flow Configuration

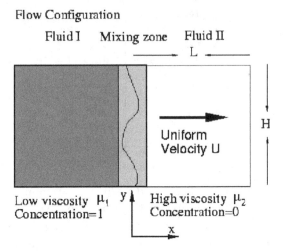

Figure 11.   Fluid II (more viscous, concentration $c = 0$) being displaced by Fluid I (less viscous, concentration $c = 1$) in a porous medium with superficial velocity U. The mixing zone is the region of diffusive mixing and viscous finger formation.

operators, the FEM model accurately reproduces the predictions of linear stability theory.

3. Viscous fingering instabilities

The Benard problem is a paradigm for instabilities of a stationary state. Viscous fingering is an instability of a nonhomogeneous state in motion — a less viscous fluid displacing a more viscous fluid. Figure 11 shows the flow configuration for *miscible* viscous fingering, where diffusion tends to spread out viscous fingers and oppose their formation. Nevertheless, viscous fingering is a long wave instability — broad channels originally form along any displacement front, and then subsequently nonlinear interactions force fluid along these paths, leading to narrow channels.

The phenomenon is a recurring fundamental instability in many realms. Enhanced oil recovery, for instance the injection of dilute detergents into oil sands or flooding with CO_2 gas, as well as the remediation of contaminated acquifers are common geophysical applications. Miscible displacement and the concomitant pesky viscous fingering instability recur as well in regeneration processes. Of special interest to chemical engineers is the flushing of catalytic systems with solvents or oxidants that remove the impurities fouling the catalysts or liquid chromatography columns. It was in the context of a regeneration process, the "sweetening off" of sugar liquors displaced by water from a charcoal packed column, that Hill [12] recognized and first analyzed the channelling instability. Homsy [13] gives the best review of the early work in this area. The standard venues for miscible viscous fingering are porous media, which are well described by Darcy's Law, which is a simpler momentum equation than the Navier-Stokes equations, typically semi-empirically based on measures of pressure drop and superficial velocity in porous media:

$$\nabla p = -\frac{\mu}{k}\mathbf{u}\,. \tag{12}$$

COMSOL Multiphysics has a Darcy's Law application mode built into the Chemical Engineering Module. p is the pressure; u is the velocity vector; μ is the viscosity, and k is the permeability of the medium. Along with (12), it imposes the conservation of mass for an incompressible fluid as

$$\nabla \cdot \mathbf{u} = 0\,. \tag{13}$$

The mixing as depicted in Figure 11 is due to convection and diffusion, also a built-in application mode in the Chemical Engineering Module, with concentration satisfying

$$\frac{\partial c}{\partial t} + u\frac{\partial c}{\partial x} + v\frac{\partial c}{\partial y} = D\nabla^2 c\,, \tag{14}$$

where D is the molecular diffusivity. Additionally, in order to couple the mixing with the momentum transport realisitically, the fluid viscosity must depend on the concentration. The simplest model is monotonic dependence, which is a good model for glycerol-water, a common laboratory model system for the blending of viscous fluids:

$$\mu = \exp(R(1 - c))\,. \tag{15}$$

Armed with these equations, we are now ready to simulate viscous fingering using the built-in application modes. Launch COMSOL Multiphysics, bring up the Model Navigator and select the Multiphysics tab.

By judicious use of the Solver Manager and swapping initial values, the no mass transport, flow only model was used to compute the initial pressure profile. Analytically, the pressure profile takes the form of a definite integral:

$$pinit(X - x) = \int_0^x \exp(R(1 - erf(\xi)))d\xi\,, \tag{16}$$

where X is the domain length, taken here as $X = 10$. Here, $erf(x)$ is the error function of statistics. $erfc(x) = 1 - erf(x)$ is called the complementary

Table 6. Viscous fingering in miscible displacement model.   File name darcy_fingers.mph.

Model Navigator	Select 2-D space dimension Select Chemical Engineering Module|Momentum Balance|Darcy's Law (mode chdl) Mutliphyics button|Add Select Chemical Engineering Module|Mass balance|Convection and Diffusion (mode chcd) Add. OK
Draw Menu	Specify objects. Rectangle with width 10 and height 1 with corner at the origin
Physics Menu: Boundary conditions	Mode chdl Select boundary 1 Condition type: Inflow/Outflow Enter $u = 1$ (unit inflow) Select boundary 4 Condition type: Inflow/Outflow Enter $u = -1$ (unit outflow) Leave boundary conditions for boundaries 2 and 3 at the default insulation/symmetry conditions Mode chcd Select boundary 1 Condition type: Concentration Enter $c = 1$ Select boundary 4 Condition type: Convective flux Leave boundary conditions for boundaries 2 and 3 at the default insulation/symmetry conditions. OK
Physics Menu: Sub-domain settings	Mode chdl Select subdomain 1. Set $k = 1$; $\eta = \exp(3*(1-c))$; $F = 0$; $\rho = 1$. Init tab: $p(t_0) = 1$. OK Mode chcd Set D(isotropic) $= 0.01$; $u = u\_chdl$; $v = v\_chdl$ Init tab. Set $c(t_0) = \mathrm{erfc}(x)*(1. + 0.05*\sin(10*pi*y))$
Physics Menu| Equation System| Point settings	Select point 3. Weak tab. Set the constraint entry to p 0 This sets the weak constraint that $p = 0$ at this point for a pressure datum
Mesh	Click on the toolbar triangle for the default mesh Refine the mesh twice
Solve Menu	Solver Manager|Solve For tab|Darcy's Law (chdl) only Solver parameters|Set the stationary linear solver Click on the solve (=) tool on the toolbar Solver parameters|Set the time dependent solver Set output times to 0:0.05:5 Solver Manager|Solve For tab|select both modes. OK Click on the restart solver

error function. In my first attempt at this model, $pinit(x)$ was introduced as an m-file function for the initial condition. The Solver Manager eliminates the need for manually coding initial conditions that can be found from a single active mode or combinations of active application modes.

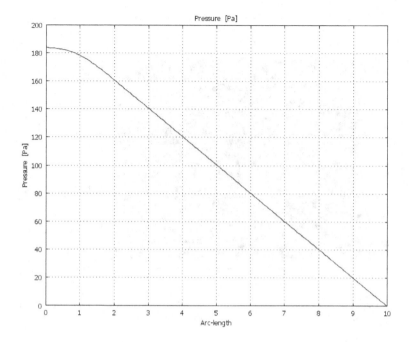

Figure 12. Initial pressure profile for concentration given by $1 - erf(x)$.

Figure 13 shows the initial condition for the concentration profile with a sinusoidal concentration profile perturbing the front, but with the pressure profile of Figure 12. Since there is no vertical variation of the pressure profile, this initial condition is not fully "relaxed" and would clearly have a transient while the pressure adjusts and develops a vertical dependency. Figure 14 shows the subsequent development of the pressure isobars and concentration profile out to time $t = 5$. The formation of a viscous finger along the bottom wall is clearly evident in the animation.

The animation of the viscous finger development demonstrates the dynamics of the process very well, but cannot be well represented in print. Figure 15 presents the transverse average concentration profile, produced according to the steps in Table 7. These profiles, separated by time increments 0.05, demonstrate that the displacement front changes from a flat profile to a narrow channel in a relatively short time. The front appears diffusive until time $t \sim 0.5$, but then rapidly switches character. How it becomes unstable can be explored, as in the previous section, by linear systems analysis. The results here are in line with the simulations of Tan and

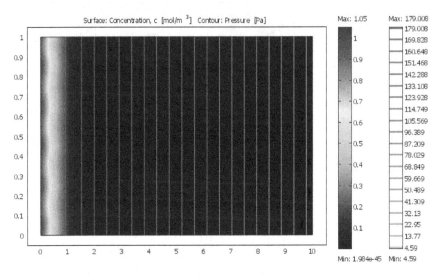

Figure 13.   Concentration and pressure profiles for Darcy's Law and Convection/Diffusion model with sinusoidal concentration perturbation.

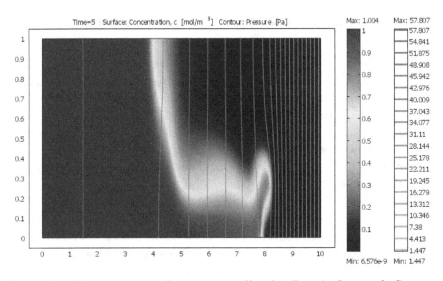

Figure 14.   Concentration and pressure profiles for Darcy's Law and Convection/Diffusion model with initial sinusoidal concentration perturbation developed until $t = 5$.

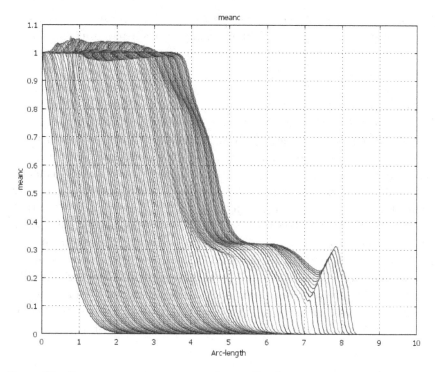

Figure 15. Transverse average concentration profiles for the viscous fingering model with time step 0.05 until $t = 5$.

Table 7. Transverse average profiles computed by convection coupling variables.

Model Navigator	Open darcy_fingers.mph
Options Menu|Projection Coupling Variables	Select subdomain variables
	Select subdomain 1
	Enter name: meanc
	Enter expression: c
	Check general transformation
	Destination tab|set destination boundary 2
Solve Menu	Select update model
Post-processing menu|Domain plot parameters |Line Plot	Select boundary 2 Enter expression: meanc
	Hit the OK button

Homsy [14] [15], who found broad unstable fingers forming in the troughs of the sinusoidal initial condition. However, the linear stability analysis was conducted analytically, and is a tour-de-force of differential operators and secular equations. The analytic calculation has the advantage of providing estimates of the length scale and growth rate of forming viscous fingers at the early stages in terms of the dimensionless parameters of the problem, such as viscosity contrast and Peclet number. The numerical linear systems analysis is not limited to short times and always has the interpretation of time scales at which the dynamics are currently evolving. The previous section demonstrated the ease at which femeig in MATLAB acting on a FEM structure or the eigenvalue solver in COMSOL Multiphysics could be used to generate the eigenvalues and eigenvectors of the FEM matrices. Little more need be added here. Figure 16 shows the eigenvector in pressure and concentration associated with the unstable most dangerous mode $\lambda = -0.6076$ at the initial state. The eigenvector is clearly not vertically symmetric with respect to the centreline, which is unexpected given that in the linear stability theory of Tan and Homsy, the normal modes are vertically sines and cosines. The only source of vertical asymmetry in this model is the unstructured mesh which is not constrained to be symmetric. Indeed, over a range of initial perturbation amplitudes, the resultant viscous fingers are reproducible in shape, but appear earlier or later depending on "incubation time." The unstructure mesh is the only selector of this reproducible behavior. Zimmerman and Homsy [16] found that if

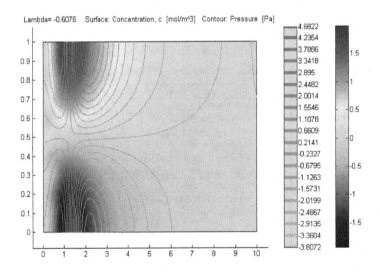

Figure 16.   Isobars of pressure eigenvector and shading of concentration eigenvector associated with the growing most dangerous mode with eigenvalue $\lambda = -0.6076$ at the initial state.

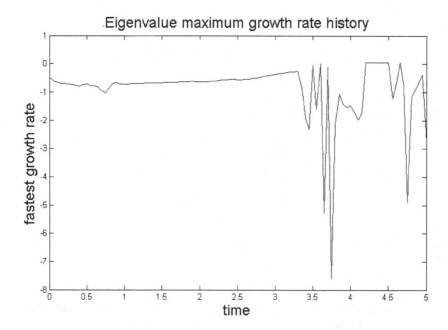

Figure 17. Eigenvalue maximum growth rate history.

the initial noise in their simulations was the same, the fingers were similar, but delayed or accelerated in growth depending on the amplitude of the initial noise. Thus, the "noise" in finite element methods that selects the formation path of viscous fingers is the placement of the grid.

Figure 17 shows the value of the maximum growth rate over the whole temporal development. When computing the model itself, I observed that there we periods when the time dependent solver took very few steps (especially in the time range $0 < t < 3.5$), but in some intervals an order of magnitude more time steps were needed by the adaptive solver to resolve the dynamics. Given that maximum growth rate is practical constant in the linear growth range $0 < t < 3.5$, and then fluctuates rapidly during the nonlinear development of the single narrow channel, there is a consistency in interpretation of the growth rate as the most rapid time scale in the dynamics.

Figure 17 was generated in a similar fashion to the eigensystem analysis of Rayleigh convection. The former was parametric analysis, which required a MATLAB script for efficient computation. The viscous fingering example is temporal analysis, using the state at each time step and linearizing the operator about that state. The scripting code to carry out this transient eigensystem analysis is given here, working on the exported fem structure (saved as vf_fem.mat) of the whole transient darcy_fingers.mph model:

```
load vf_fem.mat
fem times=fem.sol.tlist;
output=zeros(length(times),21);
forj=1:length(times)
        [K,L,M,N,D]=assemble(fem,'T',times(j),'U',fem.sol.u(:,j));
        sol2=femeig('In',{'D',D,'K',K,'N',N},'Eigpar',20);
        output(j,1)=times(j);
        for k=1:20
                output(j,k+1)=sol2.lambda(k);
        end
end
dlmwrite('vf_eig.dat',output,',');
quit
```

The script for the transient eigensystem analysis is shorter than that for the parametric analysis because we do not need to conduct the parametric continuation simultaneously when working on the FEM structure. Nonetheless, substituting the FEM structure from the parametric solution of the Rayleigh convection model would parallel the above. The above was computed in the MATLAB GUI (using COMSOL Multiphysics with MATLAB). The code is remarkably compact by comparison to its FEMLAB predecessor.

femeig sometimes has difficulty finding large decay rates. Even though I requested twenty "SM" eigenpairs, after $t = 0.04$, it can only find 19, and after $t = 0.42$, only 18. femeig uses the sparse eigenanalysis routines of ARPACK, which is essentially iterative, to compute eigenvalues and eigenvectors. This package has difficulty in finding and distinguishing zero eigenvalues (associated with singular systems). Since [13] and [15] show theoretically and numerically that the linear stability theory has a neutral mode at zero wavenumber and at a finite cut-off wavenumber of the longwave unstable wave packet, k_c, the linear system is nearly singular and will have difficulty resolving these neutral (or numerically near-neutral) modes.

4. Summary

In this chapter, we explored how COMSOL Multiphysics can be used to set up simulations and study nonlinear dynamics and stability. For stationary nonlinear problems, stability studies through eigensystem analysis give the growth rates and eigenmodes which are equivalent to the modes found in transient analysis of the same problem from initially noisy conditions. The most dangerous mode is expected to be observed asymptotically as long as it is smaller than nonlinear interactions. If the operator is nonself-adjoint, however, this is not necessarily the case (see [11]). So interestingly, eigensystem analysis informs about the results of *simulations*, even with stationary solutions. In the case of the Benard problem, the stationary solution returns the no motion base state, even in the situation that the eigensystem analysis identifies critical or growing modes. The second example, viscous fingering, is a paradigm for simulation of evolving instabilities — the base state is moving and changing with time, and the instabilities formed have complex nonlinear interactions. The eigensystem analysis in the uniform outflow Darcy's Law model showed an asymmetric instability. Typically, noisy initial conditions are introduced directly in the simulation by a random number (normally distributed) modulating the concentration base state. As we claimed in the introduction, by simulation we normally expect some element of randomness is modelled. However, the asymmetric feature that introduces this noise is the unstructured mesh. This case is the least controversial use of randomness in a simulation — noisy or uncertain initial conditions due to mesh discretization. Thereafter, the simulation is a completely deterministic model. In general, COMSOL Multiphysics can be used for simulating more complicated stochastic processes by alternating random processes and deterministic ones.

It should be noted that noisy initial conditions may not be necessary in such simulations simply due to the approximation error in FEM analysis and roundoff errors in truncation of fixed precision arithmetic. Since the user has control over error tolerances, stochasticity can be simulated by using unconverged or unresolved analysis, but this is a dangerous practice as the statistics of the noise so introduced may be unquantifiable, and the "simulation" may just be numerical instability. A more controlled simulation with quantifiable levels of noise is preferable.

As averse to classical linear stability theory, the application of FEM analysis and subsequent interrogation of the eigensystem analysis of the FEM operator is not limited to a specific type of basis functions — typically "normal modes." The advantage of normal modes is that the transform space that is dual to the physical space has useful measures as coordinates — wavenumber, for instance, specifies the lengthscale characterizing the associated eigenmode. With FEM eigensystem analysis, the growth rates are elucidated for whatever the natural growing mode(s) turns out to be, but the eigenmode does not have an unequivocal length scale, for instance. Where the normal modes are eigenmodes, the FEM methodology usually shows this qualitatively with regard to the patterns in the eigenmode. Normal modes do not necessarily get excited in systems that have FEM operators that are nonself-adjoint. I would speculate that this methodology for numerical computation of stability is far more likely to capture the pseudomodes of [11] for a nonself-adjoint problem than the linear stability theory.

This chapter introduces several new aspects of eigensystem analysis that can be done by using COMSOL Multiphysics and MATLAB tools and a little user defined programming. The ease by which this can be done is a major advantage of the pde engine and programming language of COMSOL Multiphysics. It is now common practice in stability theory, for instance of viscoelastic flows [17], across many disciplines [18], to compute via numerical methods the eigenvalues and eigenmodes of instabilities in transient conditions. Smith *et al.* [17] use the Arnoldi iterative method implemented in ARPACK [19] for their computation. The eigs() sparse eigensolver of MATLAB/COMSOL Multiphysics does as well. The eigenvalue solver, based on the Krylov subspace decomposition, becomes computationally cost effective with larger, sparse systems; the MATLAB/COMSOL Multiphysics implementation of the ARPACK routines is robust and highly accurate.

References

[1] D. Coca and S. A. Billings, A direct approach to identification of differential models from discrete data, *J. Mechanical Systems and Signal Processing* **13** (1999) 739–755; D. Coca, Y. Zheng, J. E. M. Mayhew and S. A. Billings, Nonlinear system identification and analysis of complex dynamical behaviour in reflected light measurements of vasomotion, *Int. J. Bifurcation and Chaos* **10** (2000) 461–476.

[2] S. Wolfram, *A New Kind of Science* (Wolfram Media, Inc., 2002).

[3] S. C. Roberts, D. Howard and J. R. Koza, Evolving modules in genetic programming by subtree encapsulation, *Genetic Programming, Proceedings Lecture Notes in Computer Science*, Vol. 2038 (2001), pp. 160–175.

[4] C. Sabbah, R. Pasquetti, R. Peyret, V. Levitsky and Y. D. Chashechkin, Numerical and laboratory experiments of sidewall heating thermohaline convection, *International Journal of Heat and Mass Transfer* **44**(14) (2001) 2681–2697; Y. D. Chashechkin and V. V. Mitkin, High gradient interfaces in a continuously stratified fluid in field of 2D adjoined internal waves, *Doklady Akademii Nauk* **362**(5) (1998) 625–629.

[5] P. G. Drazin and W. H. Reid, *Hydrodynamic Stability* (Cambridge University Press, Cambridge, 1981).

[6] G. W. Haarlemmer and W. B. Zimmerman, Advection of pollutants by internal solitary waves in oceanic and atmospheric stratifications, *Nonlinear Processes in Geophysics* **5** (1999) 209–217.

[7] L. Rayleigh, Investigation of the character of the equilibrium of an incompressible heavy fluid of variable density, *Proc. London Math. Soc.* **14** (1883) 170–177.

[8] W. H. Reid and D. L. Harris, Some further results on the Benard problem, *Phys. Fluids* **1** (1958) 102–110.

[9] D. G. Hassell and W. B. Zimmerman, Investigation of the convective motion through a staggered herringbone J micromixer at low Reynolds number conditions, *Chemical Engineering Science* (2006).

[10] S. H. Davis, Convection in a box: Linear theory, *J. Fluid Mech.* **30** (1967) 465–478.

[11] L. N. Trefethen, A. E. Trefethen, S. C. Reddy and T. A. Driscoll, Hydrodynamic stability without eigenvalues, *Science* **261** (1993) 578–583.

[12] S. Hill, Channelling in packed columns, *Chem. Eng. Sci.* **1** (1952) 247.

[13] G. M. Homsy, Viscous fingering in porous media, *Ann. Rev. Fluid Mech.* **19** (1987) 271.

[14] C. T. Tan and G. M. Homsy, Stability of miscible displacements in porous media: Rectilinear Flow, *Phys. Fluids* **29** (1986) 3549.

[15] C. T. Tan and G. M. Homsy, Simulation of nonlinear viscous fingering in miscible displacement, *Phys. Fluids* **30** (1987) 1239.

[16] W. B. Zimmerman and G. M. Homsy, Nonlinear viscous fingering in miscible displacement with anisotropic dispersion, *Physics of Fluids A* **3**(8) (1991) 1859.

[17] M. D. Smith, Y. L. Joo, R. C. Armstrong and R. A. Brown, Linear stability analysis of flow of an Oldroyd-B fluid through a linear array of cylinders, *J. Non-Newtonian Fluid Mech.*, to appear.

[18] R. T. Goodwin and W. R. Schowalter, Interactions of two jets in a channel: Solution multiplicity and linear stability, *J. Fluid Mech.* **313** (1996) 55–82.

[19] R. B. Lehoucq, D. C. Sorensen and C. Yang, ARPACK users guide: Solution of large scale eigenvalue problems by implicitly restarted Arnoldi methods, ftp://ftp.caam.rice.edu/pub/software/ARPACK.

Chapter Six

CHANGING GEOMETRY: CONTINUATION AND MOVING BOUNDARIES

V.R. GUNDABALA[1], W.B.J. ZIMMERMAN[1] and A.F. ROUTH[2]

[1]*Department of Chemical and Process Engineering, University of Sheffield, Newcastle Street, Sheffield S1 3JD United Kingdom*
[2]*Department of Chemical Engineering, Cambridge University, Pembroke Street, Cambridge, CB2 3RA*
E-mail: w.zimmerman@shef.ac.uk

Geometric continuation occurs if the mesh of the domain must change from one solution to the next due to variation of the geometry model. In this chapter, we take two examples as paradigmatic — the additional pressure loss in a channel due to various size orifice plates is an example of steady state geometric continuation. Conceptually, this problem is little different from the parametric continuation by Rayleigh number in the Benard problem of Chapter 5. The second example is a drying film with latex particles embedded in the fluid. In this example, two moving fronts are treated; one by coordinate transformation and the second by a smoothed interface model. This technique has been found to be an improvement on the moving weak term previously used to model this system.

1. Introduction

1.1. *Geometric continuation*

We have already seen several examples of parametric continuation — the traversing in small steps of a range in a parameter, using the previous solution of a nearby parameter value as the initial guess for the solution at the new value of the parameter. As long as the parameter does not pass through a bifurcation point, the new solution should be smoothly connected to the old one if the step in parameter is small enough. Even if there is a bifurcation, however, the old branch may still be a solution, as we found with Benard convection in Chapter 5.

Geometric continuation is qualitatively different from parametric continuation in one important respect. In geometric continuation, the geometrical change of the domain leads to the requirement of re-meshing with each geometric parameter value. We should be careful to class as geometric continuation changes in a parameter that do not lead to a similar geometry. For instance, in pipe flow, it is well known that the flow is characterized by a Reynolds number:

$$Re = \frac{\rho U D}{\mu}.\tag{1}$$

This dimensionless parameter rolls the influences of the fluid density ρ, inlet velocity U, the diameter D, and viscosity μ into one parameter that describes the dynamic similarity of the flow. Thus changes in the pipe diameter for fully developed flow are not classed as geometric variation, but rather the more common parametric variation.

Just as in the last chapter where examples of simulations were given for stationary models and for transient models, in this chapter we will give examples of stationary geometric continuation and transient geometric continuation. In the former, distinct models are solved with slightly different domains and therefore different meshes. Therefore the solutions are incompatible (different degrees of freedom) from one geometric parameter value to the next. If the old solution is to be taken as the initial guess for new geometry, then mapping the old solution to the new domain in a consistent fashion must be done. In the examples given here, the system of PDEs for the stationary models are linear, so the solution can be determined directly in one FEM step. Thus mapping old solutions onto the new geometry has no additional value. The transient problem, however, involves a problem in a shrinking domain with a moving front. The domain changes after each time step, so the mapping of the solution at the old time step onto the new domain is essential to the model. Consequently, after each time step, remeshing must be done as well. One class of problem where this is a crucial step is the free boundary problem. Film flows and jet flows, for instance, are cases where the position of the boundary is intimately related to the solution of the velocity field. The boundaries should be located wherever the stress balances are satisfied.

The 2-D incompressible, laminar Navier-Stokes equations can be solved by several standard means (finite difference, finite element, spectral element, lattice Boltzmann, and multigrid techniques) and have been implemented in standard simulation engines commercially with fixed boundary conditions and complex geometries. Standard computational fluid dynamics packages have two standard engines: (1) the grid generator to cater for complex geometry, and (2) the PDE engine, which can solve more general systems of transport equations that include the pressure as in the

Navier-Stokes equations as a Lagrange multiplier for the continuity equation. These two steps are typically conducted separately. The grid is generated initially, and thereafter many simulations are conducted. FEMLAB is no different in this respect.

1.2. *Free boundary problems and the arbitrary Lagrange-Eulerian transformation*

This paradigm for computational fluid dynamics does not deal particularly well with free boundary problems. An iterative scheme for coupling the flow solution to grid generation could be envisaged, but automation with standard packages is difficult to implement. Ruschak [1] described the now standard method of implementing boundary stress conditions with grid adaptation. Goodwin and Schowalter [2] have successfully implemented their simultaneous solution for the position of the mesh with the solution of the flow equations and boundary conditions using Newton iteration in the treatment of a capillary-viscous jet using finite element methods. FI-DAP, which does treat free boundary flows, uses the iterative flow solution/elliptic mesh regeneration methodology, rather than the simultaneous Newton iteration. In our transient model in a shrinking domain §3, we adopt a coordinate transformation and smoothed interface approach to the variation of the geometric domain over time with two moving boundaries. In general with multiphase flows, smoothed interface models are preferable to moving boundary models since they permit topological changes. Chapter eight of this book illustrates the level set method for two phase flows which accommodates coalescence of droplets.

In principle, COMSOL Multiphysics could also do the latter by using the ALE mode (arbitrary Largrange-Eulerian transformation. The ALE mode implements the equations for the COMSOL Multiphysics application modes, augmented with the residuals for the movement of the grid positions. Standardizing the methodology for including these extra terms in all application modes whenever the grid is "active," i.e. there is a free boundary, is a substantial advancement on previous versions of FEMLAB. Given that the number of models that require free boundary computations, even in surface tension dominated flows, is rather few, such a general alteration to the package was not driven by this application mode. Typically, the need for linking of mechanical action (impellers and turbines) to transport phenomena is one extreme driver; another is the linkage of structural mechanics to transport phenomena. I wish I had a simple model which would illustrate ALE moving boundary coupling, but the reader is recommended to read the COMSOL Multiphysics Model Library entries on peristaltic pump and ALE cantilever beam models.

2. Stationary geometric continuation: Pressure drop in a channel with an orifice plate

In this section, we consider two related models that require geometric continuation. They are the orifice plate and the platelet in a duct filled with viscous fluid. They are related, as in fact there is only a slight change in the model from one case to the other. Figure 1 shows the orifice plate geometry. We would like to estimate the additional pressure drop in this channel required to maintain a constant volumetric flow rate for extreme cases of the plate blockage factor ε. This figure shows the typical mesh generated for a 40% blockage factor. Since it is necessary to resolve small features around the orifice plate, the mesh must be very densely packed as the blockage factor approaches unity. The mesh resolution therefore becomes problematic. Rather than solve the high blockage factor problem directly, we will extrapolate the required additional pressure drop from geometric parameter variation.

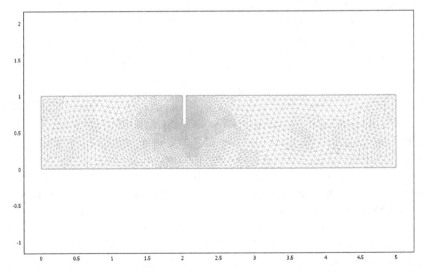

Figure 1.   Mesh generated for the orifice plate in a duct filled with viscous fluid. The parameter representing the percentage of blockage is $\varepsilon = 0.4$ (aspect ratio).

Although it is possible to consider the calculation of the flow around the orifice plate at arbitrary Reynolds number, the major effects in laminar flow are similar to those with artificially vanishing Reynolds number — the Stokes equations. The fundamental reason for this is that most of the dissipation occurs in the opening of the orifice plate, where flow is accelerated yet the small gap leads to strong viscous friction dominating the flow. So to a first approximation, we will model the momentum transport by the Stokes equations:

$$\nabla p = \mu \nabla^2 \mathbf{u} + \rho g \,,$$
$$\nabla \cdot \mathbf{u} = 0 \,, \tag{2}$$

where μ is the fluid viscosity and ρ its density, and all other symbols have their usual fluid flow interpretation. Equation (2) are dimensional, pseudo-stationary, and inertia-free. As they are also linear, they have been the subject of exhaustive analysis. Ockendon and Ockendon [3] is a good reference for the area. Homsy *et al.* [4] provides several excellent visualizations of the "pathologies" of viscous flow with vanishingly small Reynolds number. My attention to the problem of an orifice plate was drawn by Professor Dugdale [5], who arrived at the solution to (2) in the vicinity of a sharp-edged orifice by requiring the condition of optimum energy dissipation within the orifice itself, ignoring the dissipation on all other boundaries of the vessel. His argument is that since the orifice is so small, and all of the flow is forced through the orifice, nearly all of the energy must be dissipated through it, gives a dimensional argument that for a three-dimensional orifice with characteristic opening length a, the energy dissipation rate E must satisfy

$$E = c\frac{\mu Q^2}{a^3} = W = Q\Delta p \,, \tag{3}$$

where Q is the volumetric flow rate and W is the rate of working. c is an unknown constant of proportionality that Dugdale calculates theoretically on the basis of the extremum of the energy absorption or can be found experimentally by measuring Q and pressure loss. In a two dimensional system, the analogous dimensional argument makes E' the dissipation loss per unit length and Q' the cross-sectional area flow rate, giving rise to the scaling argument

$$E' = c\frac{\mu Q'^2}{a^2} = W = Q'\Delta p \,. \tag{4}$$

Dugdale reports experiments with molasses determining c in the range of 3.17 to 3.30. His theoretical result was 3.0. Bond [6] gives an argument of the similarity of orifice plates to Hagen-Poiseuille pipeflow in a pipe of length $2ka$, where a is the orifice radius, and his pressure drop equated to $k = 0.631$, implying $c = 3.21$.

One of us has been interested for some time in the drag on close fitting particles in tubes. For the same rationale leading to (3) or (4), close fitting particles in tubes have drag controlled by the gap width. Zimmerman for thin discs [7] (broadside motion) and for spheres [8] sedimenting in cylindrical tubes, reports on the rapid growth of drag as the particle is taken as having larger radius (smaller gap width a). By using perturbation methods in small particle radius $(1 - a)$ and summing the series expansion, it is possible to determine the nature of the singularity as the particle approaches scraping the duct wall. (4) would suggest a second order singularity, $O(a^{-2})$, on dimensional analysis alone for the thin disc in broadside motion by analogy with pressure loss and drag for 2-D or axisymmetric gaps. The sphere problem is not amenable to dimensional analysis, as the gap width changes with polar angle relative to the sphere's center. Bungay and Brenner [9] computed that the singularity for the drag on the sphere is $O(a^{-5/2})$. Using finite element methods, Harlen [10] found convergence difficulties with close-fitting spheres in a cylindrical duct, indicating the extreme difficulty in resolving large scale differences in numerical computations, even with linear models, when small length scales dominate the dynamics of the flow. It is my guess that much of the dynamics of close fitted particles with small gap width can be found by extrapolation of solutions for larger gap width.

In this section, we have proposed first solving for the additional pressure drop Δp due to the presence of the orifice plate with blockage factor ε obstructing the flow over the pressure drop for laminar flow in a channel without the orifice plate. The gap radius is related to the blockage factor, $a = 1 - \varepsilon$. The difference between this problem and the drag on a sedimenting particle is conceptually very small. For instance Shail and Norton [11] calculated both for the thin disc in broadside motion in a cylindrical duct, as well as the couple — the induced force that opposes rotation of a stationary disc. As these quantities are linearly related due to the linearity of (2), it is expected that the singular behavior of one mirrors that of the other as the gap width is squeezed.

Table 1. Orifice plate inserted into a 2-D channel model. File name orifice05.m.

Model Navigator	Select 2-D space dimension Select COMSOL Multiphysics|Fluid Dynamics| Incompressible Navier-Stokes (mode ns) OK
Draw Menu	Specify objects Specify line. Style: polyline Coordinates x: 0 0 2 2 2.05 2.05 5 5 0 y: 0 1 1 0.95 0.95 1 1 0 0 OK Draw Menu|Coerce to|Solid which creates composite object CO1
Options menu|Constants	Name of constant: rho0 Expression: 0 Name of constant: mu0 Expression: 1 Name of constant: U_{mean} Expression: 1 OK
Physics Menu: Boundary settings	Select boundary 1 Set inflow/outlflow BC with $u_0 = 6 *$ $U_{\mathrm{mean}} * s * (1 - s)$; $v_0 = 0$ Select boundary 8. Set to normal flow/pressure BC with $p = 0$ (default) Accept default no slip BCs on all other boundaries. OK
Physics Menu: Sub-domain settings	Select subdomain 1 Set $\rho = $ rho0; $\eta = $ mu0; $F_x = 0$; $F_y = 0$ Select the init tab and give $u(t_0) = U_{\mathrm{mean}}$, $v(t_0) = 0$; $p(t_0) = 0$. OK
Mesh	Click on the toolbar triangle for the default mesh Refine the mesh twice
Solve Menu	Solver parameters|Set the Stationary nonlinear solver Advanced tab: set scaling of variables to none Solve with the (=) button on the button bar Save as orifice05.m

2.1. *Model of an orifice plate inserted in a 2-D channel*

Table 1 has the instructions for setting up the orifice plate in a channel model. In the drawing section, it would be possible to set up the small orifice "bump" with blockage factor ε easily by creating a composite geometry object that subtracts one rectangle from another. The more roundabout method of creating a closed polyline and coercing it to a solid is used here because it creates a geometry object that is editable. Our intention is to automate the geometric parameter change in a MATLAB script, because automatically changing geometric parameters is difficult in the COMSOL Multiphysics GUI.

The inlet boundary condition is fully developed Hagen-Poiseuille flow in a 2-D channel, with U_{mean} as the single parameter characterizing the inlet condition. The arc length parameter s is used on the boundary to specify the vertical dependence. s is available on every boundary and spans the range from 0 to the total arc length in the direction that defines the positive sense of the boundary. Since the parabolic velocity profile is independent of the directional sense of the boundary, there is no need to account for the directional sense.

By accepting the standard mesh parameters and hit the remesh twice, the mesh in Figure 1 is generated. Note that the output specification gives a pressure datum, so we would expect the pressure to be well conditioned. There are multiple choices for the output condition, but the "straight out" condition, labelled normal flow/pressure reproduces the Hagen-Poiseuille pressure drop at the end of the channel with no orifice plate. The question of which boundary conditions to employ is always a modelling question and involves a degree of idealization and estimation. In this case, inlet conditions that are fully developed suggest there is a long straight section of pipe upstream and outlet conditions that are "straight out" suggest the same downstream.

Under solver parameters, the "scaling of variables" is set to the None option as a matter of course in problems where the mean flow is well conditioned. Furthermore, as our selection of density (rho0 = 0) forces this to be a linear problem, there is no point in complicating matters with scaling the variables to improve convergence. Linear problems are well-posed in terms of convergence — a single matrix inversion step. Nevertheless, we use the stationary nonlinear solver — because not only is this the default for the incompressible Navier-Stokes equations, the linear solver is not an option. It actually takes two iterations for COMSOL Multiphysics to be convinced it has converged, although the initial error being 10^{-12} might have been a good clue! Figure 2 gives the arrow plot of velocity vectors for an $\varepsilon = 0.05$ notch.

Figure 3 shows the pressure profile along the centerline of the channel generated by cross-section plot parameters on the Post-Processing menu. The boundary integral on boundary 1 of the pressure gives the average pressure across the boundary, as it has unit length. This prints in the message dialogue box as 60.462. A quick calculation with fully developed laminar Hagen-Poiseuille flow gives $\Delta p = 60$. The fully developed u-velocity profile is

$$u = 6U_{\mathrm{mean}}y(1 - y).\qquad(5)$$

Substitution into (2) yields the constant pressure gradient as $-12U_{\mathrm{mean}}$. Over five unit lengths downstream, one would expect $p_{\mathrm{inlet}} = 60U_{\mathrm{mean}}$ on

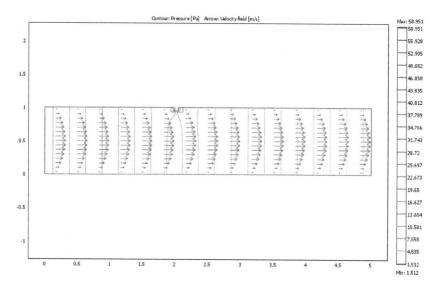

Figure 2.   Velocity arrow and isobars plot for an $\varepsilon = 0.05$ notch.

the inflow plane to achieve $p = 0$ at the outflow. So the additional pressure drop over Hagen-Poiseuille flow is 0.462 (unitless due to scaling of viscosity and velocity).

Exercise 6.1.

Refine the mesh and compute the additional pressure drop. Use the standard refinement on the toolbar, and restart with the old solution as the initial guess. Comment on the uniformity of the mesh and the variation in the additional pressure drop. Is it worth refining the mesh yet again?

Now go to Draw Mode, and double click on the vertices at the bottom of the notch. Edit them to place the orifice plate across to 40% blockage of the gap, but with the same width (0.05). Solve. Figure 4 shows the arrow plot of velocity vectors aqnd isobars of pressure. Clearly the velocity profile must "turn the corner," which causes substantially more disruption and by implication more dissipation of energy.

Boundary integration gives a pressure loss of $\Delta p = 84.85$ required to achieve uniform outflow with $p = 0$. Note that boundary integration along the outflow boundary of the x-velocity gives 1, the value of U_{mean}. Figure 4 shows the isobars which clearly show rapid dissipation of pressure in the orifice. Also, just upstream of the plate, the maximum pressure occurs, due to the need to force flow "around the corner."

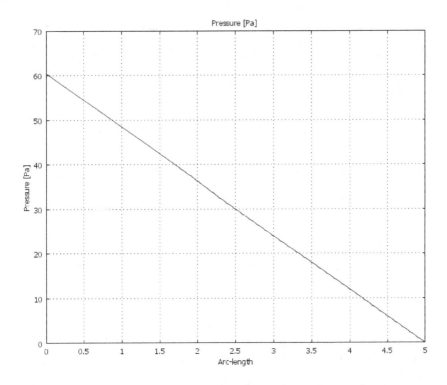

Figure 3.   Centerline pressure profile for the $\varepsilon = 0.05$ notch.

So how do we implement geometric continuation? In this case, all models, even without the use of nearby geometric parameters (blockage factor) converge in one iteration, since the problem is linear. However, the grid refinement studies are required to ensure resolution. First, export a model m-file. Then edit it to set up geometric parameters. The first part of my MATLAB m-file script reads as follows:

```
%%%%%%%%%%%%%%%%%orifice.m%%%%%%%%%%%%%%%%%%%%%%%%%%%%%%%
slot=[0.05:0.05:0.75];
% WZ: Set up storage
output=zeros(length(slot),5);
% WZ: Now loop around the whole FEMLAB model m-file with j
for j=1:length(slot) eps=slot(j);
% Constants
fem.const = {'rho0','0','mu0','1','Umean','1'};

% Geometry
carr={curve2([0,0],[0,1]), curve2([0,2],[1,1]),
```

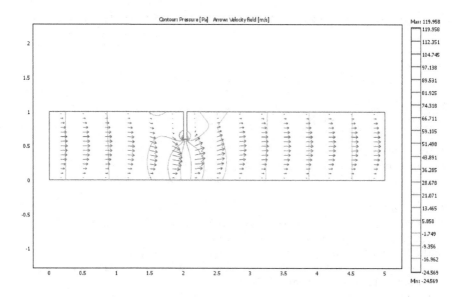

Figure 4.   Velocity arrow and isobars plot for an $\varepsilon = 0.05$ notch.

```
curve2([2,2],[1,1-eps]), ...
  curve2([2,2.05],[1-eps,1-eps]), curve2([2.05,2.05],[1-eps,1]), ...
  curve2([2.05,5],[1,1]), curve2([5,5],[1,0]), curve2([5,0],[0,0])};
g1=geomcoerce('curve',carr); g2=geomcoerce('solid',{g1}); clear s
s.objs={g2}; s.name={'C01'}; s.tags={'g2'};

fem.draw=struct('s',s); fem.geom=geomcsg(fem);

% Initialize mesh
fem.mesh=meshinit(fem,'report','off');

% Refine mesh
fem.mesh=meshrefine(fem,'mcase',0,'rmethod','regular');
fem.mesh=meshrefine(fem,'mcase',0,'rmethod','regular');
fem.mesh=meshrefine(fem,'mcase',0,'rmethod','regular');

% Application mode 1
clear appl appl.mode.class = 'FlNavierStokes'; appl.gporder = {4,2};
appl.cporder = {2,1}; appl.assignsuffix = '_ns'; clear prop
prop.analysis='static'; appl.prop = prop; clear bnd bnd.u0 =
{0,'6*Umean*s*(1-s)',0}; bnd.type = {'noslip','uv','strout'};
bnd.ind = [2,1,1,1,1,1,1,3]; appl.bnd = bnd; clear equ equ.init =
{{'Umean';0;0}}; equ.rho = 'rho0'; equ.eta = 'mu0'; equ.cporder =
{{1;1;2}}; equ.gporder = {{1;1;2}}; equ.ind = [1]; appl.equ = equ;
```

```
fem.appl{1} = appl; fem.frame = {'ref'}; fem.border = 1; fem.units =
'SI';

% Multiphysics
fem=multiphysics(fem);

% Extend mesh
fem.xmesh=meshextend(fem);

% Solve problem
fem.sol=femnlin(fem,'solcomp',{'u','p','v'},...
'outcomp',{'u','p','v'}, 'uscale','none','report','off');
fem0=fem;
% Solve problem
fem=adaption(fem, 'init',fem0.sol, 'solcomp',{'u','p','v'}, ...
'outcomp',{'u','p','v'}, 'ntol',1e-006, 'nonlin','on', ...
'solver','stationary','l2scale',[1],'l2staborder',[2], ...
'eigselect',[1],'maxt',10000000, 'ngen',2,'resorder',[0], ...
'rmethod','longest','tppar',1.7, 'uscale','none', ...
'geomnum',1,'report','off');

% Integrate
I1=postint(fem,'u','dl',[1],'edim',1);
I2=postint(fem,'K_x_ns','dl',[4,5,6],'edim',1);
I3=postint(fem,'p','dl',[1],'edim',1);
% WZ: write out output to the array output
output(j,1)=slot(j); output(j,2)=I1; output(j,3)=I2; output(j,4)=I3;
output(j,5)=length(fem.mesh.p);
% WZ: end our j-loop
end
% WZ: record results to file
save orifice.mat fem
dlmwrite('orifice.dat',output,',');
```

This m-file script is perhaps the longest of the book and therefore deserves some commentary. The first point to notice is that the geometry is constructed in the script (as exported to the m-file) with substitution of $1 - \epsilon$ on the description of the polyline for the depth of the orifice plate. By creating a geometry in the exercise with $\varepsilon = 0.4$. Most of the programming in this script is generated by the COMSOL Multiphysics GUI by the Save As m-file command under the file menu. Manual contributions are about program control (FOR loop) and storage of the desired results. Note that the adaptive solver was used in this script. This can also be used in the GUI (check the adaptive solver check box on the Solver Parameters General

Table 2. Global properties produced by orifice plate model. Average exit x-velocity, integrated viscous stress on the orifice plate, average pressure loss, and the number of degrees of freedom used in arriving at the COMSOL Multiphysics solution as a function of the variation of blockage factor ε.

ε	exit velocity	viscous stress	Δp	dof
0.05	1	−0.78522	60.463	90511
0.1	1	−1.0336	61.551	94235
0.15	1	−1.2105	63.248	1.0034e+05
0.2	1	−1.4211	65.605	99151
0.25	1	−1.5548	68.729	1.0326e+05
0.3	1	−1.8241	72.788	1.0092e+05
0.35	1	−1.9963	78.043	96579
0.4	1	−2.3204	84.864	1.0645e+05
0.45	1	−2.6338	93.817	1.03e+05
0.5	1	−3.1446	105.78	1.0411e+05
0.55	1	−3.6672	122.21	1.0388e+05
0.6	1	−4.383	145.46	1.0534e+05
0.65	1	−5.2545	179.9	99807
0.7	1	−6.933	233.9	99231
0.75	1	−9.2868	325.69	92935

tab). The solution to the whole script takes about 10 CPU minutes on a conventional linux workstation. The estimate of DOF in Table 2 is given by the length of fem.sol.u.

Eventually, Table 2 is generated by the results of this geometric continuation study. Clearly, the additional pressure loss rises rapidly with increasing blockage factor. The viscous stress along the orifice plate shows a similar rapid rise (factor of 12, approximately) from $\varepsilon = 0.05$ to $\varepsilon = 0.75$.

$$\frac{\Delta P}{\Delta P_0} = 1.0 + 0.6155\varepsilon - 1.472\varepsilon^2 + 6.955\varepsilon^3 . \tag{6}$$

Although (6) fits the data well in the range shown, over the whole range, a Laurent series in inverse powers of $(1 - \varepsilon)$ gives a better fit to the $\varepsilon = 0.75$ model than the cubic of (6), when the fit is only done on the range $\varepsilon \in [0, 0.5]$.

$$\frac{\Delta p}{\Delta p_0} - 1 = \frac{0.395652\varepsilon - 0.3832\varepsilon^2}{(1 - \varepsilon)^3} . \tag{7}$$

Figures 5 and 6 shows the plots for equations (5) and (7) against the data (dots). The prediction by (7) is $\Delta p = 371$, by (6.6) 214, and the model gives 309. The key feature of (7) is that, if you account for the coefficients in the numerator nearly being equal, it arrives at the predicted dependency by dimensional analysis only, equation (4).

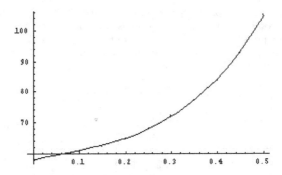

Figure 5. Δp versus ε for the orifice plate of thickness 0.05. The curve is the best cubic fit for ε in the range [0,0.5].

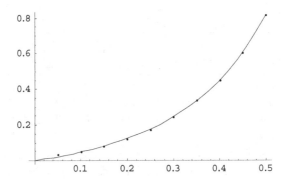

Figure 6.   Fit to Laurent series in inverse powers of $1 - \varepsilon$.

Exercise 6.2: Sharpness effects

Dugdale's orifice plate was sharp. Ours has a thickness of 5% of the channel width. Try making the orifice plate sharper: 4%, 3%, 2%. What effect does this have on the additional pressure drop? According to [12], the detailed shape of the particle has a considerable effect on the drag force as the gap width becomes smaller. If the gap is flat, then Dugdale's dimensional analysis is correct, equation (3), but if the particle has finite curvature, then Bungay and Brenner's $O(a^{-5/2})$ result is recovered.

Exercise 6.3: Platelet geometric continuation

(a) Change the top boundary condition of the orifice plate to be a symmetry boundary condition. This models a two-dimensional platelet with viscous flow past it. Try geometric continuation.

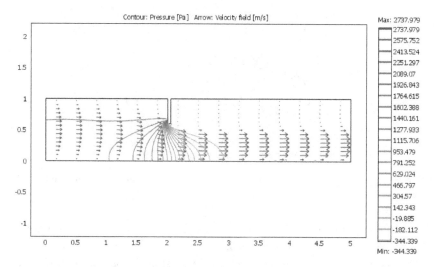

Figure 7. $Re = 1500$ with $\varepsilon = 0.4$. Notice the leeside stagnant flow area with separated flow behind the orifice plate. Fluid inertia causes the stream to largely bypass this stagnant zone. In viscous flow, the fluid "turns the corner" and does not separate.

(b) Alter your m-file to use the solution to the last geometric configuration as the initial condition for the next. Does your m-file finish executing faster?

(c) This example isn't really multiphysics. Try adding the streamfunction-vorticity equation as in the buoyant convection example so as to compute streamlines.

The platelet problem was studied by Kim [13] with an analytically determined long perturbation series that was summed to yield the singular behavior of the drag force as the gap width becomes small.

Exercise 6.4: Geometric continuation with inertia effects

These examples do not actually implement geometric continuation since each geometric change is met by a linear problem since the Reynolds number $Re = 0$, which causes the nonlinearity caused by inertial forces to vanish. Suppose the $Re = 1500$. Given the constants defined in fem.consts, this is achieved by setting rho0 = 1500. In Figure 7, notice the leeside stagnant flow area with separated flow behind the orifice plate. Fluid inertia causes the stream to largely bypass this stagnant zone. In viscous flow, the fluid "turns the corner" and does not separate. This flow structure is significantly different from linear flow that as the Re increases, finding a solution from

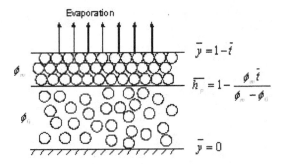

Figure 8.   Schematic of the two fronts in film drying: Evaporation front at the top and compaction front in the interior.

"scratch" is problematic. There are essentially two strategies — parametric continuation in Re from the linear flow solution or geometric continuation in the geometry. The latter method involves using the asseminit command to interpolate the old solution onto the new mesh/geometry:

fem.sol = asseminit(fem,'init',fem0) transfers the solution fem0.sol in the source FEM structure fem0 to the mesh in fem, using interpolation.

This is necessary since the new geometry and mesh are incompatible with the old solution, old mesh, and old geometry. Alter the m-file script orifice.m to include inertial effects and true geometric interpolation using asseminit.

3. Transient geometric continuation: Film drying

In the previous section, geometric continuation did not require using the previous solution with a different geometry, varied slightly, as an initial condition for the new solution, until nonlinear effects are considered (Exercise 6.4). Geometric continuation was carried out for the obvious reason of exploring the model for a range of geometric parameters that alter the domain. In this section, the solution of a transient problem is posed in the case that domain is changing over time, so the solution at the previous time is essential for the prediction of the solution at the current time. The application is to film drying. The model here is an idealization of experiments on film drying reported by Mallegol *et al.* [14].

A thin film of liquid containing particles at an initial volume fraction of ϕ_0 is subject to evaporation from the top surface at a constant rate. If diffusion of particles throughout the film is small an accumulation at the top surface is observed, with particles packing at a volume fraction ϕ_m. Over

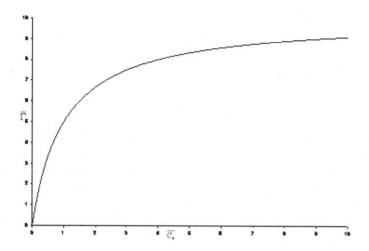

Figure 9. Langmuir adsorption isotherm with $k = 10$ and $\bar{A} = 1$.

time the thickness of the packed layer above the still fluid layer increases. The overall film thickness decreases linearly with time. Scaling vertical distance and time with initial film thickness and the characteristic drying time respectively, allows the film surface to be described by $\bar{y} = 1 - \bar{t}$. A simple mass balance gives that the compaction front moves at a velocity α, given by $\alpha = \frac{\phi_m}{\phi_m - \phi_0}$. From this, the position of the particle front is given by $\overline{h_p} = 1 - \frac{\phi_m}{\phi_m - \phi_0}\bar{t}$. The schematic of a propagating particle front is shown in Figure 8. It follows that no further compaction can take place after time $\bar{t} = 1/\alpha$, in which case a steady film thickness is reached. The function for particle volume fraction, reflecting the abrupt change at the particle front, is easily represented using a step function. Thus,

$$\phi = \phi_0 + (\phi_m - \phi_0)H(\bar{y} - \overline{h_p}),$$

where $H(\bar{y} - \overline{h_p})$ is a heaviside step function.

There is also surfactant present in the film. As the solvent evaporates from the film the nonvolatile surfactant is trapped. This surfactant can either be in solution or stuck to the particles. The surfactant in solution is taken to be initially uniformly distributed at some concentration C_{S0}, a known *a priori*. The surfactant concentrations in solution and on particles are scaled using this initial concentration C_{S0}. In the context of these packing dynamics, the surfactant concentration is also changing due to adsorption on to the packed particles. We note the following conditions on $\overline{C_s}$, the scaled surfactant concentration in solution:

Initial condition: $\overline{C_s} = 1$.

Boundary condition at $\bar{y} = 1$: $\frac{\partial \overline{C_s}}{\partial \bar{y}} = 0$, no surfactant flux across impermeable surface.

Boundary condition at $\bar{y} = 1 - \bar{t}$:

$$(1 - \phi)\frac{\partial(\overline{C_s})}{\partial \bar{y}} = Pe_S((1 - \phi)\overline{C_s} + \phi\bar{\Gamma}),$$

no surfactant across material surface — nonvolatile surfactant is trapped.

Here, Pe_S, the surfactant Peclet number is given by $Pe_S = \frac{H_0 E}{D_S}$, H_0 is initial film thickness, E is top surface velocity, and D_S is the surfactant diffusion coefficient. $\bar{\Gamma}$ is the scaled surfactant concentration on particles, which is a function of $\overline{C_s}$. The function relating $\bar{\Gamma}$ and $\overline{C_s}$ is called the adsorption isotherm. Assuming the surfactant adsorption onto latex particles to be of Langmuir type, the isotherm is given by $\Gamma = \frac{\Gamma_\infty C_S}{A + C_S}$, where Γ is in moles/m^2. Hence in terms of concentration, we obtain a dimensionless isotherm $\bar{\Gamma}$ as

$$\bar{\Gamma} = \frac{\frac{3\Gamma_\infty}{RC_{S0}}\overline{C_S}}{\frac{A}{C_{S0}} + \overline{C_S}} \equiv \frac{k\overline{C_S}}{\bar{A} + \overline{C_S}}, \tag{8}$$

where Γ_∞ and A are the Langmuir isotherm parameters and, k and $\bar{A}$ are the scaled Langmuir isotherm parameters. Figure 9 shows the Langmuir isotherm represented by Equation (8).

The dynamic adsorption of surfactants in miscible displacement is a fundamental, recurring situation in the chemical and petrochemical industries. Enhanced oil recovery by detergent flooding has been practiced for more than twenty years. Liquid chromatography, where the adsorption-desorption isotherm is key to separation processes, is another common example. The desorption of the isotherm forced by the compaction front, however, is a unique feature of the model presented here. Trogus *et al.* [15], in the context of enhanced oil recovery, proposed a kinetic model for adsorption/desorption rates, and Ramirez *et al.* [16] developed a two-equation (concentration and surfactant loading), 1-D spatio-temporal model for dynamic adsorption. Nevertheless, their transport model is still for a homogeneous porous media, where in ours, given below, the compaction front between the close packed and looser packed layers, serves as an impetus for desorption, and thus as a propagating point source of surfactant.

The transport equation for the surfactant is given by [18]:

$$\frac{\partial(\overline{C_s}(1 - \phi) + \bar{\Gamma}\phi)}{\partial \bar{t}} = \frac{1}{Pe_s}\frac{\partial((1 - \phi)\frac{\partial \overline{C_s}}{\partial \bar{y}})}{\partial \bar{y}}, \tag{9}$$

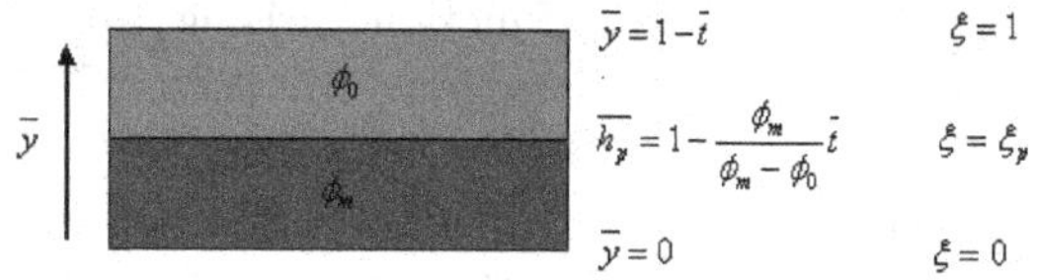

Figure 10. Coordinate transformation: One front.

where the term on the LHS represents accumulation of surfactant, the second part of the factor being due to accumulation in the adsorbed phase. The RHS represents a diffusion term. Representative values of packings are: $\phi_m = 0.64$ $\phi_0 = 0.4$.

It is rather difficult to analyze a two front problem, especially with one an effective point source that is moving. It makes sense to transform our coordinate system to remove one, if not both, of the moving fronts. We experimented with nonlinear coordinate transforms in time to remove both fronts to a fixed domain. Surprisingly, this was possible, adding some greater complexity to the PDE (9), but it is not physical, as the variation is not monotonic for the transform coordinate. Better to stick to one front (the internal compaction front) and transform away the top front to a fixed domain.

The tranformation that achieves this is simple:

$$\xi = \frac{\bar{y}}{1 - \bar{t}}; \qquad \tau = \bar{t}, \tag{10}$$

and results in a new specification for the compaction front:

$$\xi_p = \frac{1 - \alpha\tau}{1 - \tau}. \tag{11}$$

Differentials are expressed in the new coordinates according to the chain rule:

$$\frac{\partial}{\partial \bar{y}} = \frac{1}{1 - \bar{t}} \frac{\partial}{\partial \xi}; \qquad \frac{\partial}{\partial \bar{t}} = \frac{1}{(1 - \bar{t})} \frac{\partial}{\partial \xi} + \frac{\partial}{\partial \tau}, \tag{12}$$

which results in a transformed PDE for surfactant transport:

$$\left(1 - \phi + \phi \frac{d\bar{\Gamma}}{d\overline{C}_S}\right)\left(\frac{\partial \overline{C}_s}{\partial \tau} + \frac{\xi}{1 - \tau} \frac{\partial \overline{C}_s}{\partial \xi}\right)$$

$$- \frac{1}{Pe_S(1 - \tau)^2} \frac{\partial\left((1 - \phi)\frac{\partial \overline{C}_s}{\partial \xi}\right)}{\partial \xi} + \frac{(\bar{\Gamma} - \overline{C}_S)\phi_m}{(1 - \tau)}\delta(\xi - \xi_p) = 0. \tag{13}$$

Here $\delta(\xi - \xi_p)$ is a dirac-delta function, arising from the heaviside step function used for the particle volume fraction. COMSOL Multiphysics provides an easy way to use heaviside step functions and dirac-delta functions. Two m-files, flc1hs.m for smoothed heaviside step function and flc1hs.m for smoothed dirac-delta function, need to be added to the workspace. The expression $Y = $ flc1hs(X, SCALE) approximates the logical expression $Y = (X > 0)$ by smoothing the transition within the interval $-\text{SCALE} < X < \text{SCALE}$. The solution is "sensible" for $0 \leq \tau < \frac{1}{\alpha}$. The boundary and initial conditions are now expressed as: initial condition: $\overline{C_S}(\tau = 0) = 1$ BCs:

$$\frac{\partial \overline{C_s}}{\partial \bar{\xi}}\bigg|_{\xi=0} = 0,$$

and

$$(1 - \phi)\frac{\partial \overline{C_S}}{\partial \bar{\xi}}\bigg|_{\xi=1} = Pe_S((1 - \phi)\overline{C_S} + \phi\bar{\Gamma})(1 - \tau).$$

Table 3 contains the COMSOL Multiphysics instructions to set up the simulation of this double moving front model. It should be noted that the COMSOL Multiphysics modeling strategy used here provides a simple way of dealing with moving boundary problem without having to export it as a m-file. The current strategy avoids the hassle involved in modifying the m-file for the interpolation of the solution from previous time step to a new mesh. Here we deal with the moving front by defining it as a variable (frontpos) and then making use of the built-in functions available for smoothed versions of the Heaviside step function and Dirac delta function.

Figure 11 shows the time evolution of surfactant distribution. The solution from the COMSOL Multiphysics model gives us $\overline{C_s}$ as a function of ξ (the distance from the substrate). But in Figure 11 we plot the % surfactant excess/depletion as a function of ξ. This is the % excess/depletion of surfactant over the amount that would be expected if the surfactant were uniformly distributed. This is given by

$$\frac{((\phi_m\frac{k\overline{C_S}}{\bar{A}+C_S} + (1 - \phi_m)\overline{C_S})(1 - \tau) - (\phi_0\frac{k\overline{C_S}}{\bar{A}+C_S} + (1 - \phi_0)\overline{C_S})|_{\bar{t}=0})100}{(\phi_0\frac{k\overline{C_S}}{\bar{A}+C_S} + (1 - \phi_0)\overline{C_S})|_{\bar{t}=0}}.$$

$$(14)$$

Now that the model is solved, the question is, "Is the solution right?" This can be answered in two ways. First, to see, if the solution makes sense (physically). The four curves in Figure 11 represent the surfactant

Table 3. Film drying model.

Model Navigator	Select 1-D dimension Select COMSOL Multiphysics|Diffusion|Convection and Diffusion, transient analysis OK
Draw Menu	Specify objects Specify line from 0 to 1. OK
Options menu|Constants	Define constants as below: theta_m Expression: 0.64 theta_0 Expression: 0.40 alpha Expression: theta_m/(theta_m–theta_0) Pe Expression: 1 k Expression: 2 A Expression: 10 scale Expression: 0.001
Options menu|Scalar expressions	Define expressions as below: theta Expression: theta_0 + (theta_m − theta_0) * flc1hs(x-frontpos, scale) frontpos Expression: $(1 - alpha * t)/(1 - t)$ y Expression: fldc1hs(x-frontpos, scale) ads_part Expression: $k * u/(A + u)$ dtherm Expression: $k * A/((A + u)^2)$
Physics Menu: Boundary settings	Select boundary 1 Set flux BC with $N_0 = -u*x*(1 -$ theta $+$ theta*dtherm$)(1 - t)$ Select boundary 2. Set to flux BC with $N_0 = ((1 -$ theta$)*u+$ theta*ads_part$)(1 - t) - u*x*(1 -$ theta $+$ theta*dtherm$)(1 - t)$
Physics Menu: Subdomain settings	Select subdomain 1 Set $\delta_{ts} = (1 -$ theta $+$ theta*dtherm$)(1 - t)^2$ Set $D = (1 -$ theta$)/Pe$ Set $R = -$theta_m*(ads_part-u)$y(1 - t)$ Set $U = x(1 -$ theta $+$ theta*dtherm$)(1 - t)^2$ init tab: set $u(t_0) = 1.0$
Mesh	Select subdomain 1; enter maximum element size: 0.001
Solve Menu	Solver parameters|Set the time dependent solver Set output times to 0:0.001:0.374 Advanced tab: weak form Solve with the (=) button on the button bar

distribution profiles for four different times. It can be noted that during the solvent evaporation stage, the surfactant distribution profiles are not continuous across the entire film thickness, but have a discontinuity at the particle front position. This discontinuity is due to the sharp change in particle volume fraction at the particle front position. When the particle volume fraction is finally uniform ($\tau = 0.374$; $\xi_p = 0$), the surfactant profile is a continuous curve. Throughout the drying process, the upper regions of the film have higher particle volume fractions compared to the

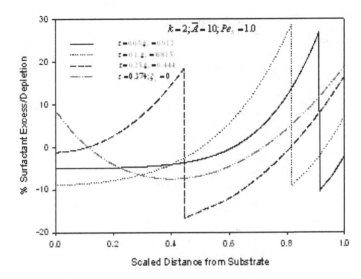

Figure 11. Percentage surfactant excess/depletion at different stages of drying as a function of scaled distance from substrate for fixed k, $\bar{A}$, Pe_S, and values (τ is time and ξ_p is position of particle front).

lower regions of the film. As drying proceeds, the surfactant concentration increases in the bulk due to water evaporation, and as a result, there is more adsorption onto the particles. As a result of the increasing particle volume fraction and increasing adsorption onto particles, the upper regions of the film grow laden with surfactant and hence the excess seen towards the air-water interface at $\tau = 0.374$.

The second approach to verifying the accuracy of the model is to obtain the total surfactant mass. The surfactant being nonvolatile, it should be conserved. The expression, $((1 - \phi)\overline{C_S} + \phi k \overline{C_S}/(A + \overline{C_S}))(1 - \tau)$ gives the total mass of surfactant at any time. So, to obtain the total mass, click on post processing tab and choose subdomain integration, pull down expression to integration and enter $((1-\phi)u+\phi ku/(A+u))(1-\tau)$. Evaluate the integral at $\tau = 0$ and $\tau = 0.374$, the initial and final times. COMSOL Multiphysics gives the values for the expression (surfactant mass) at the two times. The values are almost identical with about 0.5% error. The error stays below 1% for a wide range of operating parameters (k, $\bar{A}$, and Pe_S), indicating the effectiveness of COMSOL Multiphysics in solving moving boundary problems. For a full discussion of the particle and surfactant distribution problems, see our previous publications [17][18].

4. Conclusions

In this chapter, we explored how COMSOL Multiphysics can be used to set up simulations where the geometry model changes smoothly over either a parametric range or smoothly due to transient evolution of a front. The groundwork for these two situations was laid with previous discussion of parametric continuation. In particular, in the previous version of this chapter, an operator splitting technique to deal with transient geometric continuation, with geometry modification occurring during the first part of the time step, and a PDE being solved during the second part was introduced. This cumbersome technique, which requires MATLAB programming for time stepping about COMSOL Multiphysics GUI generated subroutines for PDE solution, is avoided here by using a smoothed interface approach for the internal moving boundary. The new technique was shown to be self-consistent by mass conservation.

As an aside, in many cases the two stage approach for treating moving boundaries can be avoided by specifying the moving boundary in the new ALE mode of COMSOL Multiphysics. The two stage approach required re-starting the solution at one time step with the old solution at the last time increment. Yet, in order to do this, the old solution must be interpolated onto the new mesh, with potentially different numbers (and relevance) of the degrees of freedom. asseminit() was found to have sufficient power to do this, with the code supplied by the COMSOL Multiphysics GUI programming interpreter, through a model m-file translation. This common technique for programming MATLAB routines calling COMSOL Multiphysics functions can sometimes be replaced by an ALE mode. In the case studied here, the coordinate transformation and smoothed interface methods also avoid the two stage time stepping method.

References

[1] K. J. Ruschak, A method for incorporating free boundaries with surface tension in finite element fluid-flow simulators, *Int. J. Num. Methods Eng.* **15** (1980) 639.

[2] R. T. Goodwin and W. R. Schowalter, Arbitrarily oriented capillary-viscous planar jets in the presence of gravity, *Phys. Fluids* **7**(5) (1995) 954–963.

[3] H. Ockendon and J. R. Ockendon, *Viscous Flow* (Cambridge University Press, Cambridge, 1995).

[4] G. M. Homsy *et al.*, *Multimedia Fluid Mechanics* (*CD-ROM*) (Cambridge University Press, Cambridge, 2000).

[5] D. S. Dugdale, Viscous flow through a sharp-edged orifice, *Int. J. Eng. Sci.* **8** (1997) 725–729.

[6] W. N. Bond, *Proc. Roy. Soc.* **34** (1922) 139.

[7] W. B. Zimmerman, The drag on sedimenting discs in broadside motion in tubes, *Int. J. Eng. Sci.* **40** (2002) 7–22.

[8] W. B. Zimmerman, On the resistance of a spherical particle settling in a tube of viscous fluid, *Int. J. Eng. Sci.* **42**(17–18) (2004) 1753–1778.

[9] P. M. Bungay and H. Brenner, The motion of a closely-fitting sphere in a fluid-filled tube, *Int. J. Multiphase Flow* **1** (1973) 25.

[10] O. G. Harlen, High-Deborah-number flow of a dilute polymer solution past a sphere falling along the axis of acylindrical tube, *J. Non-Newtonian Fluid Mech.* **37** (1990) 157–173.

[11] R. Shail and D. J. Norton, On the slow broadside motion of a thin disc along the axis of a fluid-filled circular duct, *Proc. Camb. Phil. Soc.* **65** (1969) 793.

[12] H. A. Stone, On lubrication flows in geometries with zero local curvature, private communication.

[13] M. U. Kim, On the slow broadside motion of a flat plate along the centerline of a fluid-filled two-dimensional channel, *J. Phys. Soc. Jpn.* **53**(1) (1984) 139.

[14] J. Mallegol, J.-P. Gorce, O. Dupont, C. Jeynes, P. J. McDonald and J. L. Keddie, Origins and effects of a surfactant excess near the surface of waterborne acrylic pressure-sensitive adhesives (2002), preprint.

[15] F. Trogus, T. Sophany, R. S. Schechter and W. H. Wade, Static and dynamic adsorption of anions and nonionic surfactants, *SPE J.* **17** (1977) 337–344.

[16] W. F. Ramirez, P. J. Shuler and F. Friedman, Convection, dispersion, and adsorption of surfactants in porous media, *SPE J.* **20**(6) (1980) 430–438.

[17] A. F. Routh and W. B. Zimmerman, Distribution of particles during solvent evaporation from films, *Chem. Eng. Sci.* **59** (2004) 2961–2968.

[18] V. R. Gundabala, W. B. Zimmerman and A. F. Routh, A model for surfactant distribution in latex coatings, *Langmuir* **20** (2004) 8721–8727.

Chapter Seven

COUPLING VARIABLES REVISITED: INVERSE PROBLEMS, LINE INTEGRALS, INTEGRAL EQUATIONS, AND INTEGRO-DIFFERENTIAL EQUATIONS

W.B.J. ZIMMERMAN

Department of Chemical and Process Engineering, University of Sheffield,
Newcastle Street, Sheffield S1 3JD United Kingdom
E-mail: w.zimmerman@shef.ac.uk

In this chapter, coupling variables are explored in great depth with regard to their role in solving inverse equations and integral equations of various types. Four important applications are taken as example studies — using lidar to detect position and spread of dense gas contaminant clouds, the inverse problem in electrical capacitance tomography, the computation of nonlocal heat transfer in a fiber composite medium, and the population balance equations in particle processing. En route, we encounter several features of COMSOL Multiphysics not previously explored — coupling to optimization tools through MATLAB, extended meshes, using the time-dependent solver as an iterative tool for stationary nonlinear models, and the ability to selectively activate/deactivate multiphysics modes in coupled models. The latter is particularly useful if there is only one-way coupling (as in the hydrodynamics around the catalyst supported on the pellet in Chapter 3). In the case of the integral equations treated here, a fictitious dependent variable on an auxiliary domain is set up. The domain is used by coupling variables for various operations, but the dependent variable is never needed itself. So deactivating it results in better conditioning the FEM approximation to the integral equation.

1. Introduction

We are already familiar with boundary and subdomain integration — options available on COMSOL Multiphysics's post processing menu. Boundary integration is useful for computing all manner of surface quantities: the charge on a body in electrostatics and the drag on a body in hydrodynamics, for example. Subdomain integration is typically used for averages and higher moments of combinations of the degrees of freedom defined in the domain. These features are reliable, and given the nature of the finite element method expressed through an integral property, the Galerkin method (see Chapter 2), COMSOL Multiphysics naturally incorporates efficient and accurate integration schemes. Yet if the reader is interested in numerical integration of arbitrary integrands or ODEs, the built-in MATLAB schemes are generally sufficient (see Chapter 1) and do not warrant further discussion here. In this chapter, more complicated applications of integral equations and theory are explored with an eye to computation within COMSOL Multiphysics. Line integrals, integral equations, and integro-differential equations are the target applications in applied mathematics. These are all treatable with recourse to COMSOL Multiphysics's coupling variable capability. Perhaps that is reason to have titled this chapter "Extended multiphysics II." The title selected, however, is probably more descriptive. As ever, we target illustrations in chemical engineering of the use of the COMSOL Multiphysics features. The most important treated here are using lidar to detect position and spread of dense gas contaminant clouds, population balance equations which are exemplary of IDEs, the inverse problem in electrical capacitance tomography, and the computation of nonlocal heat transfer in a fiber composite medium.

Extended multiphysics

At first approach, I must admit to being skeptical of extended multiphysics as something that I was likely to use. Eventually, the utility of scalar coupling variables dawned on me, and provided the impetus for Chapter 4. It also spawned our interest in a new adventure for our research team, modeling microfluidics networks. Yet COMSOL Multiphysics provides two other conceptual constructs for coupling variables — extruded and projected coupling variables. The examples of their use in the Model Library are nearly all about post-processing, i.e. to express solutions in cross domain functionals to analyze particular features.

Rarely, however, coupling variables (extruded and projection) have been incorporated in the model and solved for simultaneously with the independent field variables. Multi-domain, multiple scale, and multiple process models are not common in engineering mathematics and mathematical physics. Typically, models are local in character — conceived of as a set of (partial) differential equations and boundary and initial conditions that are well posed. These are termed continuum models. Historically, this development has been predicated on the use of analysis techniques that have some scope for treating this class of models in closed form. Computational models, even in situations that are treatable by continuum methods, are approximated by discrete interaction rules that need not be local. Smooth particle hydrodynamics [1] and discrete element methods [2] are growing in popularity, but older methods like molecular dynamics simulations [3], Monte Carlo methods [4], microhydrodynamics [5], cellular automata [6] and exact numerical simulation in gas/plasma dynamics [7] bridge the continuum/discrete system gap in modelling distributed systems. Another set of techniques is based on optimization theory to satisfy PDE constraints — penalizing the degree to which constraints are not satisfied. Mixed integer nonlinear programming [8], genetic algorithms [9] and genetic programming [10] are all suitable for treating models of mixed discrete/continuum systems. COMSOL Multiphysics was formulated with a strong bias towards continuum systems with pde constraints. Yet, conceptually, extended multiphysics is not an afterthought for dealing with awkward situations. It permits treating discrete systems on an even footing with continuum systems characterized not only by PDE constraints, but by integral constraints as well. Essentially, coupling variables permit nonlocal and discrete modelling.

In sections §3 (scalar), §4 (projection) and §5 (extrusion) we revisit coupling variables to explore the COMSOL Multiphysics treatment of inverse problems, line integrals, and integral/integro-differential equations, respectively.

Scalar coupling variables

Undoubtedly, scalar coupling variables are the conceptually easiest to grasp. In Chapter 4, scalar coupling variables were used to link up a recycle stream in a flowsheet for a heterogeneous chemical reactor — the output of the reactor, suitably scaled, re-enters with the feed stream. An abstract 0-D element in a second geometry was created for the purpose of modeling a buffer tank that achieved the algebraic relationship between the recycle stream and the reactor outlet. Very simply, a scalar coupling is a single value passed to the destination domain, subdomain, boundary, or edge, where it is used anywhere in the description of the domain FEM residuals. The scalar coupling variable is created by an integration on the source domain. Since in our example, sources were 0-D (endpoints or the single element construct), the integrations were trivially the same as the integrand. Furthermore, that buffer tank model was artificial since the recycle relations could have been more readily incorporated in a weak boundary constraint without recourse to the second domain. So we have yet to see an example of scalar coupling variables where the source integration was nontrivial and the coupling itself essential. In the next subsection, we tackle an inverse problem where coupling is essential and intricate. An inverse problem has the connotation that there is an associated forward problem that is well-posed, but that the inverse problem is ill-posed. Our selected inverse problem is a tomographic inversion for electrical capacitance tomography.

Electrical capacitance

Process tomography has matured as an engineering science in the past decade. One of the most common configurations is electrical capacitance tomography, frequently used for imaging processes with multiphase flows in cylindrical pipelines. Sensing of multiphase pipeline flows with information about the distributed flow of dispersed phases can be crucial to tight control of chemical and processing unit operations. Noninvasive and nonintrusive measurements of two-phase flow are notoriously difficult to obtain. The difficulty is often exacerbated by the highly time-varying flows some times encountered in gas-liquid flows in the oil and gas production industry. Accurate measurements of transients in the flow and instantaneous phase distributions cannot be achieved. One possible way of obtaining such data is to measure the spatial electrical permittivity distribution of a flowing gas-liquid mixture using Electrical Capacitance Tomography (ECT). This will give information regarding the phase distribution about the pipe cross-section.

Tomographic instrumentation can provide images, noninvasively, of the distribution of components within a process vessel or pipeline. Electrical Capacitance Tomography (ECT) provides 2-D images of the *dielectric distribution* of the components within a process pipe. Noninvasive measurements of capacitance by electrodes — excited by a charge-discharge principle [11] — are used in a mathematical reconstruction algorithm to create images of materials having different permittivities. This procedure allows different phases to be determined. To date, process engineering studies involving ECT have been sparse, but some areas of application include fluidised beds and pneumatic conveying. McKee *et al.* [12] reported the use of capacitance tomography for imaging pneumatic conveying processes in two industrial pilot scale rigs. This work pioneered the application of ECT to *dense-phase* pneumatic conveying and demonstrated the potential of capacitance tomography as an aid to on-line process control. A good review of this area can be found in [13].

The tomographic imaging device involves three main sub-units: An array of sensors (typically 12 electrodes; 66 independent measurements), a data acquisition system and an image reconstruction system. Measurements of capacitance are obtained for all possible combinations of electrodes. For each electrode pair the following charge-discharge procedure is adopted: the active electrode is charged to a given voltage (15 volts) while the detecting electrode is earthed; the active electrode then discharges to earth while the detecting electrode connects to the input of a current detector. This detector then averages the resultant oscillating current from the detecting electrode, creating a voltage directly proportional to the unknown capacitance value.

The basic capacitance data acquisition system is based on the charge transfer principle. The discharging current flows out of the current detector producing a positive voltage output. The typical charge/discharge cycle repeats at a frequency of 1 MHZ, and the successive charging and discharging current pulses are averaged in the two current detectors, producing two DC output voltages.

Calibration of the instrument is performed before use of the electrode arrangement and involves the sensor device being filled with the material of lower permittivity. This procedure provides a reference value of permittivity. A change in the measurement sensitivity of the circuit then occurs when the pipe is filled with the material having the higher permittivity. A calibration procedure is needed for each type of material studied.

The image reconstruction process yields an image of the concentration distribution within the pipe by the use of a back-projection algorithm. Existing algorithm techniques for ECT are capable of producing images at a frame rate of 100 images per second and can, virtually, provide almost real-time information about the process. However, a limiting feature of the existing ECT system is the modest spatial resolution (about one tenth of the pipe radius). The major reason for this constraint is that the surface area of each electrode is large enough that, for all practical purposes, the electric field lines are parallel between the electrode pairs in the charge/discharge cycle. This convenience permits an easy image reconstruction by the back-projection algorithm. If more and smaller electrodes are used, there is the possibility of greater spatial resolution, but at the cost of a more complicated reconstruction algorithm. This algorithm would need to solve a Poisson equation with boundary data to find the internal permittivity field.

In this subsection, we give a flavor of the image reconstruction inverse problem with a toy model of a sparse system with large electrodes and distinct, rod-like inclusions in a 12-gon duct (see Figure 1 for the mesh).

The electric charge density within the duct is related to the potential by the appropriate simplification to Maxwell's equations where there is no

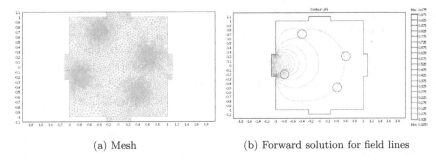

(a) Mesh (b) Forward solution for field lines

Figure 1. Left: mesh for four rod-like inclusions in a rectangular cross-sectional duct, each with dielectric constant $\varepsilon_x = \varepsilon_y = \varepsilon_z = \varepsilon_4 = 0.005$ in a medium with unit dielectric constant ε_0. Right: Steady state contours of potential (voltage) when the boundary segments on electrode 1(west) are held at unit voltage, and all other electrode segments are held at ground, zero voltage, with all other boundary segments electrically grounded.

magnetic coupling [14]:

$$\nabla \cdot \varepsilon \nabla \Phi = -\rho^{(e)} , \qquad (1)$$

where $\rho^{(e)}$ is the total electric charge per unit volume, which is clearly zero within the bulk fluid and the inclusions, but nonzero on the electrode surfaces only, ε is the dielectric constant or permittivity of the medium, depending on the choice of scaling, and Φ is the electric potential (voltage). Using equation (1) and applying over a thin control volume incorporating the interface between the electrode and the bulk fluid leads to this electric flux boundary condition:

$$-\varepsilon_s \frac{\partial \Phi}{\partial n} + \varepsilon_0 \frac{\partial \Phi}{\partial n} = q , \qquad (2)$$

where ε_s is dielectric constant of the solid constituent of the electrodes and ε_w is the dielectric constant of the bulk fluid medium. The LHS represents the electric flux out of the interface from the electrode side, the electric flux into the electrode from the bulk fluid, the difference balanced by the accumulated charge on the electrode at steady state. Rearranging (2) leads to the boundary status as

$$\frac{\partial \Phi}{\partial n} = \frac{q}{\varepsilon_0 - \varepsilon_s} = q' , \qquad (3)$$

where we shall term q' as the charge on the electrode.

With these governing equations, we can define two related tomographic mathematical problems.

1.1. *The forward problem*

Suppose the firing electrode is held at unit voltage (see Figure 1) and the sensing electrodes are held at ground (zero voltage). The solution Φ to (1) computing the total charge on the electrodes i is then

$$\langle q' \rangle_i = \int_{\partial\Omega} \frac{\partial\Phi}{\partial n} d\Omega\,, \tag{4}$$

with known dielectric constants for the inclusions, is termed the forward problem. Figure 1(right) shows the solution to the forward problem that we will shortly formulate in COMSOL Multiphysics.

The inverse problem

Our goal in modelling is to solve an *inverse problem* — finding the unknown parameters of the problem from measured outputs. To do that, we first set up the forward problem and use it in finding the inverse methodology. Rarely can one restate the forward model to "solve in reverse" for the unknown parameters (as in high school algebra). A further complication is defining which inverse problem to solve. For instance, suppose the same experiment is conducted, but that the dielectric field in the duct is not known *a priori*. The charges q'_i are measured on the electrodes and the permittivity field ε in the duct consistent (since Φ is a solution of (1)) with the measurements through (4) is sought. This is termed an inverse problem. In the next subsection, we will use our forward model to approach the solution to the inverse problem.

Modelling the forward ECT problem in COMSOL Multiphysics

Launch COMSOL Multiphysics and in the Model Navigator follow the instructions in Table 4 in setting up the forward ECT problem. Isn't the PDE system, with equations (1), BCs described in the caption of Figure 1, and outputs measured as boundary integrals (4), *stationary* and *linear*? Shouldn't we be using the stationary linear solver? Potentially, we will need the time-dependent solver, so specifying the *da* coefficients is useful now. The nonlinear solver takes two Newton iteration steps, which in practice is a more accurate solution to linear problems than the linear solver. If you don't believe me, check the Solver log with the View log option on the Solve Menu. In this problem, the first Newton step achieves 10^{-14} error estimate, and the second Newton step achieves 10^{-16}. But there are other more subtle reasons to use the nonlinear solver — to be certain that all the contributions to the solution are identified. If some of the dependencies are

Table 1. Electrostatic field lines model. Filename: ectrect.mph.

Model Navigator	Select 2-D dimension
	Select COMSOL Multiphysics|PDE Modes|General Mode
	Name the dependent variable phi
Options Menu| Axes/Settings	Set the x-axis to -1 to 1; set the grid spacing to 0.1 for both x and y
Draw Menu	Specify a rectangle R_1 with corner $(-0.25, 1)$, height 0.1, width 0.5
	Specify a rectangle R_2 with corner $(1, -0.25)$, height 0.5, width 0.1
	Specify a rectangle R_3 with corner $(-0.25, -1.1)$, height 0.1, width 0.5
	Specify a rectangle R_4 with corner $(-1.1, -0.25)$, height 0.5, width 0.1
	Specify a square SQ1 with corner $(-1, -1)$ and width 2
	Specify a circle with center $(0.6, 0.2)$ and radius 0.1
	Specify a circle with center $(-0.4, 0.7)$ and radius 0.1
	Specify a circle with center $(0.4, -0.5)$ and radius 0.1
	Specify a circle with center $(-0.8, -0.2)$ and radius 0.1
	Select Create Composite object. Put in formula
	SQ1 $+R_1+R_2+R_3+R_4$; uncheck "keep internal boundaries"
	OK

Options Menu| Constants					
	e_0	e_1	e_2	e_3	e_4
	1	0.05	0.05	0.05	0.05
	phi1	phi2	phi3	phi4	
	1	0	0	0	

Physics Menu: Boundary settings	Set boundary 4, 5, 6, 7, 13, 15, 16, 18 keep Dirichlet (R $=-$phi) default
	Set boundary 1, 2, 3 Select Dirichlet, $R = $ phi1 $-$ phi
	Set boundary 8, 9, 12 Select Dirichlet, $R = $ phi2 $-$ phi
	Set boundary 10, 11, 14 Select Dirichlet, $R = $ phi3 $-$ phi
	Set boundary 17, 19, 20 Select Dirichlet, $R = $ phi4 $-$ phi

Physics Menu: Subdomain settings	Subdomain	Γ	F	da	ea
	1	e_0 * phi$x$ e_0 * phi$y$	0	1	0
	2	e_2 * phi$x$ e_2 * phi$y$	0	1	0
	3	e_3 * phi$x$ e_2 * phi$y$	0	1	0
	4	e_4 * phi$x$ e_2 * phi$y$	0	1	0
	5	e_1 * phi$x$ e_1 * phi$y$	0	1	0

Mesh	Mesh with standard mesh and refine once
Solve Menu	Set the stationary nonlinear solver
	Click on the solve (=) tool on the toolbar

hidden in the "F" side of a general PDE mode, then subsequent Newton iterations will draw these out. This is not a broadside against the linear solver, however, since it has the very useful feature that the linear solver always generates an output if the problem is well-posed. The nonlinear solver may not converge.

Table 2. Boundary integration for each "electrode".

Boundary	integrand	Q
1, 2, 3	nx * phix + ny * phiy	2.050065
8, 9, 12	nx * phix + ny * phiy	−0.030689
10, 11, 14	nx * phix + ny * phiy	−0.033562
17, 19, 20	nx * phix + ny * phiy	−0.015282

Table 2 contains the boundary integrations of the normal flux. Note that (nx, ny) are reserved geometric variables in COMSOL Multiphysics. These are the components of the outward pointing normal vector to the boundary, which permits the computation of the normal flux according to the standard formula

$$\frac{\partial \Phi}{\partial n} = \hat{n} \cdot \nabla \Phi. \tag{5}$$

In an electrostatic problem, this has the interpretation of the normal electric field component. It should be noted that there is an exact analogy between electrostatics and steady heat transfer. Φ could be interpreted as the temperature, the fluxes q are heat fluxes, and the medium is a "rod composite." So the forward model here predicts the fluxes due to imposed potentials at specific boundaries (electrodes) in an otherwise grounded medium. This field leakage through the electrodes is measurable. So the inverse problem is to use the measured field leakage to determine the material properties of the medium in the domain. For instance, could we, knowing the permittivity of the bulk medium and the positions of the rods, determine the permittivity of the rods? The major issue is whether the measurables, i.e. the fluxes q_i are sensitive to the parameters, e_j, of the problem.

Sometimes, the inverse problem is solvable in the COMSOL Multiphysics GUI. Coupling variables play a major role in the GUI inverse methodology. The simplest problem to pose is that the flux q across electrode 1, boundaries 1, 2, 3, is measured and used to infer the permittivity e_2 in subdomain 2 (circle $C4$, closest to electrode 1). We would imagine that the flux q_1 is sensitive to variation in e_2, but can we demonstrate this sensitivity? This is precisely the sort of problem that the parametric solver was built to address. Follow the steps in Table 3 to create an integration boundary variable to compute q_1 and use the parametric solver to show the sensitivity to e_2. Figure 2 shows the results of the sensitivity study. First, the good news. There is a monotonic variation of q_1 with e_2, so the "inverse problem" to find e_2 given a measure of q_1 is uniquely solvable. The bad news is the sensitivity level — a one percent variation from the q_1 from

Table 3. Parametric sensitivity modifications to ECT forward model. Filename: ec-
trect_param.mph.

Model Navigator	Open ectrect.mph
Options Menu| Integration coupling variables| Boundary variables	Select boundaries 1, 2, 3 Name: q_1 Expression: nx *phi$x$ $+ ny$*phi$y$ Keep global destination
Solve Menu	Select the parametric nonlinear solver. General tab Name of parameter: e_2 List of parameter values: 0.01:0.01:0.2 Click on the solve ($=$) tool on the toolbar
Post-processing menu	Selection cross-section plot parameters Point tab. Set expression $q_1/2.050065$ General tab. Select point plot

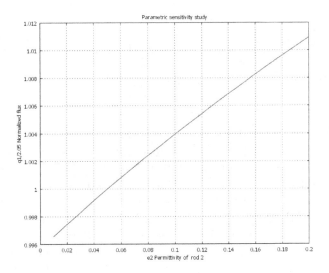

Figure 2. Parametric sensitivity plot of normal flux q_1 with variation of e_2.

$e_2 = 0.05$ to $e_2 = 0.2$, a 400 percent change. This level of sensitivity makes
identifying the parameter e_2 in the presence of experimental noise problematic. It also makes determination of other parameters difficult if there
is more than one unknown permittivity and more than one measurement.
As an aside, note that this model had 21645 degrees of freedom.

1.2. *Inverse methodology in the GUI*

Now let's use the GUI to solve an inverse problem. Suppose we measure
$q_1 = 2.07$ and wish to know the permittivity e_2. The quick witted will spot

that we already know the answer from the parametric solver — we can interpolate within Figure 2 and find the requisite value. This works if we have a low dimensional parameter space and low dimensional measurable space. We can use a graphical method (traditional) or table look-up and interpolation. Since we have one measurement and one unknown, this is essentially a root finding problem. One strategy would be to package the COMSOL Multiphysics forward problem, i.e. $q_1 = f(e_2)$ as a matlab function and use matlab's built-in rootfinding algorithm. But as we saw in Chapter one, the nonlinear solver of COMSOL Multiphysics is a Newton's method root-finding solver. So how do we convince the COMSOL Multiphysics GUI to find our root for us? The simple answer is that we need another degree of freedom — a variable to solve for, say E_2, and another constraint to solve, i.e. $q_1 = q_{1,m}$, where the latter is the measured flux and the former is the predicted flux from the model. One way to add another degree of freedom is to set up a 0-D domain as we did to solve root-finding problems in Chapter one. Another is to use ODE settings. The state variables are global and can be used anywhere in the model set up. Recall in chapter four, the concentrations u_c and v_c in the buffer tank were used as state variables and coupled to the boundary conditions of the reactor inlet and outlet.

The third way to set up additional degrees of freedom is the weak point mode. The 0-D construct of chapter one is somewhat tedious to set up, but it's usage is straight forward. The ODE settings for state variables are very easy to set up, but not as flexible for back coupling as the other two methods. The weak point mode is easy to set up, but requires a basic understanding of the weak form and how COMSOL Multiphysics sets it up. Only trained numericists in finite element analysis are really comfortable with the weak form and weak form programming. We will use it here and again in chapter eight so as to make the reader aware of the alternative ways of adding additional independent variables and degrees of freedom.

Table 4 gives the description of setting up the inverse problem solution in the GUI. Note that since there are 36 vertices, a weak point mode creates 36 new degrees of freedom. By deactivating all but vertex 1, we have now the desired one extra degree of freedom. The weak field expression entered, E_2_test*$(q_1 - 2.07)$, might require some explanation. E_2_test is the "test function" conjugate to the dependent variable E_2 which was defined in Chapter two. In this expression, it serves mainly as the format for entering the constraint desired, i.e. $q_1 = 2.07$. In order to make the variable E_2

Table 4. Modifications to the ECT model for inverse problem in the GUI. Filename: ectrect_inv.mph.

Model Navigator	Open ectrect_param.mph
	Click the multiphysics button. COMSOL Multiphysics|PDE modes|weak point form
	Name the dependent variable: E_2
	Name the application mode: error
Physics Menu|Point settings Mode error	Select points 2–36 and uncheck the "Active in this domain" check box
	Select point 1. In the weak value/expression field enter: E_2_test*$(q_1 - 2.07)$
	Init tab. Enter $E_2(t_0) = e_2$
Options Menu| Integration Coupling Variables| Point variables	Select point 1
	Enter variable name: P_2
	Enter expression: E_2
	Keep global destination
Physics Menu|Subdomain settings Mode g	Select domain 2
	Change to $\Gamma = -P_2*$phi$x - P_2*$ phi$y$
	OK
Solve Menu Solver Manager	Initial value tab: Set initial value expression
	Solve For tab. Select only mode g
Solve Menu	Select Get initial value
Solve Menu Solver Parameters	Select the nonlinear solver
	Advanced tab. Set solution form to weak
	Click on the solve (=) tool on the toolbar
Solve Menu Solver Manager	Solve For tab. Select both modes g and error
Post-processing menu	Data Display|Global
	Enter P_2 and q_1 to find
	Value: 0.167904, Expression: P_2
	Value: 2.07, Expression: q_1

a "state variable," an integration coupling variable (P_2 here) needs to be defined, since E_2 is only defined on vertex 1.

Note under mesh statistics that the final model does have 21646 (one more) degrees of freedom. You might wonder why the Solver Manager in invoke to turn off and turn on the solution for the error mode. Actually, this is a mystery to me. If I try to solve for both modes from the initial conditions simultaneously, the error message "singular matrix" rears its ugly head, identifying a single DOF that does not contribute to the linearized matrix. Since we have built this model up step-by-step, we know exactly which DOF is the culprit — E_2, which does not appear directly in any equation, but is indirectly constrained by the error mode. By staging the solution, the Solver becomes alerted to the indirect E_2 dependence through all the coupling variables. As I have become a bigger user of coupling variables, this anomalous behaviour becomes more common, and the staging of the solution to bring out the dependencies in the coupling variables is a necessary step.

Exercise 7.1.

It should be noted that in the above example, because the number of unknowns is equal to the number of measurements, the inverse problem is essentially a root-finding exercise. The GUI method can be used in the case of more than one measurement. Suppose that $q_1 = 2.061215$ and $q_2 = -0.031372$. Given this additional degree of freedom, the inverse problem becomes one of minimizing the error subject to variation of e_2. The typical error is a sum of squared errors, i.e. error $= (q_{1_p} - q_{1_m})^2 + (q_{2_p} - q_{2_m})^2$. The subscripts p and m refer to the model prediction and measurements, respectively. Routinely, however, modellers choose to scale the error term components by the measurements, i.e. error $= (\frac{q_{1_p} - q_{1_m}}{q_{1_m}})^2 + (\frac{q_{2_p} - q_{2_m}}{q_{2_m}})^2$, so that the relative error is computed. Each error term makes a comparable contribution. We can "trick" the GUI into solving the minimization problem by entering the above error formula into the expression solved in the weak point mode error, i.e. expression $E_2\_test*error$. We are asking the Newton solver of COMSOL Multiphysics to do the impossible — satisfy two constraints by varying one degree of freedom. In practice, it will seek the smallest value of error, and when it cannot drive that value to zero, it will not converge. But the nonconvergent solution is an excellent approximation to the solution to the inverse problem, i.e. finding the minimum error at the best value of E_2. Try this out in the GUI. You may find that it is preferable to use the linear solver to get started, and restart solver from the nonconverged solutions to drive towards zero value of the error formula.

1.3. *Inverse method programming in MATLAB*

The COMSOL Multiphyics GUI could be "tricked" into solving the inverse problem very easily (see Exercise 7.1). However, as the complexity of the inverse problem rises, the flexibility of MATLAB programming arises. As we saw in the parametric sensitivity study, it is important to select measurements which are sensitive to the parameter that is desired to be identified. As important, however, is to select experimental conditions that lead to sensitivity. In our forward problem, we deliberately created constants for the electrode voltages phi1, phi2, phi3, phi4, but "fired" only one, while the others remained at ground. In many tomographic devices, the electrodes are fired sequentially, and the normal field leakage is measured at each of the other electrodes. The result is that substantially information is collected about the permittivity distribution internal to the domain. Where one sensor might be insensitive to some of the permittivity distribution, another might be sensitive. Using this information effectively to identify the parameters desired requires a strategy, but if there is a forward model, the inverse problem can be expressed across several experimental conditions more readily by using MATLAB programming wrapped around COMSOL Multiphysics functions/subroutines.

The easiest way to code this is to save the fem structure within a file and load it into a MATLAB function. Load the first example for the forward problem, ectrect.mph, and export the fem structure to MATLAB. Use the matlab command save to store it:

```
save forward.mat fem
```

The following MATLAB function computes the solution to the forward problems of firing the first two electrodes and predicting the normal flux across those electrodes:

```
function [Q1,Q2]=forward(E2)
% Computes normal fluxes when electrodes 1 and 2 fire sequentially.
load forward.mat fem
% Constants
fem.const = {'e0','1','e1','0.05', ...
   'e2',E2, ...
   'e3','0.05','e4','0.05', ...
   'phi1','1', ...
   'phi2','0', ...
   'phi3','0','phi4','0'};
% Fire first electrode
fem=multiphysics(fem); fem.xmesh=meshextend(fem);
fem.sol=femnlin(fem,'solcomp',{'phi'},'outcomp',{'phi'},'report','off');
% Compute normal flux
Q1=postint(fem,'nx*phix+ny*phiy','dl',[1,2,3],'edim',1);
% Constants
fem.const = {'e0','1','e1','0.05', ...
   'e2',E2, ...
   'e3','0.05','e4','0.05', ...
   'phi1','0', ...
   'phi2','1', ...
   'phi3','0','phi4','0'};
% Fire second electrode
fem=multiphysics(fem); fem.xmesh=meshextend(fem);
fem.sol=femnlin(fem,'solcomp',{'phi'},'outcomp',{'phi'},'report','off');
% Compute normal flux
Q2=postint(fem,'nx*phix+ny*phiy','dl',[8,9,12],'edim',1);
```

The matlab call is simple:

```
>> [Q1,Q2]=forward(0.143)

Q1 =

    2.0646

Q2 =

    2.0035
```

The two calls to the nonlinear solver compute the solution to two separate models — differing only in the operating conditions. The next step is to create the error function. It is a simple wrapper around the forward function call(s). The error function must return a single value, the estimated error, and take a single vector input, the vector of parameters to be called:

```
function b=errornm(v);
[q1,q2]=forward(v);
Q1=2.0646; Q2=2.0035;
x=(q1-Q1)/Q1;
y=(q2-Q2)/Q2;
[v,q1,q2,sqrt(x^2+y^2)]
b=sqrt(x^2+y^2);
```

This function returns the error in the model if the guess $v = E_2$ does not give the measured values $Q_1 = 2.0646$; $Q_2 = 2.0035$; For instance,

```
>> errornm(0.05);

ans =

    0.0500    2.0501    2.0034    0.0070
```

By trial and error, it is possible to get Matlab to minimize the error for us. In practice, it is better to automate the search. Matlab has a built-in optimization routine called fminsearch which implements the directed search for the minimum automatically. The command in this case is:

```
v=0.05;fminsearch(@errornm,v);
```

The @ preceeding the function name treats it as a pure function argument. The second argument represents the initial condition. fminsearch() provides a simple algorithm for minimizing a scalar function of several variables. It implements the Nelder-Mead simplex search algorithm, which modifies the input arguments "v" to find the minimum of $f(v)$. This is not as efficient on smooth functions as some other algorithms, especially those that compute the derivatives, but on the other hand, costly gradient calculations are not made either. It tends to be robust on functions that are not smooth. If the function to be minimized is inexpensive to compute, the Nelder-Mead algorithm usually works very well.

It takes 28 steps for fminsearch to home in on the minimum error at $E_2 = 0.143$. Given that each step computes two 26000 DOF finite element method solutions, this is an expensive method for solving inverse problems.

We have seen two different methods — GUI based and Matlab based — for approaching inverse problems. I know of one other COMSOL Multiphysics method for solving inverse problems which is potentially useful. Niklas Rom of COMSOL published in the Knowledge Base a methodology for using calculus of variations to fit an unknown diffusivity with a known concentration profile. The method involves computing the stationary form of the error function and requiring it to vanish at the parameter value (diffusivity) generating the minimum error. This method is useful if the parametric dependence can be stated explicitly so that the variational form can be symbolically computed. The two methods presented here are "black box" in that gradient information is not explicitly computed.

If greater resolution in the composition of the inclusions in the domain or their positions or sizes are desired than in the examples given here, then the better quality boundary data is diluted across the domain, possibly obscuring the "image" of the included data. Image reconstruction is a complicated problem for capacitance tomography. A good review of applications can be found by Dyakowski *et al.* [15]. The work of Lionheart and coworkers [16], especially the EIDORS MATLAB based software package, is the best source of novel inversion techniques.

Exercise 7.2.

Write the matlab scripts and functions to solve the inverse problem of Exercise 7.1. Compare the timings for the GUI solution and the MATLAB optimization solution. Which method is computationally cheaper (CPU time)?

2. Projection coupling variables and line integrals

The projection coupling variable performs a line integral across a 2-D domain according to a specified coordinate dependent transformation, i.e. a path integral. This useful concept is seen, for example, in formulating quantum electrodynamics [14]. In its simplest form, the path is taken as one of the coordinates (a simple grid line) and thus achieves a reduction in the order of the domain or variable dependence:

$$I(x) = \int_{y_1(x)}^{y_2(x)} f(x,y)dy \quad \text{on} \quad D_2 \,. \tag{6}$$

$I(x)$ is the coupling variable, which must be defined on a domain D_1 of dimension one less than $D_2 = (x_1, x_2) \times (y_1(x), y_2(x))$ in the case shown above. A more complicated projection can be achieved by local mesh transformation using either the space coordinates (dependent variables x, y, z, ...) or local mesh parameters, e.g. s, s_1, or s_2, which are then used to make a new source mesh either for interpolation or directing the curves on which the line/projection integrals are to be computed. For example

$$I(x) = \int_{C(x)} f(\xi(s), \eta(s))ds \,, \tag{7}$$

where the curve C is parametrized by x. So there is one such line integral for each point x in the destination domain. Generally, a projection coupling variable is one order of dimensionality lower than the source domain and therefore must be defined on a new domain, perhaps created explicity to receive the coupling variable as its destination domain. Inherently, a projection coupling requires two distinct domains (though the destination might be a boundary of the source domain) and thus must be planned from the start as at least a two domain (and potentially two geometry) model.

The coupling variable $I(x)$ contains more information than one line integral. So if you are interested in a particular value of the line integral, then you need only click on the point in the destination domain on the post plot of the coupling variable, and the message window will display the interpolated value at that point. Alternatively, you can export the FEM structure and use postinterp to provide a numerical value.

2.1. *Example: Lidar positioning and sizing of a dispersing pollutant cloud*

Lidar works on the same principle as several other optical devices, for instance spectrometry and spectroscopy, where light received of a given wavelength is of lower intensity due to absorption by a chemical species. For dilute chemical species, the signal received is proportional to the integrated concentration along the optical path, i.e.

$$proj = \int_C c(x(s), y(s))ds, \tag{8}$$

where $\underline{C}$ is the curve $(x(s), y(s))$ and s is the coordinate along the length of the arc. $c(x, y)$ is the concentration field of the chemical species. Suppose the domain is quasi-2D and an array of lidar are arranged along the x-axis which is the lower bound of the domain which is mapped to $[0, 1] \times [0, 1]$. Then the lidar array receives the discrete equivalent of the projection coupling variable $proj_1$:

$$proj_1(x) - \int_0^1 c(x, y)dy. \tag{9}$$

The curves C in (8) are taken here to be vertical lines. This is the standard action for COMSOL Multiphysics projection coupling variables on a 2-D domain. The projection coupling variable is only a function defined on a 1-D independent variable and the default choices of "local mesh transformation" $(x \leftarrow x, y \leftarrow y)$ for the source domain and of "evaluation point" for the destination domain $(x \leftarrow x)$, produces (9) on a unit square. The choices for nonrectangular domains make more sense if one uses local coordinates: (s_1, s_2) in 2D for domains with curving boundaries, s in 1D for nonlinear curves.

Now for the example. The standard model for an instantaneous release of a dense pollutant gas in the atmosphere is a cloud with an average profile of a Gaussian in 2D. Zimmerman and Chatwin [17] analyze wind tunnel data of such dense gas releases, showing the instantaneous structure of fluctuations is highly intermittent. Yet the ensemble average or windowed time averages approach the 2-D Gaussian profile as the cloud becomes dilute.

So let's suppose that we have an initial profile of concentration of

$$c_0(x, y) = \exp\left(-\frac{(x - x_0)^2}{l_x^2} - \frac{(y - y_0)^2}{l_y^2}\right). \tag{10}$$

Suppose that this profile is subjected to a uniform velocity field $\mathbf{u} = (u_0, v_0)$. It follows that the projections for horizontal and vertical arrays, respectively, of lidar would initially measure:

$$proj_1(x) = c_1 \exp\left(-\frac{(x - x_0)^2}{l_x^2}\right),$$

$$proj_2(y) = c_2 \exp\left(-\frac{(y - y_0)^2}{l_y^2}\right). \tag{11}$$

Let M be the total concentration dose in the domain

$$M = \int \int_{\Omega} c(x,y)d\Omega . \tag{12}$$

Then these projection coupling variables can be treated as normalized conditional probability density functions, with moments, in the case of $proj_1$:

$$m_{x,1} = \int_0^1 x \frac{proj_1(x)}{M} dx ,$$
$$m_{x,2} = \int_0^1 x^2 \frac{proj_1(x)}{M} dx . \tag{13}$$

For $c = c_0(x,y)$, the initial Gaussian, one can show that the first moment locates the x-coordinate, $m_{x,1} = x_0$. The second moment permits the computation of the standard deviation according to

$$s_x = \sqrt{m_{x,2} - m_{x,1}^2} = \frac{l_x}{\sqrt{2}} . \tag{14}$$

Both conclusions obviously hold for x-projections onto the y-axis. So projections can locate and size a cloud of Gaussian shape. For non-Gaussian clouds, they provide at least a notion of centrality and degree of spread.

Now we demonstrate this example in action in COMSOL Multiphysics. Launch COMSOL Multiphysics and in the Model Navigator follow the instructions in Table 5. Tables 6 and 7 demonstrate the post-processing stages necessary to compute the first moments and standard deviations. The computation of (13) follows as below in Table 6 for $t = 0.06$.

The latter two integrations give $s_x = 0.2738$ and $s_y = 0.2765$, nearly identical spread, but this is expected given the nearly diffused final state. For time $t = 0$, the same contributions result in Table 7. The latter two integrations give $s_x = 0.0707$ and $s_y = 0.0872$, consistent with the settings of l_x and l_y as expected.

Figure 3 shows the extent to which the initial condition diffuses very rapidly. Although $Pe = 1$ in this simulation, the numerical diffusivity is strong on this mesh resolution. Likely the result is less rapid diffusion on a finer mesh. Figure 4 gives the projection coupling variables demonstrating the near Gaussian profiles captured by our synthetic "lidar." Clearly, even the late stage evolution where periodic boundary conditions obscure the usual "long tails" of the Gaussians, exhibit a central peak and spread captured by the central moments according to (13) and (14).

Table 5.   Projection coupling variables mimicking lidar model.

Model Navigator	Select 2-D space dimension Select Chemical Engineering|Mass balance|Convection and diffusion application mode (chcd) Accept defaults
Options Menu|Constants	Name of constant: x_0 Expression: 0.4 Name of constant: y_0 Expression: 0.6 Name of constant: l_x Expression: 0.1 Name of constant: l_y Expression: 0.12 Name of constant: u_0 Expression: 1 Name of constant: v_0 Expression: 0 Name of constant: Pe Expression: sqrt(u_0^2 + v_0^2)
Options Menu|Axes/Grid settings	Set the grid to $(-0.1, 1.1) \times (-0.1, 1.1)$ and the grid spacing to 0.1, 0.1
Draw Menu|Specify Objects	Select square and accept the default width 1 and corner at the origin
Physics Menu|Subdomain settings	Select domain 1 $D = 1/Pe$; $u = u_0$; $v = v_0$ Init tab: $c(t_0) = \exp(-(x - x_0)^2/l_x^2 - (y - y_0)^2/l_y^2)$
Physics Menu|Periodic Conditions|Periodic boundary conditions	Select boundary 1. Source tab: Enter expression c Click in constraint name field Destination tab: Select boundary 4. Enter expression c Source vertices: 1, 2. Destination vertices 3, 4 Source tab: Select boundary 2. Enter expression c Click in constraint name field Destination tab: Select boundary 3. Enter expression c Source vertices: 1, 3. Destination vertices: 2, 4. OK
Options Menu|Projection Coupling Variables|Subdomain variables	Source tab. Select subdomain 1. Enter name $proj1$ Expression: c. Select general transformation $(x \leftarrow x, y \leftarrow y)$ Destination tab. Select boundary 2 Evaluation point $(x \leftarrow x)$ Source tab. Select subdomain 1. Enter name $proj2$ Expression: c. Select general transformation $(x \leftarrow x, y \leftarrow y)$ Destination tab. Select boundary 1. Evaluation point $(x \leftarrow x)$
Solve Menu|Solver Parameters	Select time dependent solver General tab: Set .output times to 0:0.001:0.06 Solve with the (=) button on the toolbar
Post-processing Subdomain integration	Compute total mass (any time) by integrating c
Post-processing Boundary integration	Compute boundary integrals of $proj2*x/0.037699$ and $proj2*y/0.037699$ (first moments) and $proj2 * x^2/0.037699$ and $proj2 * y^2/0.037699$ (second cumulants) at the first and last times

Table 6. Moment and cumulant calculations at $t = 0.06$.

Post Mode $t = 0.06$
Subdomain integration: domain 1 c (at any time) $I_1 = 0.037702$
Boundary integration: bnd 2 $proj1 * x/0.037702$ $I_2 = 0.49325$
Boundary integration: bnd 1 $proj2 * y/0.037702$ $I_3 = 0.51522$
Boundary integration: bnd 2 $proj1 * x\hat{\ }2/0.037702$ $I_4 = 0.31825$
Boundary integration: bnd 1 $proj2 * y\hat{\ }2/0.037702$ $I_5 = 0.34188$

Table 7. Moment and cumulant calculations at $t = 0$.

Post Mode
Boundary integration: bnd 2 $proj1 * x/0.037702$ $I_2 = 0.39999$
Boundary integration: bnd 1 $proj2 * y/0.037702$ $I_3 = 0.60012$
Boundary integration: bnd 2 $proj1 * x\hat{\ }2/0.037702$ $I_4 = 0.165$
Boundary integration: bnd 1 $proj2 * y\hat{\ }2/0.037702$ $I_5 = 0.36726$

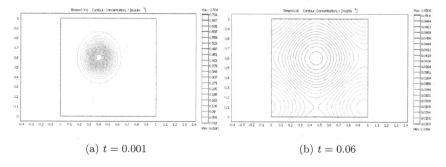

(a) $t = 0.001$ (b) $t = 0.06$

Figure 3. Isopycnals for times $t = 0.001$ (left) and $t = 0.06$ (right) for the time evolution of the concentration field from c_0 (10) according to the convective-diffusion model with $Pe = 1$ and uniform horizontal flow under doubly periodic boundary conditions.

Exercise 7.3: Artificial (numerical) diffusivity

Repeat the lidar example with a refined mesh. Does the Gaussian cloud disperse slower (less spread) with a refined mesh? How could you use this computation to quantify the numerical diffusivity that is artificially created?

3. Extrusion coupling variables

An extrusion coupling variable was named after one of its most common uses; it maps information from a domain of dimension n to one of higher dimension $n + 1$. Yet extrusion is only one of its potential uses, which are generalized as interpolation, projection, or mapping, depending on the

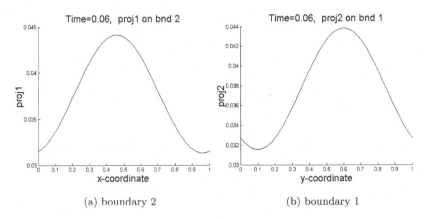

(a) boundary 2   (b) boundary 1

Figure 4.   Line integral projections for $t = 0.06$ onto the horizontal boundary (left) and vertical boundary (right) for the model of Figure 3.

information passed. The other two coupling variable types — scalar and projection — perform integrations over their source domains (or subdomains) and are thus able to be incorporated in integral equations. Extrusion coupling variables map detailed or distributed information from one domain to another, with the destination position selected by the local mesh transformation. So extrusion variables are useful intermediaries in models with multi-domain coupling. Yet they need not be defined on domains of different geometries. In COMSOL Multiphysics seminars, the common example given of extrusion coupling variables is for aesthetic reasons. Frequently, given the symmetry in a physical configuration, the model can be solved over only part of the domain or even a lower dimension, yet the real physical configuration is required to visualize the solution. So, for instance, in a cylindrical duct, axi-symmetry may only require solution in the r–z plane, yet visualization on the cylinder may be desirable. Extrusion over the θ-coordinate of the r–z solution will permit the desired visualization. Suppose placement of baffles with hexagonal symmetry in the domain permitted solution over a wedge of $\theta \in [0, \pi/3]$ with r and z bounded. Yet, if visualization is required over the whole duct, extrusion of the wedge to the other fiver wedges would permit this. So extrusion coupling variables may merely extend information for postprocessing into other domains.

3.1. *Integral equations*

Integral equations are distinguished by containing an unknown function within an integral. As with differential equations, linear systems are the best characterized and therefore most commonly occurring. The classification system is straightforward.

If the integration limits are fixed, the equation is termed of Fredholm type. If one limit is a variable, it is termed of Volterra type.

If the unknown function appears only under the integral sign, it is labeled as of the "first kind." If it appears both inside and outside the integral, it is labeled of the "second kind."

Here are the four combinations symbolically:

Fredholm integral equations of the first kind:

$$f(x) = \int_a^b K(x,t)g(t)dt\,. \tag{15}$$

Fredholm integral equations of the second kind:

$$g(x) = f(x) + \lambda \int_a^b K(x,t)g(t)dt\,. \tag{16}$$

Volterra integral equations of the first kind:

$$f(x) = \int_a^x K(x,t)g(t)dt\,. \tag{17}$$

Volterra integral equations of the second kind:

$$g(x) = f(x) + \lambda \int_a^x K(x,t)g(t)dt\,. \tag{18}$$

In all four cases, g is the unknown function. $K(x,t)$, called the kernel, and $f(x)$ are assumed to be known. When $f(x) = 0$, the equation is said to be homogeneous.

One might reasonably ask why we bother with integral equations. The answer is the theme of this chapter — integral equations are fundamentally nonlocal. Some physical phenomena are inherently nonlocal in character, so their description leads to integral or integro-differential equations. For instance, Shaqfeh [18] derived a theory for transport properties of composite materials that naturally leads to a nonlocal description of effective properties. Many systems are "elliptical" in nature — the boundary data diffuses everywhere, say steady state heat transfer or mass transfer — which results in the solution at a point depending on the solution everywhere. Such nonlocal systems can be conveniently described in terms of a Green's function, which then leads to an integral equation description for inhomogeneous systems. Finally, some processes are conveniently described in a phase space (Fourier space, Laplace space, size, volume or mass distribution) that involve nonlinear coupling of the variables in phase space. When described in physical space, these phase space couplings manifest as convolution integrals which are both nonlocal and nonlinear. Rarely, transform methods, for instance the Abel transform, through a clever change of variables, permits the restatement of an integral equation as an equivalent differential equation, at least for smooth functions. Howison *et al.* [19] give an example that was cited with regard to film drying in Chapter 6. Otherwise, either discretization or power series expansion are the preferred analysis techniques.

It is not the intention of this chapter to teach integral equation theory. An introduction worth reading is given in Arfken's book [20] and a thorough grounding can be found in Stakgold [21] or Lovitt [22]. Here we intend only to explore some aspects of COMSOL Multiphysics's ability to compute solutions to integral and integro-differential equations.

3.2. *Solving a Fredholm integral equation of the second kind*

Zimmerman [23] gives the derivation of a Fredholm integral equation of the second kind as an intermediate in the solution for the drag on a thin disk in broadside motion in a cylindrical duct. The variation on (16) is slight:

$$g(x) = 1 + \varepsilon \int_0^1 K(x,t)g(t)dt\,, \tag{19}$$

where $\varepsilon < 1$ is a small parameter. The kernel K was bounded, so a theorem in integral equation theory [21] ensures that a solution for $g(x)$ can be found by iteration, with each iterate improving in accuracy by at least one order of correction in ε. Zimmerman [23] demonstrated a solution by series expansion in powers of x and ε, albeit relying on numerical computation of the series coefficients. As the kernel of that problem is not particularly tractable (it too was expanded in powers of x and t), a simpler kernel will be selected here for demonstration.

As alluded to in §4, projection variables are the variables of choice for a line integration that returns a function. Although we wish to achieve a line integration of the form

$$f_1(x_1) = \int_0^1 K(x_1, x_2)g_2(x_2)dx_2\,, \tag{20}$$

it is easier to achieve

$$f_1(x_1) = \int_0^1 K(x_1, x_2)g_2(x_1, x_2)dx_2\,, \tag{21}$$

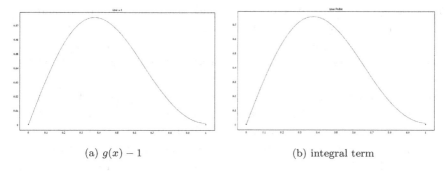

(a) $g(x) - 1$ (b) integral term

Figure 5. Solution $g(x)$ to Fredholm integral equation with $K(x, t) = \sin(2\pi xt)$.

Table 8. Fredholm integral equation solution using the dest() command.

Model Navigator	Select 1-D dimension Select COMSOL Multiphysics|PDE Modes|General Mode OK
Draw Menu	Specify geometry|line. Define interval $[0, 1]$
Options Menu| Constants	Define eps = 0.1
Options Menu| Integration Coupling Variables| Subdomain variables	Name: fredint Expression: sin(2*pi*x*dest(x))*u Keep 4th order integration and global destination defaults OK
Physics Menu: Boundary settings	Set boundary 1, 2 Select Neumann, $G = 0$
Physics Menu: Sub- domain settings	Set $\Gamma = 0$, $da = 0$, $F = u - 1 -$ eps*fredint OK
Mesh	Global tab: set maximum element size: 0.01 Remesh. OK
Solve Menu	Set the stationary nonlinear solver Click on the solve (=) tool on the toolbar

where $g_2(x_1, x_2) = g_1(x)$, with the mapping $(x_2 \leftarrow g)$ and extruded along the x_2 coordinate. The combination of projection and extrusion variables will be used to compute an integral equation term in the population balance equations at the end of the chapter. In this example and the next, integration coupling variables alone are used, replacing the combination of extrusion and projection. This can be achieved due to a new feature of COMSOL Multiphysics called the dest() operator.

Launch COMSOL Multiphysics and in the Model Navigator, follow the instructions in Table 8 to set up the solution to a simple Fredholm integral equation. This calculation uses the dest() command in the integrand of an integration coupling variable to reference the value of the coordinate in the destination mapping, $dest(x)$, whereas the plain use of x refers to the coordinate in the source mapping. When the kernel of the integral equation depends on the destination operator in this way, the recipe for computing integral terms that depend on the destination coordinate (as in integral equations) follows as in Table 8. I am indebted to Dan Smith of MKS Instruments who co-authored the only integral equation solution currently in the Model Library (see under COMSOL Multiphysics — Equation Based Models — An Integro-Differential Equation) for pointing out the utility of the dest() operator in solving integral equations. I should point out, however, if the dest() operator is not used, COMSOL Multiphysics does not make the integration coupling variable a function of the destination coordinate — it's simply a scalar. So there is a health warning in using this otherwise simple technique. If your kernel is independent of the destination coordinate system, my original technique which uses both extrusion and projection coupling variables should be used. An example of the original technique is given in the model for population balance equations at the end of this chapter for the "death" term in the integro-differential equation.

Briefly, the reason why the kernel must be destination dependent is as follows: In the new example, scalar coupling variables are "destination aware" because they use dest operator in them.There are two "scalar" coupling variables, the old **elcplscalar** and the new **elcplgenint** which is "destination aware" — the whole of the integrand is computed at the destination point, otherwise you are not computing the spatial contribution of the integral equation. **elcplgenint** should compute these spatial contributions (destination aware) even if the kernel does not include the dest() operator.**elcplscalar** should only compute a scalar value, not a function of position. However, in scripting (Matlab or Comsol Script) your model, it is possible to select which type of coupling variable you want. Using the GUI, COMSOL Multiphysics selects according to the presence or absence of the dest() operator in the integrand.

3.3. *Solving a Volterra integral equation of the second kind*

In searching for a Fredholm integral equation of the second kind as an example from the literature for the last section, I hit upon Shaqfeh's [18] equation (22) for the edge effect near an impermeable wall for characterization of effective boundary conditions for thermal conduction in a fiber composite medium, where the fibers are better conductors than the fluid matrix:

$$g(z) + 1 + N \int_{-2}^{2} K(z,x)g(z+x)dx = 0. \tag{22}$$

Here, $g(z)$ is the gradient of the ensemble average temperature at a distance z from the edge of the wall in scaled coordinates. Shaqfeh's theory derives the nonlocal contributions for average extra flux due to the presence of randomly positioned fibers. N is the dimensionless parameter expressing number density and slenderness of the fibers. The kernel is given here in MATLAB notation

```
K(z,x)=((x>0)*(z^2*(6-3*x-2*z)+(2*z-x+2)*(x+z-2)^2*(x+z>2))+
(x<0)*((2*z+3*x+2)*(z-2)^2*(z>2)+(x-2*z+6)*(z+x)^2*(z+x>0)))/12;
```

It should be noted that this expression corrects equation (83b) of [18] for a typographical error. The proof of this is that with the correction, Shaqfeh's assertion that the kernel is homogeneous for $z \geq 2$ is borne out. Figure 6 (left frame) shows the invariant kernel profile in this regime. The contours of $K(z,x)$ are shown in Figure 6 (bottom frame), with the regime of parallel lines at the top consistent with this assertion. Essentially, in this regime, the heat flux sees the same environment whether in the direction of the edge or away from it, statistically, and therefore, the driving force for nonlocal heat flux is lost. Furthermore, with the correction, we find that $K(z,x)$ is continuous at the origin, but has a discontinuous slope, typical of Green's functions in 1-D [21]. Finally, we shall see that we reproduce results consistent with Shaqfeh's finite difference solution of (22).

It turns out that (22), however, is not a Fredholm integral equation at all. Why? Because the $g(z+x)$ dependency in the integrand is not the standard form for a Fredholm equation. Change of variable leads to a Volterra integral equation. Let $x_2 = z + x$. Then $dx_2 = dx$. Re-writing (22) yields

$$g(z) + 1 + N \int_{z-2}^{z+2} K(z, x_2 - z)g(x_2)dx_2 = 0. \tag{23}$$

Since how one writes the kernel is not at issue, (23) clearly has the dependent variable in the limits of integration, so can be identified as a Volterra integral equation. Nevertheless, with a second alteration to the kernel, we can re-write (23) in a form that is treatable by our recipe for Fredholm integral equations in COMSOL Multiphysics, using extrusion and projection coupling variables on an abstract intermediate domain with coordinates (z, x_2):

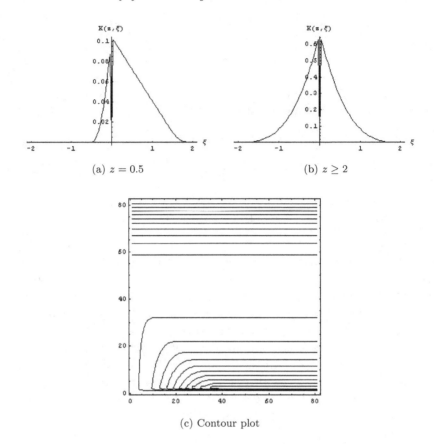

(a) $z = 0.5$ (b) $z \geq 2$

(c) Contour plot

Figure 6.   Left: $z = 0.5$. Right: $z \geq 2$. Lower: Contour plot of the Kernel $K(z, x)$, of the integral equation (22). Kernel $K(z, x)$, of the integral equation (22). Abcissa and ordinate are in the range $(x, z) \in [0, 4] \times [0, 4]$. Homogeneity of K for $z \geq 2$ is apparent.

$$g(z) + 1 + N \int_a^b K'(z, x_2) g(x_2) dx_2 = 0, \tag{24}$$

where $K' = (x_2 > z - 2) * (x_2 < z + 2) * K(z, x_2 - z)$. With this kernel, (24) is of the same form as (19), so the same strategy should suffice to a large extent. The one major modification is that (22) is ill-posed as it stands for any finite interval in z. Simply, for $z \in [0, l]$ (24) shows that g must be defined for $x \in [-2, l + 2]$. Shaqfeh posited that to regularize the problem, the homogeneous behavior of K for $z > 2$ leads to the asymptotic solution that $g \to -1/(1 + 2/3N)$. This can be taken as the solution in the regime $x \in [-2, l + 2]$. For $x \in [-2, 0]$, the position is within the wall, so the

Table 9. Shaqfeh's integral equation model for the nonlocal heat transfer in a fibre composite medium.

Model Navigator	Select 1-D dimension Select COMSOL Multiphysics|PDE Modes|General Mode OK
Draw Menu	Specify geometry|line. Define three domains: bottom:$[-2, 0]$, middle:$[0, 5]$, top:$[5, 7]$
Options Menu| Constants	Define epsilon $= 1$
Options Menu| Integration Coupling Variables| Subdomain variables	Select domains 1, 2, 3. Check group box Name: voltint Expression: $u * (x > \text{dest}(x) - 2)*(x < \text{dest}(x) + 2)*((x > \text{dest}(x))*$ $(\text{dest}(x)^2*(6-3*(x-\text{dest}(x))-2*\text{dest}(x))+(\text{dest}(x)+(x-$ $\text{dest}(x)) > 2)*(2*\text{dest}(x)-(x-\text{dest}(x))+2)*(\text{dest}(x)+(x-$ $\text{dest}(x)) - 2)^2)/12 + (x < \text{dest}(x))*((2*\text{dest}(x) + 3*(x-$ $\text{dest}(x))+2)*(\text{dest}(x)-2)^2*(\text{dest}(x) > 2)+((x-\text{dest}(x))-$ $2*\text{dest}(x) + 6)*(\text{dest}(x) + (x-\text{dest}(x)))^2*(\text{dest}(x) + (x-$ $\text{dest}(x)) > 0))/12);$ Keep 4th order integration Uncheck global destination box Destination tab: set level subdomain, select subdomain 2 only. OK
Physics Menu: Boundary settings	Set boundary 1,2 Select Neumann, $G = 0$
Physics Menu: Sub- domain settings	Select subdomain 1 Set $\Gamma = 0$, $da = 0$, $F = u + 1$ Select subdomain 2 Set $\Gamma = 0$, $da = 0$, $F = u + 1 + $ epsilon*voltint Select subdomain 3 Set $\Gamma = 0$, $da = 0$, $F = u + 1/(1 + 2*$epsilon$/3)$ OK
Mesh	Global tab: set maximum element size: 0.01 Remesh. OK
Solve Menu	Set the stationary nonlinear solver Click on the solve ($=$) tool on the toolbar

homogeneous conductivity there must match the flux at the wall, $g = -1$. So in our abstract 2-D domain of coupling variables, we impose these two limiting behaviors outside the solution domain $z \in [0, l]$.

Launch COMSOL Multiphysics and in the Model Navigator follow the instructions in Table 9.

Figure 7 shows that the physically important region over which the temperature gradient moves from edge value (-1) to asymptotic value $-1/(1 + 2N/3)$, is not more than about a dimensionless length of 2.5. This limiting behavior matches the theoretical predictions of Shaqfeh [18], and

is a consistency check on the kernel. An interesting feature of this profile
is the internal maximum of temperature *gradient*. Shaqfeh graphed the
temperature profile itself, so given the relatively smooth transition, the integration of g might have such a modest internal maximum discernable, but
here, in the gradient, it clearly manifests. I am curious if this feature is an
intrinsic aspect of the edge conduction near an impermeable (to the fibers)
wall bounding a composite. From the description of the calculation in [18],
it is not clear which conditions are applied for the region $x_2 \in [-2, 0]$. The
original integral equation (22) clearly shows that the integral is computed
over that region. The last term of the kernel, with the factor $(x + z > 0)$,
clearly has a contribution from $x \in [-2, 0]$ when $z \in [0, 2]$. So what is the
consistent value of the temperature gradient? I argued that $g(x) = -1$ for
$x \in [-2, 0]$. Perhaps this choice influences the prediction of an internal
maximum in temperature gradient.

Figure 8 gives a parametric study over N (epsilon in our COMSOL
Multiphysics model) for the same eight values given for the profiles of temperature in Figure 6 of [18]. The internal maximum in temperature gradient
seems to be a persistent feature for $N > 1$, but is not apparent for $N < 0.5$.
The computation given here is self-consistent, leading to confidence in using COMSOL Multiphysics to compute the solutions to canonical linear 1-D
integral equations of either the Fredholm or Volterra type, of either kind.
Although not particularly envisaged by the software developers themselves,

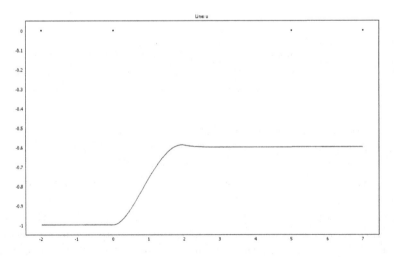

Figure 7.   Temperature gradient, solution $g(z)$ to (22) for $N = 1$.

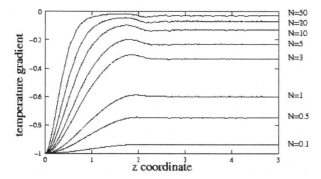

Figure 8. Solution $g(z)$ to (22) of chapter 7 for parameters $N = 0.1$, 0.5, 1, 3, 5, 10, 20, 50. Shaqfeh's Figure 6 solves for the integral $\langle T \rangle = \int_0^z g(z')dz'$ where the reference temperature is at the wall.

this feature has its own niche among software packages for general engineering/mathematical physics productivity. As an experienced, Mathematica, Macsyma, and Matlab user, I can confidently claim that solution to integral equations by other means is a *tour de force* in difference equations, matrix assembly, and sparse matrix solvers for linear integral equations. As we will see in the next subsection, our FEMLAB recipe for integral equations extends to nonlinear integral equations, even of the convolution type, in a straightforward manner.

As a coda to this subsection, one notes that the problem considered here is a variant on the electrical capacitance models of §3.2 and §3.3 in this chapter, particularly as there is a direct analogue to heat conductance in a fluid medium with solid inclusions. The difference is that the ECT models were of nonhomogeneously placed rods and thus the relative positions dominated the flux calculations. Here, the homogeneity of the fibrous inclusions simplifies the conductance model.

3.4. *Convolution integrals and integral equations*

Convolution integrals are typically nonlinear and nonlocal, viz.

$$I(z) = \int_a^b K(z, x)g(z)g(z + x)dx.\qquad(25)$$

They arise naturally in turbulence theory as two point correlation functions — statistics of the turbulence [24]. There is also a well known duality with nearly all linear transforms — convolutions in physical space transform to quadratic products of the individual transforms in transform space, and vice versa — known as the convolution theorem [25]. Since quadratic nonlinearity is fairly common in transport phenomena (inertia and convective terms), convolutions in transform space are just as common. Another important class where convolutions occur is in phase space descriptions of combination processes. In liquid-liquid (droplets) and gas-liquid (bubble) flows, the population changes due to coalescence [26] are expressible as convolutions. Fragmentation mechanisms can be partially treated by collision rules. The kinetics of some mechanisms, like vibration breakup, bag breakup, bag-and-stamen breakup, sheet stripping, wave crest stripping, and catastrophic breakup can only be estimated by rate and probability laws for isolated bubbles/droplets for given local conditions. Nevertheless, collision-based processes are inherently represented in a size phase space as a convolution integral for the population change.

In solid particle dynamics, the population balance equations also are expressible in terms of some convolution integrals. Nicmanis and Hounslow [27] used FEM to describe an integro-differential equation with convolution-type integral terms, where the major processes of this type are aggregation and breakage. The collision rules for bubbles and particles depend substantially on the physicochemical properties of the liquid medium. Traces of flocculent and coagulent effect the probability of bubble-particle agglomeration and floc formation. However, in terms of particle dynamics, the collision rules can be formulated to match observed kinetic rates. Thus, a semi-empirical approach to population balance equations, fitting the coefficients of the aggregation, breakage, and growth models, is a successful technique in characterizing particle processes. Randolph and Larson [28] cite the change in number density $n(\nu)$ of particles with volume ν in the product stream of a continuous mixed-suspension, mixed-product removal crystallizer in which these three processes are occurring from an inlet stream with feed population $n_{\text{in}}(\nu)$:

$$\frac{n(\nu) - n_{\text{in}}(\nu)}{\tau} + \frac{d}{d\nu}(G(\nu)n(\nu)) = b(\nu) - d(\nu)\,, \qquad (26)$$

where τ is the residence time in the crystallizer, $G(\nu)$ is a volume-dependent growth function and the number density of nuclei is incorporated into the equation as a boundary condition $n(0) = n_0$. $b(\nu)$ and $d(\nu)$ are suggestively denoted as the birth and death terms for the volume fraction of size ν. In general, there are contributions to both terms from both aggregation and breakage. Case 1 considered by [27] is a purely aggregation model, so we will cite only the forms derived by Hulburt and Katz [29] for aggregation.

$$b(\nu) = \int_a^{\nu/2} \beta(\nu - w, w)n(\nu - w)n(w)dw \,,$$

$$d(\nu) = n(\nu) \int_0^{\infty} \beta(\nu, w)n(w)dw \,. \tag{27}$$

Succinctly, birth by aggregation is due to the probability of combining particles with volumes which sum to ν. Since these particles are presumed to stick together with some probability, a new particle with a given volume is born and two predecessors "die." The death rate is by the probability of a particle of volume ν participating in a collision (and sticking).

COMSOL Multiphysics model

Case 1 of Nicmanis and Hounslow is for aggregation only, characterized by the assignments $\beta(v, w) = $ beta0, $G = 0$, and no breakage contributions to b and d source/sink terms. The idealized feed is the exponential inlet condition

$$n_{\text{in}}(\nu) = \exp(-\nu) \,. \tag{28}$$

Launch COMSOL Multiphysics and in the Model Navigator and follow the instructions in Table 12. Nicmanis and Hounslow [27] suggest that the domain $v \in [0, 2500]$ has a suitable ceiling for convergence. The major action in this model is the computation of the extrusion and the projection coupling variables in the convolution integral term. By suitable selection of the transformations defining the auxiliary 2-D domain, it is even possible to cope with the difference of the two coordinates ν and w which appears in the convolution integral.

It should be noted that the projection coupling variable ba computes the convolution integral for the birth term in (27), with the awkward offset coordinate $(v_1 - v_2)$ treated neatly by the evaluation point transformation in the extrusion variable N_2. The independent variable in the limits of integration are catered for by the MATLAB binary logic factors $(v_2 > 0) *$ $(v_2 < v_1/2)$, in the same fashion as the treatment of the Volterra integration limits in the last section. da is far more pedestrian, only requiring the projection coupling variable for the line integral to be computed. Although da is the same for all v (a constant), it must be computed by a coupling variable. On reflection, its source could be Geom 1, subdomain 1, with integrand n_1 to save computer labor in this case due to the assignment of $\beta(v, w) = $ beta0. The treatment here is more general to accommodate potentially greater complexity of β.

Exponentially scaled mesh

Nicmanis and Hounslow [27] also employed a nonuniform mesh, with smaller elements for small volumes, and larger elements for larger volumes. COM-SOL Multiphysics will permit this as well. Those authors specified a mesh where the upper bound of element e is given by

$$
x_e = v_b \left(\frac{v_{0\,\mathrm{max}}}{v_b} \right)^{e-1/N-1} , \tag{29}
$$

where N is the number of elements and v_b is the bin volume size for the first element. Femlab 2.x used to permit the specification of graduated meshes, so the exponentially scaled mesh of Nicmanis and Hounslow was straight forward to implement. In COMSOL Multiphysics, the only possibility for specifying an exponentially graduated mesh is using the mapped mesh command, or programming in mesh primitives. The mapped mesh creates a cartesian quad mesh, but this mesh type is not supported for projection coupling variables. So the mesh produced in Table 12 is the best compromise mesh using domain growth rates and size constraints in the GUI mesh tools. Figure 9 shows the mesh generated for the convolution integral computation in the auxiliary 2-D domain.

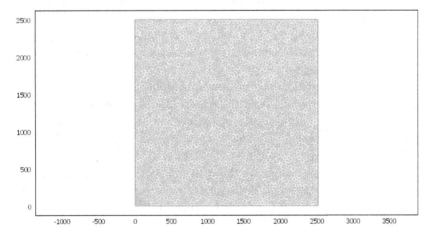

Figure 9.   Mesh in abstract 2-D extrusion domain used for computing convolution integrals.

Activating/deactivating variables with Solve for Variables

Perhaps you noticed that we are solving in the abstract domain for N, which at steady state should be the trivial solution of $F = 0$, i.e. $N = N_1 * N_2$. N is pretty useless to us, but as it is a diagonal system at steady state, it should not be hard to solve, right? And we do have to solve for something in our fictitious domain, don't we? Wrong! Even the trivial diagonal solution for $F = 0$ uses sparse matrix solvers with somewhere around 25000 back substitutions. Eventually, this work will lead to an ill-conditioned numerical solution due to round-off error alone. Furthermore, we do not need to *solve* for anything in our fictitious 2-D domain. We can disable the solution for N. Interestingly, the deactivation of variables in the 2-D using Solve For tab of the Solution Manager leaves only 605 degrees of freedom solved for. However, my most recent computation took 300s on a P4 3 GHz processor. The matrix equation is full.

To test how good the solution is now, we will compare the analytic and FEM computed moments. Moments are defined on the distribution as

$$m_j = \int_0^\infty \nu^j n(\nu) d\nu \,. \tag{30}$$

Moments are computed on our truncated domain by subdomain integrations as shown in Table 10.

Hounslow [30] gives the analytic values for $\tau = 200$ to three significant figures as shown in Table 11.

In the above computations, parameter space continuation was done with old solutions taken as the new guess. This is a more complex version of iteration than used in [27], since the COMSOL Multiphysics standard

Table 10. Moments of the population distribution.

tau	m_0	m_1	m_2
1	0.73082	0.99701	2.9864
2	0.35748	0.99379	11.867
30	0.22653	0.99573	31.884
70	0.15503	0.99768	72.533
140	0.11237	1	143.11
200	0.094934	1.0001	201.72

Table 11. Moments for $\tau = 200$ found by analysis.

200	0.0951	1	202

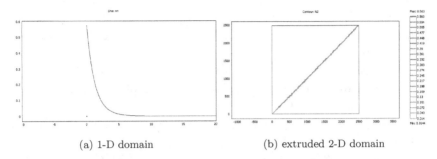

(a) 1-D domain   (b) extruded 2-D domain

Figure 10.   Solution $nn(v)$ to PBE with graduated mesh of Figure 9 for tau $= 10$. Left: 1-D solution. Right: 2-D solution for extrusion variable $N_2 = n_1(v_2, v_1 - v_2)$.

stationary nonlinear solver assembles the Jacobian matrix. Even with only 605 DOF, the solution is rather good for the moments at this level of τ. Greater refinement is necessary for higher τ values.

Exercise 7.4: An integro-differential equation

(16) is an integro-differential equation when $G \neq 0$. Set up a variation of our stationary nonlinear model for PBE with $G = 1$ and boundary condition $n_1 = 1$ at $v = 0$. Solve for the steady solution with residence time $\tau = 200$. Since the PDE is first order, only one boundary condition can be applied sensibly. The recipe with FEMLAB is to impose a "noncondition" at $v = v_{0\,max}$, i.e. the Neumann BC that the derivative of n_1 vanishes at the top volume. Since this is the natural BC in FEM, no Lagrange multiplier equation is augmented to the system. Does this natural boundary condition make physical sense in the case of PBE?

4. Summary

This chapter has explored coupling variables of all three varieties to solve an array of inverse and integral equations. We encountered several features of COMSOL Multiphysics not previously explored — coupling to optimization tools through MATLAB, extended meshes, using the time-dependent solver as an iterative tool for stationary nonlinear models, and the ability to selectively activate/deactivate multiphysics modes in coupled models. The latter is particularly useful if there is only one-way coupling (as in the hydrodynamics around the catalyst supported on the pellet in Chapter 3). In the case of the integral equations treated here, a fictitious dependent variable on an auxiliary domain is set up. The domain is used by coupling variables for various operations, but the dependent variable is never needed itself. So deactivating it results in better conditioning the FEM approximation to the integral equation. Although we implemented this procedure only with the convolution integral in our last model of the PBE, this is a generically useful technique for all the integral equations posed here.

Table 12. PBE Model of Nicmanis and Hounslow. Filename: nicmanis2.mph.

Model Navigator	Select 1-D dimension Select COMSOL Multiphysics|PDE Modes|General Form Set variable name: nn. OK Multiphysics Add Geom 2, 2-D Select COMSOL Multiphysics|PDE Modes|General Mode Set variable name: N, mode name g_2. Add. OK
Draw Menu	Geom 1: Specify geometry|line. Name: sizes Specification: [0, 2500] Geom 2: Specify geometry|square. Width 2500. Corner at $(0, 0)$
Options Menu| Constants	Define tau $= n_0 =$ beta0 $= G_0 = 1$; $v_{0\max} = 2500$
Options Menu| Extrusion Coupling Variables| Subdomain variables	Name: N_1 Source Geom 1, subdomain 1, Expression: nn Check general transformation Destination Geom 2 subdomain 1 Destination transformation: $x \leftarrow y$ Name: N_2 Source Geom 1, subdomain 1, Expression: nn Check general transformation Destination Geom 2 subdomain 1 Destination transformation: $x \leftarrow (x > y) * (x - y)$
Options Menu| Projection Coupling Variables| Subdomain variables	Source Geom 2, subdomain 1; Name: ba Expression: $(y > 0) * (y < x/2) * N_1 * N_2$ integration order 4 Select general transformation. Keep default transformation Destination tab: Geom 1, check subdomain 1 Keep default transformation Source Geom 2, subdomain 1; Name: da Expression: N_1 integration order 4 Select general transformation. Keep default transformation Destination tab: Geom 1, check subdomain 1 Keep default transformation OK
Physics Menu: Boundary settings	Mode g: geom 1 boundary 1, 2 Select Neumann, $G = 0$ Mode g_2: geom 2 boundary 1, 2, 3, 4 Select Neumann, $G = 0$

Table 12 (*Continued*)

Physics Menu: Sub-domain settings	Select Geom 1 mode g Set $\Gamma = 0$, $da = 0$, $F = (nn - \exp(-x))/\text{tau} -ba + nn * da$ Select Geom 2 mode g_2 Set $\Gamma = 0\ 0$, $da = 0$, $F = N - N_1 * N_2$ OK
Mesh Mash Parameters	Geom 1: Boundary tab: set max element size to 0.01 for boundary 1. Subdomain tab. Set maximum element size to 20. Set maximum growth rate to 1.1 Geom 2: Point tab: set max element size to 0.05 for point 1 Subdomain tab. Set maximum element size to 100. Global tab. Set element growth rate to 1.25
Solve Menu	Set the stationary nonlinear solver Advanced tab: set solution form to weak Solver Manager. Solve For tab. Select Geom 1 only Click on the solve (=) tool on the toolbar

My contacts at COMSOL have led me to believe that coupling variables and extended multiphysics were an addition to FEMLAB 2.2 "because they could" without necessarily a vision of how they might prove to be practically useful. With the wide survey of applications shown here, and earlier in Chapter 4, my impression is that coupling variables and extended multiphysics are the features of COMSOL Multiphysics most likely to lead to rapid growth in its usage. Complex system modeling and simulations that are envisaged for the biological systems, micromachines (MEMs), and generic networked systems are readily modeled by these features of COMSOL Multiphysics. Process simulation packages such as HYSYS and Aspen have long had the capability of simulating networks of coupled units comprising ODEs and nonlinear, algebraic constraints. Computational fluid dynamics packages such as FLUENT and FIDAP and finite element solvers like ANSYS contain the elements of PDE solver engines. COMSOL Multiphysics, through extended multiphysics and coupling variables, have made the combination appear seamless to the user.

Whether the major use of extended multiphysics will be network modelling or inverse methods remains to be seen. I use them both, but find that the role of inverse methods in engineering modelling is growing rapidly. In complex systems, the reductionist approach to measuring parameters directly is becoming increasingly more difficult to justify in terms of both cost and significance. The inverse method approach to parametric identification shown in this chapter is cheap and potentially effective if sensitive experiments can be designed.

References

[1] R. A. Gingold and J. J. Monaghan, Smooth particle hydrodynamics: Theory and application to nonspherical stars, *Mon. Not. Roy. Astr. Soc.* **181** (1977) 375–289.

[2] A. V. Potapov, M. L. Hunt and C. S. Campbell, Solid-liquid flows using smoothed particle hydrodynamics and the discrete element method, *Powder Technology* **116** (2001) 204–213.

[3] D. C. Rapaport, *The Art of Molecular Dynamics Simutation* (Cambridge University Press, 1995).

[4] B. J. Adler and W. G. Hoover, *Physics of Simple Liquids*, eds. H. N. V. Temperly, J. S. Rowlinson and G. S. Rushbrooke (Horth-Holland Publishing Co., Amsterdam, 1968).

[5] S. Kim and S. J. Karrila, *Microhydrodynamics. Principles and Selected Applications*, Butterworth-Heinemann Series in Chemical Engineering (Butterworth-Heinemann, Stoneham, MA, 1991).

[6] D. Coca and S. A. Billings, A direct approach to identification of differential models from discrete data, *J. Mechanical Systems and Signal Processing* **13** (1999) 739–755; D. Coca, Y. Zheng, J. E. M. Mayhew and S. A. Billings, Nonlinear system identification and analysis of complex dynamical behaviour in reflected light measurements of vasomotion, *Int. J. Bifurcation and Chaos* **10** (2000) 461–476.

[7] M. M. Mansour and F. Baras, Microscopic simulation of chemical systems, *Physica A* **188** (1992) 253.

[8] I. E. Grossman, Mixed integer nonlinear programming techniques for the synthesis of engineering systems, *Res. Eng. Des.* **1** (1990) 205.

[9] M. Mitchell, *An Introduction to Genetic Algorithms* (MIT Press, Cambridge Massachusetts, 1996).

[10] S. C. Roberts, D. Howard and J. R. Koza, Evolving modules in genetic programming by subtree encapsulation, *Genetic Programming, Proceedings Lecture Notes in Computer Science* **2038** (2001) 160–175.

[11] S. M. Huang, A. B. Plaskowski, C. G. Xie and M. S. Beck, Tomographic imaging of two-component flow using capacitance sensors, *J. Phys. E Sci. Instrum.* **22** (1989) 173–177.

[12] S. L. McKee, T. Dyakowski, R. A. Williams, T. Bell and T. Allen, Solids flow imaging and attrition studies in a pneumatic conveyor, *Powder Technology* **82** (1995) 105–113.

[13] T. Dyakowski, Process tomography for multiphase measurement, *Measurement Science and Technology* **7** (1996) 343–353.

[14] R. P. Feynman, R. B. Leighton and M. Sands, *The Feynman Lectures on Physics*, Vol. II (Addison-Wesley, 1964).

[15] T. Dyakowski, L. F. C. Jeanmeure and A. J. Jaworski, Applications of electrical tomography for gas-solids and liquid-solids flows — A review, *Powder Technology* **112**(3) (2000) 174–192.

[16] N. Kerrouche, C. N. McLeod and W. R. B. Lionheart, Time series of EIT chest images using singular value decomposition and Fourier transform, *Physiol. Meas.* **22**(1) (2001) 147–157; M. Vauhkonen, W. R. B. Lionheart,

L. M. Heikkinen, P. J. Vauhkonen and J. P. Kaipio, A MATLAB package for the EIDORS project to reconstruct two-dimensional EIT images, *Physiol. Meas.* **22**(1) (2001) 107–111, http://www.ma.umist.ac.uk/bl/eidors/.

[17] W. B. Zimmerman and P. C. Chatwin, Fluctuations in dense gas concentrations measured in a wind-tunnel, *Boundary Layer Meteorology* **75** (1995) 321–352.

[18] E. S. G. Shaqfeh, A nonlocal theory for the heat transport in composites containing highly conducting fibrous inclusions, *Phys. Fluids* **31**(9) (1988) 2405–2425.

[19] S. D. Howison, J. A. Moriarty, J. R. Ockendon, E. L. Terrill and S. K. Wilson, A mathematical model for drying paint layers, *J. Engineering Math.* **32** (1997) 377–394.

[20] G. Arfken, *Mathematical Methods for Physicists*, 3rd edn. (Academic Press, New York, 1985).

[21] I. Stakgold, *Green's Functions and Boundary Value Problems* (Wiley, New York).

[22] W. V. Lovitt, *Linear Integral Equations* (McGraw-Hill, New York, 1924), reprinted (Dover, New York, 1950).

[23] W. B. Zimmerman, The drag on sedimenting discs in broadside motion in tubes, *Int. J. Eng. Sci.* **40** (2002) 7–22.

[24] S. B. Pope, *Turbulent Flows* (Cambridge University Press, 2000).

[25] R. N. Bracewell, The fast Hartley transform, *Proc. IEEE* **72**(8) (1984) 1010.

[26] N. L. Kolev, Fragmentation and coalescence dynamics in multiphase flows, *Experimental Thermal and Fluid Science* **6** (1993) 211–251.

[27] M. Nicmanis and M. J. Hounslow, Finite-element methods for steady-state population balance equations, *AIChE J.* **44**(10) (1998) 2258–2272.

[28] A. D. Randolph and M. A. Larson, *Theory of Particulate Processes*, 2nd edn. (Academic Press, New York, 1988).

[29] H. M. Hulburt and S. Katz, Some problems in particle technology. A statistical mechanical formulation, *Chem. Eng. Sci.* **19** (1964) 555.

[30] M. J. Hounslow, A discretized population balance for continuous systems at steady state, *AIChE J.* **36**(1) (1990) 106.

Chapter Eight

MODELING OF MULTI-PHASE FLOW
USING THE LEVEL SET METHOD

K. B. DESHPANDE[1], D. SMITH[2] and W.B.J. ZIMMERMAN[1]

[1]*Department of Chemical and Process Engineering, University of Sheffield,
Newcastle Street, Sheffield S1 3JD United Kingdom*
[2]*MKS Instruments, Wilmington, Mass., USA*
E-mail: w.zimmerman@shef.ac.uk

Multiphysics, the feature of COMSOL Multiphysics that allows coupling of different types of physics, is demonstrated in this chapter for the level set method for modeling multiphase flow, illustrating various scenarios for the coalescence of drops. In the level set method for biphasic fluid systems, one fluid has strictly positive phase function ϕ, the other strictly negative ϕ, so the interface is tracked by the zero level set of ϕ. The transport of ϕ is computed by solving an advection-diffusion equation for ϕ and the incompressible Navier-Stokes equations simultaneously. The level set method is extensively applied here to study the coalescence of drops in biphasic flows for different configurations such as drops under influence of gravity, an acoustically suspended drop, drops approaching one another and the interaction among three drops. The curvature analysis here shows the power of COMSOL Multiphysics's post integration tools for statistical analysis of evolving fields, capturing the occurrence of coalescence by a distinguished feature — cusp formation. Introduced for the first time here is a new concept in the treatment of phase leakage, phase re-injection, which shows robust performance in eliminating this known shortcoming of the level set method. It is a tour-de-force in the COMSOL Multiphysics variational approach to constraint handling and extended multiphysics.

1. Introduction

Multiphase flows are often difficult to model computationally, especially because of the difficulty in tracking the fluid-fluid interface. Furthermore, there is a steep change in physical properties such as density, viscosity etc., which makes the computation yet more stiff. There are various computational methods available to solve incompressible two-phase problems such as the front tracking method [1], the boundary integral method [2], the volume of fluid method [3], the Lattice Boltzmann method [4], diffuse interface modeling [5], and the level set method [6] [11]. We use the level set method in this chapter, illustrating its use to compute the coalescence of two drops.

All the above mentioned methods have their advantages and disadvantages. Most are limited to relatively small average or slip velocity between the two phases. In the front tracking method, marker particles are explicitly introduced to keep track of the front that reduces the resolution needed to maintain the accuracy. However, re-gridding algorithms should be employed with front tracking method to prevent marker particles from coming together, especially at the points of larger curvature.

The volume of fluid method (VOF) is based on discretization of the volume fraction of one of the fluids. The motion of the interface is captured by solving a conservation law for volume fraction and the Navier-Stokes equations simultaneously. Since the interface is represented in terms of volume fraction, mass should be conserved, but the discrepancies can be large.* The VOF method needs to have accurate reconstruction algorithms to solve for the advection of volume fraction. A disadvantage of the VOF method is that it is difficult to compute accurate local curvature from volume fraction. This is due to the sharp transition in volume fraction near the interface.

The Lattice Boltzman method (LBM) is a mesoscopic approach to the numerical simulation of fluid motions based on the assumption that a fluid consists of many particles whose repeated collision, translation, and distribution converge to a state of local equilibrium, yet always remaining in flux. LBM has advantages such as implementation on a complex geometry, very efficient parallel processing, and ease of reproduction of the interface between the phases. However, LBM is not yet a widely used computational method to track the fluid motion in multiphase systems, due to its computational intensity.

The level set approach is another potential numerical method to solve incompressible two-phase flow incorporating surface tension term. In the level set method, the interface is represented as the zero level set of a smooth function. This has the effect of replacing the advection of physical properties with steep gradients at the interface with advection of level set function that is smooth in nature. Although level set method does not have the same conservation properties as of VOF method or front tracking method, the major strength of level set method lies in its ability to compute

---

*See the benchmark paper by flow3d at www.flow3d.com/pdfs/TN63.pdf where 5% error in mass conservation is reported.

curvature of the interface easily. Furthermore, level set method does not require complicated front tracking regridding algorithms or VOF reconstruction algorithms. Level set method is based on continuum approach in order to represent surface tension and local curvature at the interface as a body force. This facilitates the computations in capturing any topological change due to change in surface tension.

The diffuse interface method is a kindred notion to the level set method and VOF in that it computes the transport of another function that varies between the phases — the chemical potential. As is well known (see [7]), the surface tension between two fluids is also the excess partial molar Gibbs free energy per unit surface area, so that the change of chemical potential across an interface between immiscible fluids is treated by the notion of surface tension as infinitely steep. The diffuse interface method permits this condition to be merely relaxed to be steep, and then a field equation for chemical potential is tracked, rather than the imposition of topology and stress balance equations implied by the notion of surface tension. The latter method still requires grid adaption, which in state of the art computational models (see [8] and references therein) employ auxiliary equations for elliptic mesh diffusion, but are fragile in the face of topological change, e.g. coalescence or breakage phenomena [9]. Whether greater accuracy at the same computational intensity is available by the topological method of monitoring the free surface position with mesh re-gridding or by solving an auxiliary transport equation for a field variable (VOF, diffuse interface, or level set methods) is arguable. The latter auxiliary equation methods have ease of coding in their favor, which will be illustrated in this chapter with the level set method.

The level set method is used in this chapter to illustrate the coalescence of two co-axial drops with two different 2-D models. Computations are performed using COMSOL Multiphysics. This FEM approach simplifies the level set method by eliminating all the complexities in grid discretization required for free surface/interface tracking methods. The governing equations for the level set method are described in following section.

2. Governing equations of the level set method

In the level set method, a smooth function called a level set function is used to represent the interface between two phases. The level set function is always positive in the continuous phase and is always negative in the dispersed phase. The free surface is implicitly represented by the set of points in which level set function is always zero. Hence we have,

for the continuous phase: $\phi(x, y, t) > 0$,

for the interface: $\phi(x, y, t) = 0$,

for the dispersed phase: $\phi(x, y, t) < 0$.

From such a representation of the free surface, the unit normal on the interface pointing from dispersed phase to continuous phase and curvature of the interface can be expressed in terms of level set function as,

$$\mathbf{n} = \frac{\nabla\phi}{|\nabla\phi|}, \tag{1}$$

$$\kappa = \nabla \cdot \frac{\nabla\phi}{|\nabla\phi|}. \tag{2}$$

The motion of the interface can be captured by advection of the level set function,

$$\frac{\partial\phi}{\partial t} + \mathbf{u} \cdot \nabla\phi = 0. \tag{3}$$

The governing equation for the fluid velocity and pressure can be written in terms of the Navier-Stokes equations which is the equation of motion for incompressible flow:

$$\rho\frac{\partial\mathbf{u}}{\partial t} - \nabla \cdot \mu(\nabla\mathbf{u} + (\nabla\mathbf{u})^T) + \rho(\mathbf{u} \cdot \nabla)\mathbf{u} + \nabla p = \mathbf{F}; \tag{4}$$

$$\nabla \cdot \mathbf{u} = 0, \tag{5}$$

where F is body force which includes gravitational force and, due to the level set treatment of interfacial stresses, the surface tension term. The two components of the F term can be represented as,

$$F_x = \sigma\kappa\frac{\partial\phi}{\partial x}\delta(\phi); \tag{6}$$

$$F_y = \sigma\kappa\frac{\partial\phi}{\partial y}\delta(\phi) + \rho g. \tag{7}$$

The delta function treats the surface tension term at the interface which is determined by the position of the zero level set — which can be as many fluid-fluid interfaces as necessary to demarcate the dispersed phase. The Heaviside function, incorporated in order to describe the steep change in physical properties, is represented in terms of level set function such as,

$$\text{if } \phi < 0 \qquad H(\phi) = 0,$$

$$\text{if } \phi = 0 \qquad H(\phi) = \frac{1}{2},$$

$$\text{if } \phi > 0 \qquad H(\phi) = 1.$$

The dimensionless density and viscosity (scaled by the continuous phase) are constant in each fluid and are represented in terms of Heaviside function as,

$$\rho = H(\phi) + \frac{\rho_d}{\rho_c}(1 - H(\phi)), \tag{8}$$

$$\mu = H(\phi) + \frac{\mu_d}{\mu_c}(1 - H(\phi)). \tag{9}$$

We solve the above set of equations using COMSOL Multiphysics. Smoothed approximants to the Heaviside function are used to avoid Gibbs phenomena resulting in poor convergence.

3. Curvature analysis: Methodology

In the present simulations of multi-phase modeling, the coalescence phenomenon is demonstrated for the various scenarios where the motion of the interface is significant, particularly at the time of coalescence. The curvature analysis is an attempt to capture the rupture of the interface during the coalescence event.

In the level set method, the curvature of the interface is represented as shown in the equation (2). The mean value of the curvature can be estimated by integrating $|\kappa|$ over the interface as,

$$\kappa_{\text{mean}} = \frac{\int_\Omega |\kappa| d\Omega * (\phi = 0)}{\int_\Omega d\Omega * (\phi = 0)}. \tag{10}$$

Similarly, standard deviation of $|\kappa|$ can also be evaluated by first calculating variance as,

$$var = \frac{\int_\Omega \kappa^2 d\Omega * (\phi = 0)}{\int_\Omega d\Omega * (\phi = 0)} - (\kappa_{\text{mean}})^2. \tag{11}$$

The standard deviation, σ is,

$$\sigma = \sqrt{var}. \tag{12}$$

Thus, the first and the second moments of $|\kappa|$ can be evaluated at different time steps to study the behaviour of $|\kappa|$ at the time of coalescence.

The numerical results are shown for coalescence, followed by the curvature analysis with the associated MATLAB m-file script in the next section.

4. Results and discussion

The numerical simulation presented here demonstrates the power of COM-SOL Multiphysics in the modeling of multi-phase flow. We use multiphysics, the basic versatility of COMSOL Multiphysics that enables us to incorporate as many modes (physics) as we wish to include. The level set method, which requires two application modes: Incompressible Navier-Stokes and ChEM: Convection and Diffusion modes, respectively, has been applied extensively here to capture the coalescence of two drops in a two-phase system. Since the interface can be tracked by setting the zero level set at the interface, this permits the study of the evolution of the interface after the merging of two drops. In coalescence phenomena, the contact point between interfaces of two drops is very important and hence the approach of two drops is treated by various means in this computational study and is described in detail in following sections.

4.1. *Coalescence of two axisymmetric co-axial drops*

The simplest way to start the numerical simulation is to assume a symmetrical 2-D domain, which would significantly reduce the computational time. The system can be physically described as a rectangular domain with one boundary acting as an axis of symmetry and all other boundaries are insulated. Two equally sized drops are initially separated by axial distance equal to two times their diameter. It is a tricky task to initiate two drops of the same physical properties. This is accomplished using following initial condition which is used in the sub-domain settings.

$$\phi(t = 0) = \min(\sqrt{(x^2 + (y - 2)^2)} - 0.3, \min((6 - y), \sqrt{x^2 + (y - 1)^2} - 0.3)).$$
(13)

The above system of equations (1)–(9) with initial condition (10) in an initially quiescent fluid (no motion) can be solved using level set method as described below.

Start up COMSOL Multiphysics and enter the Model Navigator. Table 1 gives the instructions to follow to set up the droplet coalescence model using the level set method.

These application modes give us four dependent variables u, v, and p for Incompressible Navier-Stoke mode and one dependent variable called "phi"

Table 1. Co-axial droplet coalescence model using the level set method.

Model Navigator	Select 2-D space dimension Select Chemical Engineering Module|Momentum balance|Incompressible Navier-Stokes|Transient Analysis Set dependant variables: $u\ v\ p$. OK Select Multiphysics||Model Navigator Select Chemical Engineering Module|Mass balance| Convection-Diffusion|Transient Analysis Set dependant variables: phi. Add OK
Draw Menu	Specify objects|Rectangle/Square Draw a rectangle by clicking left mouse button Edit the dimensions of a rectangle by double clicking. Set Size|Width: 1 Height: 5 Set position|Base: Corner x: 0 y: 0 OK
Option Menu Zoom	Select Zoom Extents

Options Menu:
Constants

rhod	rhoc	nu	gy	dadd	n_1	n_2	sigma
1	10	1	-10	0.005	100	20	1

Options Menu:
Expressions:
Scalar Expressions

smhs	$(1 + \tanh(n_1 {*}phi))/2$
smdelta	$n_2/\text{sqrt(pi)}{*}\exp(-n_2{\char`\^}2{*}phi{\char`\^}2)$
kappa	$(phixx{*}phiy{\char`\^}2 \ - \ 2{*}phix{*}phiy{*}phixy \ +$ $phiyy{*}phix{\char`\^}2)/(phix{\char`\^}2 + \ phiy{\char`\^}2 + \text{eps}){\char`\^}(3/2)$
ro	rhod + rhoc${*}$smhs

Physics Menu: Boundary settings	Select Multiphysics|Incompressible Navier-Stokes Select boundary 1. Select Slip/Symmetry boundary condition Select boundary 2–4. Select No Slip boundary condition Apply Select Multiphysics|Convection-Diffusion Select boundary 1–4. Select Insulation/symmetry boundary condition. OK
Physics Menu: Subdomain settings	Select Multiphysics|Incompressible Navier-Stokes Select subdomain 1 Set $\rho = $ ro, $\eta = nu$ Set $F_x = $ sigma${*}$kappa${*}$smdelta${*}$phix Set $F_y = $ sigma${*}$kappa${*}$smdelta${*}$phiy + ro${*}$gy Select Init tab Set $u = 0$, $v = 0$ and $p = 0$ OK Select Multiphysics|Convection-Diffusion Select "phi" tab Set $D_i = $ dadd Set $R_i = 0$, $u = u$ and $v = v$ Init tab phi$(t_0) = \min(\text{sqrt}(x{\char`\^}2 + (y - 2){\char`\^}2) - 0.3, \min((5 - y), \text{sqrt}(x{\char`\^}2 + (y - 1){\char`\^}2) - 0.3))$ OK

Table 1 (*Continued*).

Mesh Menu	Select mesh parameters Global tab Set Predefined mesh sizes: Extra fine Remesh. OK (gives 2088 elements)
Solve Menu Solver Parameters	Select Time dependent solver General tab: Set Times: 0:0.025:3 OK

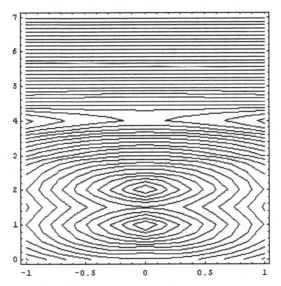

Figure 1.   Contour plot (level sets) of the initial condition for ϕ described in terms of the MATLAB function min in (10).

for ChEM: convection and diffusion mode. The 2-D space coordinates are x and y.

smhs (a hyperbolic tangent) is a common smooth approximant to the Heaviside function. **smdelta**, similarly, is a smooth approximant to the Dirac delta function. The prefactor on the Gaussian is for normalization — the quadrature over the real line must be unity. It is potentially the case that weak terms could be used to define point forces along the zero level set of phi, but the smooth approximants are easier to code. **kappa** is the major component of the curvature defined in (2). **ro** is the expression of the density as in (8).

Figure 2 captures the rise of two drops in a column that ultimately results in coalescence. The interface of the two drops is represented by the contour plot at $\phi = 0$. The velocity field is also represented in the above figure by activating the surface field for v velocity and the arrows field. Two drops are initially separated by a distance equal to two times diameter of drops and their motion under gravity is captured at different time steps. The lower drop travels faster than the upper one, although two drops are of same density and of uniform size. This can be explained by wake formation for the upper drop. The lower drop becomes entrapped into the wake region of the upper one and experiences the velocity field of the upper drop, thereby lowering the effective velocity of the upper drop. Thus, two drops suspended freely rise in a column, eventually coalesce and the subsequent coalesced drop rises again. In this way, the motion of the interface of two drops can be monitored readily using level set method in COMSOL Multiphysics. Various other configurations of the approach of the drops are discussed in the following sections.

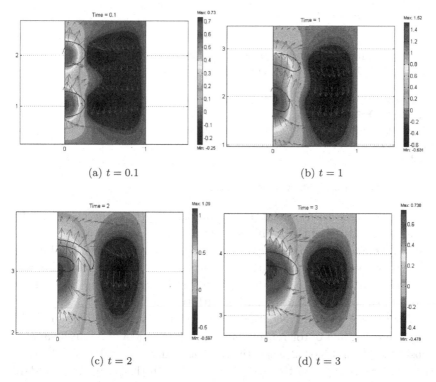

Figure 2.   Contour plot of ϕ with velocity field at different time steps.

4.2. *Curvature analysis: An application*

The results of a COMSOL Multiphysics simulation that is run in the GUI can be exported to MATLAB workspace by using "Export FEM structure as fem" from File Menu. The model m-file script shown below was saved as "analysis.m," edited, and executed from the MATLAB command line to generate the plot of standard deviation and mean of $|\kappa|$ as shown in Figure 3.

```
%%%%%%%%%%%%%%%%%%%%%%%lsm_post.m%%%%%%%%%%%%%%%%%%%%%%%%%%%%%%
t=[1:121];

m1int=postint(fem,'abs(kappa)*(phi<0.00015 & phi>-0.00015)',...
'edim',2,'solnum', t, 'dl',1);

v=postint(fem,'1*(phi<0.00015 & phi >-0.00015)',...
'edim',2,'solnum', t, 'dl',1);

m1=m1int./v;
plot(0.025*t,m1);
hold on

m2int=postint(fem,'(abs(kappa)*(phi<0.00015 & phi >-0.00015))^2',...
'edim',2,'solnum', t, 'dl',1);

m2=m2int./v;
var=m2-(m1).^2;
sd=sqrt(var);
plot(0.025*t,sd);
hold off

save collisionlong.dat t m1 sd -ascii;
```

The variables m1int, m2int and v approximate the numerators and denominator of (10) and (11). The range of tolerances surrounding the $\phi = w$ contour were selected to weight the subdomain integration by contributions in a narrow band surrounding the interface. It is to be noted that the variable "t" used in above calculation is the solution number and varies from 0 to 121, since we ran the simulation with time range 0:0.025:3.

The plots of standard deviation and mean of $|\kappa|$ are shown in Figure 3 for two different values of tolerance (commonly used in MATLAB as a variable "eps") 0.0001 and 0.0002 respectively. The smaller the value of tolerance, the greater the depth of the spike observed. Both the mean and standard deviation of the curvature as estimated are sensitive to topological changes

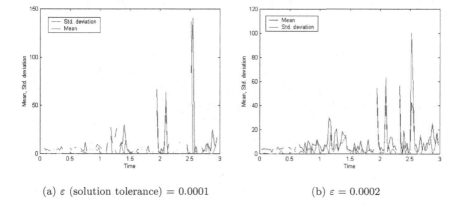

(a) ε (solution tolerance) $= 0.0001$ (b) $\varepsilon = 0.0002$

Figure 3.   Plot of standard deviation and mean value of $|\kappa|$ with respect to time for the coalescence of two drops under gravity

in the connectivity of the domains. The greatest spike was observed at the time of coalescence that can be attributed to the rupture of the interface or cusp formation.

4.3. *Coalescence of acoustically suspended drops*

The technique of acoustic levitation, using tuned sonic fields to oppose the drag force on a droplet and levitate it, has been known for many years. The group of Sadhal at USC have studied the phenomenon and its implications for droplet dynamics for years (see [10] and references therein). The coalescence of acoustically suspended drops where the lower drop is rising and upper drop is held stationary is simulated. Unlike the previous simulation where only half of the domain was considered by assuming symmetry of the domain, the present simulation is performed over the entire domain. Hence, the no-slip boundary condition is applied to all the boundaries in all the modes of Multiphysics. Change the initial condition to generate two drops as follows,

$$\phi(t = 0) = \min(\sqrt{(x^2 + (y - 2)^2)} - 0.3, \min((3 - y), \sqrt{x^2 + (y - 1)^2} - 0.3)).$$
(14)

The only other change would be in the body force (gravity term) in the Navier-Stokes equations which is modified in such a way that upper drop does not experience any gravitational force. This is the bare effect of the acoustic levitation, without consideration of capillary-gravity waves induced on the free surface by acoustic interactions. But, lower drop is rising in a column due to buoyancy. The above mentioned changes can be incorporated by changing F_y term in sub-domain settings. Table 2 details how to set up the acoustically suspended coalescence of drops model.

The new constant used yc is set to 2, i.e. y co-ordinate of the center of upper bubble. The force term used in this way applies no gravity to upper drop whereas lower drop experiences gravitational force equal to ρg.

Numerical results are shown in Figure 4 in terms of contour plot of level set function at $\phi = 0$ and surface plot of velocity field. The two drops initially separated by a distance equal to two times their diameter approach quite faster than the previous simulation where both the drops were rising. Eventually, two drops coalesce quickly and evolution of the interface of two drops after the coalescence event has been brought out through this simulation. Cusp formation is observed at time $t = 2$ s. The coalesced drop regains its original shape as it rises in a column. The different

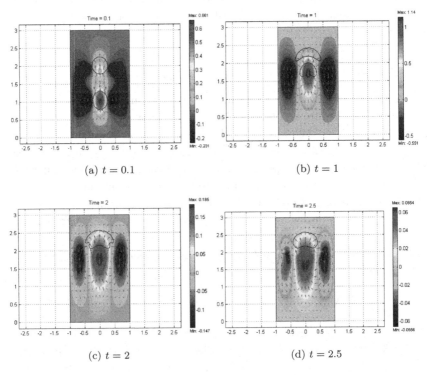

(a) $t = 0.1$ (b) $t = 1$

(c) $t = 2$ (d) $t = 2.5$

Figure 4. Contour plot of ϕ with velocity field at different time steps.

Table 2. Coalescence of acoustically suspended drops model. Filename: acoustic_noreinjection.mph.

Model Navigator	Select 2-D space dimension Select Chemical Engineering Module|Momentum balance|Incompressible Navier-Stokes|Transient Analysis Set dependant variables: $u\ v\ p$. OK Select Multiphysics||Model Navigator Select Chemical Engineering Module|Mass balance|Convection-Diffusion|Transient Analysis Set dependant variables: phi. Add OK
Draw Menu	Specify objects|Rectangle/Square Draw a rectangle by clicking left mouse button Edit the dimensions of a rectangle by double clicking. Set Size|Width: 2 Height: 3 Set position|Base: Corner x: -1 y: 0 OK
Option Menu Zoom	Select Zoom Extents
Options Menu: Constants	<table><tr><td>rhod</td><td>rhoc</td><td>nu</td><td>gy</td><td>dadd</td><td>n_1</td><td>n_2</td><td>yc</td><td>sigma</td></tr><tr><td>1</td><td>10</td><td>1</td><td>-10</td><td>0.005</td><td>100</td><td>20</td><td>2</td><td>1</td></tr></table>
Options Menu: Expressions: Scalar Expressions	<table><tr><td>smhs</td><td>$(1 + \tanh(n_1\text{*phi}))/2$</td></tr><tr><td>smdelta</td><td>$n_2/\text{sqrt}(pi)\text{*exp}(-n_2\text{^}2\text{*phi^2})$</td></tr><tr><td>kappa</td><td>(phixx*phiy^2 $-$ 2*phix*phiy*phixy $+$ phiyy*phix^2)/(phix^2 + phiy^2 + eps)^(3/2)</td></tr><tr><td>ro</td><td>rhod + rhoc*smhs</td></tr></table>
Physics Menu: Boundary settings	Select Multiphysics|Incompressible Navier-Stokes Select boundary 1–4. Select No Slip boundary condition Apply Select Multiphysics|Convection-Diffusion Select boundary 1–4. Select Insulation/symmetry boundary condition. OK
Physics Menu: Subdomain settings	Select Multiphysics|Incompressible Navier-Stokes Select subdomain 1 Set $\rho = $ ro, $\eta = nu$ Set $F_x = $ sigma*kappa*smdelta*phix Set $F_y = $ sigma*kappa*smdelta*phiy $+$ ro*gy*(tanh($-(y - yc)) > 0$) Select Init tab Set $u = 0$, $v = 0$ and $p = 0$ OK Select Multiphysics|Convection-Diffusion Select "phi" tab Set $D_i = $ dadd Set $R_i = 0$, $u = u$ and $v = v$ Init tab phi$(t_0) = $ min(sqrt(x^2 + $(y - 2)$^2) $-$ 0.3, min($(3 - y)$,sqrt(x^2 + $(y - 1)$^2) $-$ 0.3)) OK

Table 2 (*Continued*).

Mesh Menu	Select mesh parameters
	Global tab
	Set Predefined mesh sizes: Extra fine
	Remesh. OK (gives 7146 elements)
Solve Menu	Select Time dependent solver
Solver Parameters	General tab: Set Times: 0:0.025:3
	OK

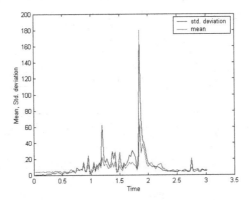

Figure 5. Plot of standard deviation and mean value of $|\kappa|$ with respect to time for the coalescence of an acoustically suspended drop.

shapes of two drops before collision can be attributed to the fact that pressure is continuously decreasing along the length of the column and hence radii of curvature would increase according to the Young-Laplace equation. This can be validated by changing the configuration so that pressure change is uniform as described in the following section.

4.4. *Curvature analysis*

The procedure outlined for the curvature analysis of the coalescence of two drops under gravity is followed for the coalescence of acoustically suspended drops. The fem structure is exported to the MATLAB workspace after the simulation is over and MATLAB model m-file analysis.m is run to study the standard deviation and mean of $|\kappa|$ as shown in the Figure 5.

Both the first and second moments of $|\kappa|$ show a sharp peak at the time of the coalescence, attributed to the rupture of the interface.

Table 3.   Coalescence of two approaching drops.

Physics Menu:	Select Multiphysics|Incompressible Navier-Stokes
Subdomain	Select subdomain 1
settings	Set F_y = sigma*kappa*smdelta*phiy + ro*gy*tanh($-(y-yc)$)
	OK

4.5. *Coalescence between two drops approaching each other*

The coalescence between two drops approaching each other in opposing directions can be achieved by defining a driving force that attracts two drops to each other. This can be simulated by incorporating a driving force term in sub-domain settings for Incompressible Navier-Stoke application mode.

Table 3 gives the GUI step to alter the acoustic droplet model for a hypothetical case of droplets approaching each other with an attractive force. The new constant used yc is a midpoint of the line of centres of two drops and set to 1.5. The force term defined in this manner applies force equal to ρg when $(y-yc) < 0$ and $-\rho g$ when $(y-yc) > 0$. Thus, upper and lower drop experiences exactly equal force but in the opposite direction.

Computational results are represented (Figure 6) in terms of a contour plot of the level set function where $\phi = 0$, a surface plot for pressure field and arrows for velocity field. Two drops separated by a distance equal to two times their diameter attract to each other, ultimately resulting in coalescence at time $t = 2$ s. Cusp formation has been clearly brought out at that time step. The coalesced drop regains its original shape at later time steps. The important feature of this simulation is that symmetry is observed at the midplane between the two drops.

The velocity field is also found to be symmetrical for both the drops which retains after the coalescence event as well. Another important feature is that both the drops are identical in their shape and size. This can be explained on the basis of a surface plot of pressure that is found to be symmetrical around the midplane between the two drops. Since the two drops experience same pressure force, they follow the same change in radii of curvature. Also, less droplet deformation is observed for the present simulation as compared to the earlier two cases. This can also be attributed to lower magnitude of the pressure force.

Curvature analysis

The curvature analysis performed for two drops approaching one another is shown in Figure 7. The peak in standard deviation and mean value of $|\kappa|$

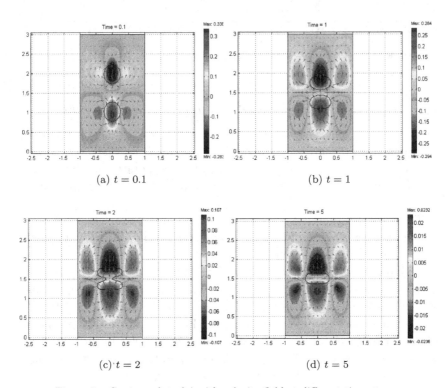

Figure 6.   Contour plot of ϕ with velocity field at different time steps.

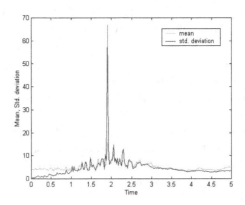

Figure 7.   Plot of standard deviation and mean value of $|\kappa|$ with respect to time for the coalescence of an acoustically suspended drop.

is confirmed at the time of the coalescence. No other spikes were observed for the present simulation because the deformation of the drops was found to be smaller than that in the earlier two cases considered. Thus, it can be concluded that peaks observed in the mean and standard deviation value of $|\kappa|$ are indeed due to the rupture of the interface.

4.6. *Multi-body coalescence*

The coalescence between three drops is simulated assuming the symmetry of the domain. Hence, only half of the domain is simulated in the present case defining one boundary as an axis of symmetry.

The above system brings out the effect of horizontal offset amongst interacting drops. In the present case, centers of drops are more than one radius apart from each other. Two drops are initiated using following initial condition,

$$\phi(t = 0) = \min(\sqrt{(x^2 + (y - 3)^2)} - 0.5, \min((4 - y),$$
$$\sqrt{(x - 0.75)^2 + (y - 1)^2} - 0.5)). \tag{15}$$

Above initial condition generates two uniform sized drops of radius 0.5 whose centers are separated by a distance equal to 0.75. The present simulation is similar to earlier one where two drops are traveling towards each other except the horizontal offset. Hence, there is no other change in the formulation of the problem than the initial condition. The contour plot of level set function at $\phi = 0$, surface plot and arrows of velocity field are represented in Figure 8.

The interaction between drops with horizontal offset is found to be very different from that without offset. The velocity field of the lower drop is found to be diverted due to the influence of the velocity field of upper drop that is traveling downwards. Hence, the shape of rising drop changes drastically. The lower drop almost skids downward traveling drop and changes the contact point between the interacting drops. Finally, coalescence takes place at 3.8 s.

Thus, it is found that the contact point of the interacting drops is very significant in the coalescence phenomenon. Different types of contacting schemes are observed for different approaches considered in the present set of simulations merely by changing the initial condition.

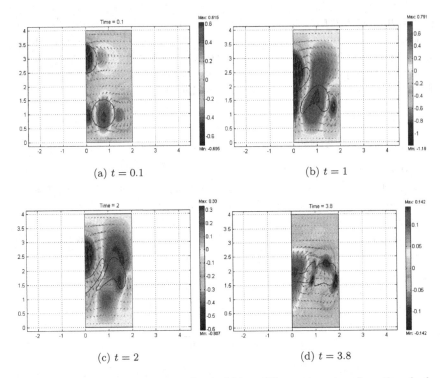

(a) $t = 0.1$ (b) $t = 1$

(c) $t = 2$ (d) $t = 3.8$

Figure 8.   Contour plot of ϕ with velocity field at different time steps for a three body collision.

5. Phase re-injection for mass conservation

A persistent problem with the level set method is that it "leaks" ϕ, i.e. the mass of one phase leaks into the other. The problem gets worse with high contrast in density and viscosity of the fluid. In this chapter, since we treat two liquid phases, phase leakage is a minor problem. Phase leakage is not an intrinsic feature of the partial differential equation system, but a numerical artifact introduced by the smoothing operators at finite grid resolution. If infinitesimal resolution is possible and true distribution functions (Dirac delta and Heaviside step) are analyzed, the level set method conserves the mass of each phase. So one straight-forward, but expensive, treatment is to use highly refined meshes and steep smoothing of the Dirac delta and Heaviside functions.

Two of us have struggled to code the standard remedy for phase leakage called re-initialization [12] [13]. In re-initialization, an auxiliary partial differential equation to condition ϕ to be close to a distance function from the interface is iterated. This has the effect of slowing the rate of phase leakage. You might wonder why, if the level set equations are a well-posed PDE system, it is possible to solve an auxiliary equation for ϕ intermediate to each time-step. The answer is that ϕ has many degrees of freedom that are not really important. We do not care how large in magnitude ϕ is away from the zero contour. The identification of the phase and its properties do not matter. Thus reinitialization poses additional constraints on the magnitude of ϕ in order to approach phase conservation. In practice, re-initialization nearly achieves its goal, but it is cumbersome to code and computationally and storage expensive.

In this section, we introduce a new concept that mimics the goals of re-initialization, but is thematically in the style of COMSOL Multiphysics, since it makes use of the ability of extended multiphysics and the weak formulation to introduce arbitrary auxiliary constraints. We have proposed to alter the level set model equations with a new source term for the transport equation for ϕ. This source term has the interpretation of a source or sink of phase function ϕ which injects precisely (up to numerical precision) the amount of phase function that leaked out of or into the other phase. We term this process "phase re-injection." The new governing equations are shown here:

$$\rho \frac{D\mathbf{u}}{Dt} = -\nabla p + \mu \nabla^2 \mathbf{u} + \mathbf{F},$$

$$\frac{D\phi}{Dt} = \nu_\phi \nabla^2 \phi + \lambda \delta(\phi). \tag{16}$$

The LHS are substantive or material derivatives (subsuming the convection and accumulation terms). ν_ϕ is an artificial diffusivity. $\rho(\phi)$ and $\mu(\phi)$ are the conventional (smoothed) Heaviside constructions that rapidly change value as the phase switches. The new source(sink) term is multiplied by an as yet undefined strength term, λ. Two integral quantities are readily computed using integration coupling variables, the total phase and the initial total phase of the bulk fluid:

$$I_\phi = \int_\Omega H(\phi) d\Omega,$$

$$I_{\phi,0} = \int_\Omega H(\phi_0) d\Omega. \tag{17}$$

The unique feature of our phase re-injection methodology is to require that the total phase is conserved, and to implement this as an auxiliary constraint in COMSOL Multiphysics:

$$I_\phi - I_{\phi,0} = 0 \,. \tag{18}$$

Recall in Chapter two, equations (41) and (43), we demonstrated that COMSOL Multiphysics satisfies auxiliary constraints by the introduction of a Lagrange multiplier in the linearization of the matrix equations for each auxiliary constraint. When these equations, shown below as block matrix equations, converge, the linearization of the constraint about the converged solution is satisfied in the least energy sense in the calculus of variations:

$$\begin{bmatrix} \mathbf{K}(U_0) & \mathbf{N}(U_0)^T \\ \mathbf{N}(U_0) & 0 \end{bmatrix} \begin{bmatrix} U - U_0 \\ \Lambda \end{bmatrix} = \begin{bmatrix} \mathbf{L}(U_0) \\ \mathbf{M}(U_0) \end{bmatrix} \,. \tag{19}$$

The upshot of this treatment of auxiliary conditions is that by requiring the auxiliary condition (18), we have acquired a Lagrange multiplier for this constraint, λ (one of the elements of Λ). It is this Lagrange multiplier which is taken as the source strength in (16). The interpretation of λ is straight-forward. It is the "shadow price" that must be paid for the degree to which the constraint is not currently satisfied in the error function that is minimized in the Newton iteration of the matrix equation system (19). It follows that when the solution (U, Λ) is found, λ should be nearly zero if it is possible to satisfy the phase conservation constraint (18). Effectively, the additional contribution for the source or sink of phase that is reinjected at the interface is numerically small if the phase is conserved.

This calculus of variations approach which is thematic to COMSOL Multiphysics constraint handling is a powerful tool to relieve phase leakage. It also parallels the concept of reinitialization — it adjusts the phi function within the timestep to ensure that phase is conserved. By solving for this additional constraint simultaneously with the residual equations for all the other degrees of freedom, there is less possibility of post-timestep iterations introducing spurious numerical oscillations. If the timestep is too large in reinitialization, the reinitialization step requires more iterations to approach the phase conservation constraint. In practice, reinitialization is implemented typically with a fixed number of iterations. As we saw in the viscous fingering transient eigenanalysis, the dynamics of nonlinear systems results in fluctuations of the important time scales. The simultaneous Newton iteration approach here is inherently more stable than staged iterations for transporting the flow and phase followed by re-initializing the phase function.

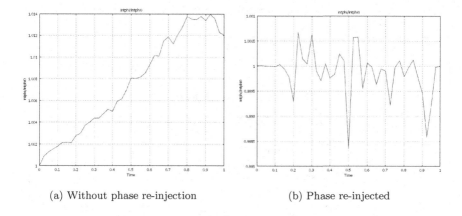

(a) Without phase re-injection   (b) Phase re-injected

Figure 9. Plot of $I_\phi/I_{\phi,0}$. Assessment of the level of phase leakage with and without phase reinjection for a unit time of the acoustic coalescence model with a fine mesh. This is deliberately coarse by comparison to the results presented in Figure 4.

But don't just take our word for it. Check out Figure 9 where it is clear that the phase leakage rate is about one percent per unit time without reinjection, and is effectively zero rate (uniformly distributed fluctuations) with a maximum magnitude of one-tenth of a percent. So at any given time, the expectation of phase leakage is negligible, and the confidence in this is very high. The COMSOL Multiphysics instructions for implementing the changes for phase re-injection are given in Table 4. The simulation without phase re-injection with the fine mesh takes 24 s on a Pentium IV PC with a 3 GHz processor. With phase re-injection, it takes 29 s. There are 980 elements in both cases, but 6598 degrees of freedom (without) and 6599 DOF (with). The speed inefficiency is, of course, due to the matrix line for phase re-injection corresponding to (18) is not sparse.

Our observation is that phase re-injection does not lead to any visual difference in the dynamics between the acoustic levitation/coalescence simulations. This section does not, however, thoroughly "road test" the new methodology. We are pursuing a careful computational study comparing the level set method including phase re-injection with conventional interfacial dynamics models to assess the degree of discrepancy. Preliminary work suggests that numerical analysis of problems with high density and/or viscosity contrast will require additional smoothing procedures for stable solution.

Table 4.   Acoustic levitation and coalescence model with phase re-injection. Filename: acoustic_reinjection.mph.

Model Navigator	Open acoustic_noreinjection.mph Select multiphysics tab COMSOL Multiphysics|PDE modes|weak form point Name the application: phase Dependent variable name: lam Add
Draw Menu	Specify objects|Line. Set x: 0 5 Set name: reactor. OK
Options Menu: Expressions	Name: phi_init Expression: $\min(\mathrm{sqrt}(x\char94 2 + (y - 2)\char94 2) - 0.3,\ \min((3 - y),\mathrm{sqrt}(x\char94 2 + (y - 1)\char94 2) - 0.3))$ Name: smhs0 Expression : $(1 + \tanh(n*\mathrm{phi\_init}))/2$
Options Menu: Integration coupling variables|Subdomain variables	Keep global destinations Name Expression intphi smhs intphi0 smhs0 OK
Physics Menu: Point settings	Mode wp Deactivate in this domain 2, 3, 4 Select point 1 Enter into the weak field: lam_test*(intphi-intphi0) OK
Options Menu: Integration coupling variables|Point vari- ables	Select point 1 Name: lamb Expression: lam Keep global destination OK
Physics Menu: Subdomain settings	Mode chcd Set $R = \mathrm{lamb}*\mathrm{smdelta}$
Mesh Menu	Global tab Select predefined mesh: fine
Solve Menu Solver Parameters	Select time dependent solver Set output times 0:0.025:1 Solve

6. Summary

The level set method is extensively used in the present simulations to study computationally the coalescence of droplets in a two-phase system. COM-SOL Multiphysics handles computationally intensive multi-phase modeling with ease using its multiphysics utility that allows the coupling of different physics into one problem as demonstrated here for the level-set method. In the present simulations, the coalescence phenomenon has been extensively studied using various configurations for the approach of drops. The curvature analysis performed captures the rupture of the interface at the time of the coalescence. Introduced for the first time here is a new concept in the treatment of phase leakage, phase re-injection, which shows robust performance in eliminating this known shortcoming of the level set method. It is a tour-de-force in the COMSOL Multiphysics variational approach to constraint handling and extended multiphysics.

Acknowledgment

We acknowledge Per-Olof Persson for use of his FEMLAB code which was modified appropriately to treat coalescence.

References

[1] S. O. Unverdi and G. Tryggvason, A front-tracking method for viscous, incompressible, multi-fluid flows, *J. Comput. Phys.* **100** (1992) 25.

[2] J. M. Boulton-Stone and Blake, Gas bubbles bursting at a free surface, *J. Fluid Mech.* **254** (1993) 437.

[3] W. J. Rider and D. B. Kothe, Reconstructing volume tracking, *J. Comput. Phys.* **141** (1998) 112.

[4] N. Takada, M. Misawa, A. Tomiyama and S. Fujiwara, Numerical simulation of two- and three-dimensional two-phase fluid motion by lattice Boltzmann method, *Comp. Phys. Comm.* **129** (2000) 233.

[5] M. Verschueren, F. N. van de Vosse and H. E. H. Meijer, Diffuse interface modeling of thermocapillary flow instabilities in a Hele-Shaw cell, *J. Fluid Mech.* **434** (2001) 153–166.

[6] M. Sussman, A. S. Almgren, J. B. Bell, P. Colella, L. H. Howell and M. L. Welcome, An adaptive level set approach for incompressible two-phase flows, *J. Comput. Phys.* **148** (1999) 81.

[7] W. Shyy, *Computational Modeling for Fluid Flow and Interfacial Transport* (Elsevier, 1994).

[8] W. B. Zimmerman, Excitation of surface waves due to thermocapillary effects on a stably stratified fluid layer, submitted to *J. Fluid Mechanics*.

[9] N. L. Kolev, Fragmentation and coalescence dynamics in multiphase flows, *Experimental Thermal and Fluid Science* **6** (1993) 211–251.

[10] H. Zhao, S. S. Sadhal and E. H. Trinh, Singular perturbation analysis of an acoustically levitated sphere: Flow about the velocity node, *J. Acoust. Soc. Am.* **106**(2) (1999) 589–595.

[11] J. A. Sethian, *Level Set Methods and Fast Marching Methods* (Cambridge University Press, 1999).

[12] K. B. Deshpande and W. B. Zimmerman, Simulation of interfacial mass transfer by droplet dynamics using the level set method, submitted to *Chem. Eng. Sci.*

[13] K. B. Deshpande and W. B. Zimmerman, Simulation of mass transfer limited reaction in a moving droplet to study transport limited characteristics, submitted to *Chem. Eng. Sci.*

Chapter Nine

MODELLING OF FREE SURFACE FLOW PROBLEMS
WITH PHASE CHANGE — THREE PHASE FLOWS

T. L. MARIN

Department of Mining Engineering, University of Chile,
Av. Tupper 2069, Santiago, Chile
E-mail: tmarin@ing.uchile.cl

A method for the solidification of a free surface liquid phase is presented using a fixed grid numerical model. The Level Set method is used to prescribe the movement of the free surface as described previously, whereas a modified version of the method presented by Voller and Prakash is used to account for the solidification of the liquid phase, including convection and conduction with mushy region phase change. In this method the properties of the fluid are dependant on the temperature, by defining a porosity function ranging from 0 to 1 (being 0 fully solid and 1 fully liquid) body forces that depend on the porosity and on temperature modify the Navier-Stokes equations in order to model the flow of a liquid or a solid. In addition, a modified heat capacity expression is used to account for the latent heat of melting into the heat equation.

1. Introduction

Solidification of fluids in which heat transfer occurs mainly by convection are difficult to model because of the problem in determining the moving interface between the liquid and the solid phases. In addition, the governing equations that are valid for the fluid are not longer valid in the solid phase. Furthermore, the problem becomes even more complicated if it involves the modelling of free surface flow that also goes under a phase change or solidification, in which case it will be necessary to track two interfaces.

There are several computational methods proposed to solve the solidification of a fluid in a close cavity under conduction and natural convection as summarized by Voller and Prakash [1]. One approach is based in a temperature formulation employing deforming grids to treat the moving liquid-solid interface. Also an enthalpy formulation together with a fixed grid methods have been employed with the advantage of a much simpler numerical problem gained from the use of the fixed grid. The problem of using a fixed grid is to account for the zero velocity of the solid in the close cavity. This can be done by simply fixing the velocity to be zero in a given computational cell based on its latent heat content [2] or by making the viscosity a function of the latent heat [3] in which case, the viscosity would increase to a very large value as the latent heat content approaches to 0 to simulate the solid phase. The method proposed by Voller and Prakash [1] considers the case in which the phase change occurs in a temperature range and models the fluid as a porous media, introducing source terms to the Navier-Stokes equations to simulate the phase change and to account for the zero velocity of the solid.

The problem of free surface flow with phase change is very challenging from both experimental and computational points of view and has been studied during the past few years. Pasandideh-Fard *et al.* presents experimental and computational studies for the solidification of tin droplets [4] and water drops on hot surface [5] using the volume of fluid method (VOF) to track the free surface together with an enthalpy formulation for the heat transfer in which the velocity is set to zero for the solid phase. In their model, they introduce the contact angle of the droplet and the heat transfer coefficient at the droplet substrate interface as boundary conditions based on experimental observations.

The method proposed here uses the Levelset method [6] to handle the free surface problem as described previously with a solidification model similar to that proposed by Voller and Prakash [1]. In this case, however, a temperature based heat equation is employed and the contribution of the latent heat during the solidification process is included in a modified heat capacity equation. It has the advantage inherited from the fixed grid methods for both, the free surface model and the solidification model and it is able to treat impure materials, in which the solidification occurs in a temperature range.

2. Governing equations

In pure materials the solidification process is determined by the melting temperature. Upon cooling of a liquid, once it reaches this temperature it will liberate the latent heat of phase change (or heat of melting) before continuing decreasing in temperature. However, in multi-component (and more widely used) systems the phase change occurs within a temperature range, starting at the liquidus temperature when the first solid crystal appears and finishing at the solidus temperature, when the last liquid portion of the system solidifies. In these cases, the latent heat of melting is released during this continuous change in temperature.

The solid fraction (F_S) during the phase change can be described as a function of temperature:

$$F_s(T) = \begin{cases} 0, & T \geq (T_m + \varepsilon), \\ \dfrac{T_m + \varepsilon - T}{2\varepsilon}, & (T_m + \varepsilon) > T \geq (T_m - \varepsilon), \\ 1, & T < (T_m - \varepsilon), \end{cases} \tag{1}$$

where T is the temperature of the system, T_m is the middle temperature between solidus and liquidus temperature and ε is half the difference in temperature between liquidus and solidus. Therefore the solidus and liquidus temperatures are:

$$T_S = T_m - \varepsilon, \tag{2}$$

$$T_L = T_m + \varepsilon. \tag{3}$$

As mentioned before, the total heat content H of the system is composed by two components, the sensible heat h and the latent heat ΔH. The sensible heat is represented by the following equation:

$$h = c_p \cdot T, \tag{4}$$

and the latent heat can be represented as a function of temperature, using the previous definition of solid fraction:

$$\Delta H(T) = \begin{cases} L, & T \geq (T_m + \varepsilon), \\ L(1 - F_s), & (T_m + \varepsilon) > T \geq (T_m - \varepsilon), \\ 0, & T < (T_m - \varepsilon). \end{cases} \tag{5}$$

The heat transfer equation by conduction and convection is expressed in terms of the temperature of the system as follows:

$$\rho \cdot c_p \frac{\partial T}{\partial t} + \nabla \cdot (-k\nabla T + \rho \cdot c_p T u) = 0, \tag{6}$$

in this case, the effect of the latent heat release can be incorporated into the heat capacity term by defining a new equation for the "effective" heat capacity:

$$c_{pT} = c_p + c_{pH}, \tag{7}$$

$$c_{pH} = L \cdot \delta^*(T) \approx L \cdot fldc2hs(T - T_m, \varepsilon), \tag{8}$$

in this case, $\delta^*(T)$ represents a smooth delta function, which is built in COMSOL multyphysiscs as $fldc2hs$. Note that it is desired that the integral of $\delta^*(T)$ be equal to one over the whole range of temperature but it has to span from $T = T_m - \varepsilon$ to $T = T_m + \varepsilon$. Only in the case of modelling the solidification of a pure material, in which case ε becomes 0, $\delta^*(T)$ will become the real Dirac's delta.

In this way, the total heat content of the system is expressed by integrating equation (7) over temperature.

$$H = \int_{T1}^{T2} c_{pT} \cdot dT = \int_{T1}^{T2} c_p \cdot dT + L \cdot \int_{T1}^{T2} \delta^*(T) \cdot dT. \qquad (9)$$

The governing equations of the velocity field and the pressure are the Navier-Stokes equations:

$$\rho\frac{\partial u}{\partial t} - \nabla \cdot \mu(\nabla u + (\nabla u)^T) + \rho(u \cdot \nabla)u + \nabla p = F, \qquad (10)$$

$$\nabla \cdot u = 0. \qquad (11)$$

The body Force F includes the gravitational force, the surface tension term as described by the levelset treatment and additionally, it includes the terms depending on the solid fraction to account for the solidification process. The two components of the body force terms are:

$$F_x = \sigma\kappa\frac{\partial\phi}{\partial x}\delta(\phi) + S_x, \qquad (12)$$

$$F_y = \sigma\kappa\frac{\partial\phi}{\partial y}\delta(\phi) + \rho g + S_y, \qquad (13)$$

where σ is the surface tension of the liquid, κ is the curvature of the interface, ϕ is the levelset function, g is the gravity acceleration and the two terms S_x and S_y are described below.

In order to account for the solidification process into the body forces, the liquid phase can be regarded as a porous medium, where the porosity λ depends on the temperature. A fully liquid phase corresponds to a porosity value of 1, and a fully solid phase corresponds to porosity equals to 0. The porosity is defined as follows:

$$\lambda = 1 - F_s(T) = \begin{cases} 1, & T \geq (T_m + \varepsilon), \\ \dfrac{T + \varepsilon - T_m}{2\varepsilon}, & (T_m + \varepsilon) > T \geq (T_m - \varepsilon), \\ 0, & T < (T_m - \varepsilon). \end{cases} \qquad (14)$$

Then the source terms are defined as:

$$S_x = -Au\,, \tag{15}$$

$$S_y = -Av\,, \tag{16}$$

where A is described by the following equation as a function of the porosity:

$$A = \frac{C(1-\lambda)^2}{(\lambda^3 + q)}\,. \tag{17}$$

The effect of these source terms can be explained as follows. When the temperature is higher than the liquidus, thus the system is fully liquid, the source terms take the value of 0 and the Navier-Stokes equations are not modified. In the mushy region, it means between liquidus and solidus temperatures, the value of A increases and begin to dominate the transient, convective and diffusive terms and the equations of fluid flow approximate the Darcy law for flow in a porous medium. When the temperature decreases further down and the porosity approaches 0 (solid), this source terms dominate all other terms forcing the predicted values of velocity to approach zero (solidified phase). The constants C and q in equation (17) are chosen arbitrarily depending on the problem being solved, C is a large value and q is small enough to avoid dividing by 0 as λ goes to 0.

The levelset function ϕ defines the interface between the two initial phases (gas-liquid, for example). However the third phase that will appear from the solidification (or melting in the opposite problem) just takes place in one of the initial two phases, so the source terms S_x and S_y need to be modified in order to act exclusively in one of the initial two phases. The solution to this is very simple, just by multiplying these terms by the Heaviside function of ϕ we limit the applicability of the effects of the source terms to the phase of our choice. Then, the final expressions for the body forces in equation (10) are:

$$F_x = \sigma\kappa\frac{\partial\phi}{\partial x}\delta(\phi) + S_x H(\phi)\,, \tag{18}$$

$$F_y = \sigma\kappa\frac{\partial\phi}{\partial y}\delta(\phi) + \rho g + S_y H(\phi)\,. \tag{19}$$

The Heaviside function is also used to define the steep change in properties between the continuous and discontinuous phases, as described in the levelset method. These properties now include density, viscosity and thermal conductivity.

3. Results and discussion

Three physics models are required for this simulation: Incompressible Navier-Stokes, mass transfer by convection and diffusion and heat transfer by convection and conduction. The solution of this problem uses the multiphysics capabilities of COMSOL Multiphysics to simultaneously handle and solve three applications modes.

3.1. *Solidification of a falling droplet*

The example presented here corresponds to a liquid droplet that spreads onto a cold surface, thus solidifying. This can be reduced to a two dimensional geometry to simplify the calculations. The whole system is represented by a square domain of unit dimension with bottom left corner at $x = y = 0$ and the bottom edge set at a fixed temperature with all the rest of the boundaries being insulated. The initial interface between the liquid and the continuous (ambient) phase is defined by the following initial levelset function:

$$\phi(t = 0) = \sqrt{x^2 + y^2} - 0.5 \,. \tag{20}$$

This will represent a quarter of circumference on the bottom left corner of the numerical domain considered for the model. Note that in this case, the liquid phase takes the negative value of ϕ, whereas the continuous phase takes the positive value of ϕ.

For simplification, in this case we will not consider the effect of the surface tension on the liquid/gas interface to focus on the solidification problem. However, the implementation of these effects is very simple, as described in the levelset methodology.

Start up COMSOL Multiphysics and enter the Model Navigator. Follow the instructions in Table 1. Note that slip/symmetry boundary condition has been select for boundary 2 (bottom edge) for Navier-Stokes model. Although the proper boundary condition is no-slip, this "assumption" will help in the advection of the levelset function "phi" on this boundary and simulate better the contact between the droplet and the bottom wall. In addition, as the droplet solidifies from the bottom edge and up, the velocity will automatically be calculated as zero velocity, since a solid phase will be solved, so the "assumption" of slip/symmetry in this case is not too unrealistic. If no-slip boundary condition is to be used, re-initialization of the levelset function will be necessary.

Table 1. Level set three-phase solidification model.

Model Navigator	Select 2-D dimension COMSOL Multiphysics|Fluid Flow/Incompressible Navier-Stokes. Click on the Multiphysics tab. Add Select Convection and Diffusion. Set dependent variable name: phi. Add Heat Transfer/Convection and Conduction. Add. OK
Draw Menu	Specify Objects/Square Enter width $= 1$; Base corner $x = y = 0$. OK
Options Menu| Constants	$x_0 = y_0 = 0$; $r = 0.5$; rhog $= 1$; rhol $= 10$; etag $=$ etal $= 1$; $g_y = -10$; $n = 0.02$; $T_i = 1$; $TC = -0.2$; $e = 0.1$; $cp = 1$; $kl = 0.5$; $kg = 0.01$; $L = 1$; $C = 1600$; $q = 0.001$; $tm = 0$
Options Menu | Expressions	phi0 $=$ sqrt$((x - x_0)^2 + (y - y_0)^2) - r$ $H_{\text{phi}} = (1 + \tanh(-\text{phi}/n))/2$ rho $=$ rhog $+$ (rhol $-$ rhog)*Hphi eta $=$ etag $+$ (etal $-$ etag)*Hphi $k = kg + (kl - kg)$*Hphi lambd $= (T - T_m + e)/(2 * e)*((T <= (T_m + e))*(T >= (T_m - e))) + (T > (T_m + e))$ $F_s = 1 -$ lambd $A = -C * (1 - \text{lambd})^2/(\text{lambd}^3 + q)$ $S_x = -A * u$; $S_y = -A * v$; $c_{pH} = L * fldc2hs(T - T_m, e)$ $c_{pT} = c_p + c_{pH}$
Physics Menu: Point settings	ns mode. Check the point constraint option for point 3 to set the pressure to 0
Physics Menu: Boundary settings	ns mode Boundary 1 and 2 to slip/symmetry and boundaries 3 and 4 to no-slip chcd mode: Set all the boundary conditions to Insulation/symmetry cc mode: boundary 2 to Temperature and enter TC into the temperature field. Set boundaries 1, 3 and 4 to Thermal insulation
Physics Menu Subdomain Settings	ns mode: $\rho =$ rho, $\eta =$ eta, $F_x = -S_x * H_{\text{phi}}$, $F_y = -S_y * H_{\text{phi}} +$ rho*gy chcd mode: D(isotropic) $= 0$; $R = 0$, $u = u$ and $v = v$ Init tab: phi$(t_0) =$ phi0 cc mode: k (isotropic) $= k$; $\rho =$ rho, $C_p = c_{pT}$; $Q = 0$, $u = u$ and $v = v$. Init tab: $T(t_0) = T_i$. OK
Mesh Menu Mapped Mesh	Select Boundary 1 and check the "Constrained edge element distribution" and enter 40 edge elements Select Boundary 2 and check the "Constrained edge element distribution" and enter 40 edge elements Remesh. OK
Solve Menu	Set the time dependent solver. Set output times 0:0.02:2 Click on the solve $(=)$ tool on the toolbar

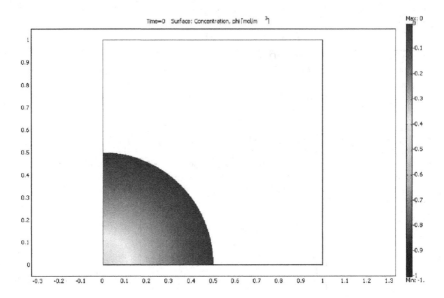

Figure 1.   Surface plot for the initial condition of the level set function.

The initial levelset function "phi0" is defined in terms of the spatial coordinates "x" and "y" according to Eq. (20), being negative for the liquid phase (or at the droplet domain) and positive for the gas phase, as seen in Figure 1. The Heaviside function "H_{phi}" is defined in terms of the negative value of phi, thus being equal to 1 inside the droplet and 0 elsewhere. The physical properties of the liquid and gas phases are defined in terms of the Heaviside function that depends on the level set function "phi". The porosity function is defined using a logical expression according to Eq. (14), being dependent on the temperature variable "T" and on the definition for the solidus and liquidus temperatures. This function is then used to define the rest of the temperature dependent properties.

The novel feature of the problem set is the use of a mapped mesh. This provides quad elements tessellating a grid.

Figure 2 shows the solution at different times. The surface plot shows the levelset function only in the negative range (droplet domain), the contours were set to show the temperature isolevels at the liquidus ($T_m + \varepsilon$), average melting temperature (T_m) and solidus temperature ($T_m - \varepsilon$) and the arrow plot shows the velocity vectors.

From Figure 2 it is observed how the solidification model works. The velocity field in the gas and liquid regions follows the expected results.

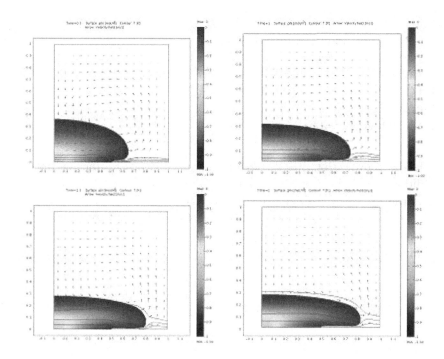

Figure 2. Results of the numerical model at times 0.5, 1. 1.5 and 2 s.

In the mushy region, where both solid and liquid phases are present, the velocity profile still exists but decreases as it approach the solid region, where the calculated velocity is completely zero, as expected for a solid phase.

The heat transfer inside the droplet is much faster than in the gas phase since the thermal conductivity in this phase is higher.

3.2. *Analysis of the solidification rate*

For the length of the simulation time, the solidification process already starts in the droplet, however not the entire droplet has solidified. By running the next scrip (called *"analysis.m"*) in either COMSOL Script or MATLAB, after exporting the fem structure as "FEM" from the "File" menu, the total area of the droplet and the area of the fully solid region can be calculated by using the Heavyside function of ϕ and integrating its value over the space domain using the *"postint"* function from COMSOL Multiphysics at different times.

```
t=1:101;
A=postint(fem,'Hphi','Solnum',t);
As=postint(fem,'Hphi*(1-flc2hs(T-Tm+e,0.001))','Solnum',t);
Fs=As./A*100;
plot(0.02*(t-1),A,0.02*(t-1),As);
figure
plot(0.02*(t-1),Fs);
```

In the above script A is the area of the droplet, which from the initial conditions should be constant (if mass is conserved) and equal to 0.196 m^2; A_s is the area of the solid which is obtained by intersecting the Heaviside function of f and another Heaviside function ($flc2hs$) around the solidus temperature $(T_m - \varepsilon)$; F_s is the percentage of the droplet that has solidified.

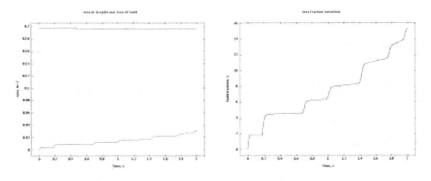

Figure 3.   Area of droplet, area of solid region and solid fraction as function of time.

It is observed in Figure 3 that the total area of the droplet is conserved as the time advances, thus mass conservation is retained, since the density of the liquid and solid are assumed to be the same and no volume change from the solidification process is expected. Also, it is observed that almost 16% of the droplet is fully solidified after 2 s of simulation.

Next, to observe the effect of the latent heat of phase change on the rate of solidification, we can turn off this by setting the value of $L = 0$ in the constants previously defined. Go to "Options" in the main menu and then select "Constants ...," set the value of L to 0 and click OK. Solve the problem again.

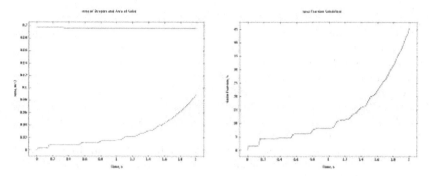

Figure 4. Area of droplet, area of solid region and solid fraction as function of time for zero latent heat of phase change.

Export the fem structure to COMSOL Script or MATLAB as "FEM" and run the "*analysis.m*" script again. Figure 4 is obtained.

After the same period of time, the fraction that has fully solidified is almost 45% of the droplet. This agrees well with the expected result, since in this case time is not consumed in liberating the latent heat during the phase change, therefore the temperature drops faster than in the previous case. This example shows the importance of considering this effect on the solidification problem.

4. Summary

The problem of free surface flow with phase change has been solved using a combination of the Level Set method and a solidification model for an example of metallurgical importance. In this example, the continuous phase is the gas phase as opposed to the examples shown for the Level Set method to model the coalescence of bubbles in liquids. The flexibility and easy of use of COMSOL Multiphysics allows to implement the model starting from the basic module, even though this model is of high modelling complexity since it includes three physics phenomena fully coupled. Finally, the example presented here sets the basis for more complex and realistic simulations.

Acknowledgment

The author acknowledges the comments and useful discussion with Professor Torstein Utigard from University of Toronto.

References

[1] V. R. Voller and C. Prakash, A fixed grid numerical modelling methodology for convection-diffusion mushy region phase-change problems, *Int. J. Heat Mass Transfer* **30** (1987) 1709.

[2] K. Morgan, A numerical analysis of freezing and melting with convection, *Comput. Meth. Appl. Eng.* **28** (1981) 275.

[3] D. K. Gartling, Finite element analysis of convective heat transfer problems with change of phase, *Computer Methods in Fluids*, eds. K. Morgan *et al.* (Pentech, London, 1980), p. 257.

[4] M. Pasandideh-Fard, R. Bole, S. Chandra and J. Mostaghimi, Deposition of tin droplets on a steel plate: Simulations and experiments, *Int. J. Heat Mass Transfer* **41** (1998) 2929.

[5] M. Pasandideh-Fard, S. D. Aziz, S. Chandra and J. Mostaghimi, Cooling effectiveness of a water drop impinging on a hot surface, *Int. J. Heat Fluid Flow* **22** (2001) 201.

[6] M. Sussman, P. Smerka and S. Osher, A Level Set approach for computing solutions to incompressible two-phase flow, *J. Comput. Phys.* **114** (1994) 146.

Chapter Ten

NEWTONIAN FLOW IN GROOVED MICROCHANNELS

D.G. HASSELL* and W.B.J. ZIMMERMAN

*Department of Chemical and Process Engineering, University of Sheffield,
Newcastle Street, Sheffield S1 3JD United Kingdom*
*Present address: Department of Chemical Engineering, Cambridge University,
Pembroke Street, Cambridge, CB2 3RA*
E-mail: w.zimmerman@shef.ac.uk

The current trend within research towards miniaturised systems and microchannel technology has offered a tantalising glimpse into a world of increased control over chemical and biological systems at scales equivalent to "lab on a chip" or "plant on a bench." At these small length scales the issue of fluid mixing proves interesting as the conventional method of mixing (turbulence with its chaotic nature) becomes unfeasible due to the prohibitively high pressures required. While molecular diffusion alone may be enough in some systems, those with large molecules (such as DNA) and low molecular diffusivity require a more elegant approach where controlled convective mixing induced either through passive or active methods increases exponentially the mixing of initially segregated fluid streams. This case study investigates the flow patterns present in the Staggered Herringbone Mixer (SHM) proposed by Stroock and co-workers [1], and looks to identify the relationship between various flow parameters that affect mixing in these devices. After comparing analytical solutions of simple parallel plate flow with COMSOL Multiphysics solutions, work is then expanded to incorporate three-dimensional modeling of two test cases at various flow conditions. Post-processing is then used to evaluate properties of interest in an attempt to gain a better understanding of the fluid mechanics involved in this complex flow.

1. Introduction

The Staggered Herringbone Mixer (SHM), shown in Figure 1, has been proposed as a method of improving mixing in microchannel devices without a correspondingly large increase in the pressure drop [1]. It uses successive cycles, consisting of two groups of asymmetric herringbone grooves, to fold the flow upon itself, such as a baker might kneed dough. This convective folding leads to increased areas over which diffusive mixing can take place,

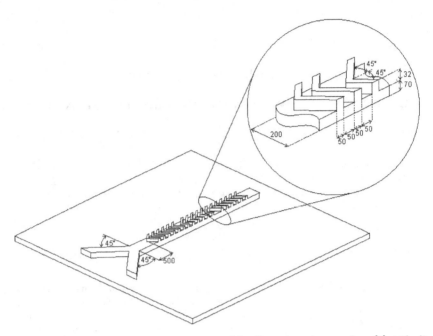

Figure 1. Example of a micromixer proposed by Stroock and co-workers [1] with dimensions given in micrometers. The chip consists of sequential cycles consisting of two units of asymmetric herringbone grooves off axis by one sixth the channel width.

and while several computational studies have been carried out that focus upon mixing [2] [3], velocity profiles and associated fluid properties have not been investigated for these systems. This work focuses on the effects the grooves have on the bulk channel flow for varying test conditions. Given the restriction that computational power places upon the resolution with which these flows can be resolved using COMSOL Multiphysics we use two different geometries. The first is a channel containing a single groove, while the second models a system of infinite symmetric grooves using periodic boundary conditions. We begin with an analysis of simple flow between parallel plates to evaluate the numerical method before moving on to more complicated three-dimensional domains with their increase complexity. The flow was modeled using nondimensional Navier-Stokes equations, where u_e was set such that the inlet velocity was equal to unity and L was set equal to the width of the channel.

$$\nabla \cdot \left[\frac{1}{Re} \right] (\nabla u' + (\nabla u')^T) + (u' \cdot \nabla)u' + \nabla p' = 0 \,, \qquad (1)$$

$$\nabla \cdot u' = 0 \,, \qquad (2)$$

$$Re = \frac{Lu_e\rho}{\mu} \,, \qquad (3)$$

$$x_i' = \frac{x_i}{L} \,, \qquad u_i' = \frac{u_i}{u_e} \,, \qquad p' = \frac{p}{\rho u_e^2} \,. \qquad (4)$$

We define the ratio of inertial to viscous forces through the Reynolds number (Re), which we will add as an additional constant. This method has the advantage of simplifying the problem for COMSOL Multiphysics by reducing the problem variable values within the problem to order unity and thus making the problem easier to solve.

2. Newtonian flow in 2-D

We begin by illustrating a simple 2-D simulation of Poiseuille flow using COMSOL Multiphysics, shown in Table 1. The velocity profile is shown in Figure 2 along with velocity profiles along the channel. The profiles illustrate that it takes time for the flow to develop from initial plug flow at the inlet to Poiseuille flow, and that the outlet condition also affects the profile near the outlet. Hence only the flow near the centre is actually Poiseuille flow.

Another method of solving this problem is by using periodic boundary conditions such that the velocities at the inlet and outlet are set equal to one another, theoretically modeling flow in channel of infinite length.

$$u_i(x_{\text{inlet}}, y) = u_i(x_{\text{outlet}}, y) \,.$$

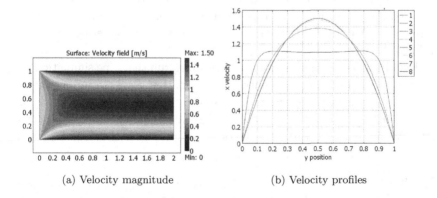

(a) Velocity magnitude   (b) Velocity profiles

Figure 2. (a) Velocity field along the 2-D channel and (b) the corresponding cross channel profiles for the x component of velocity.

Table 1. Newtonian flow in 2-D.

Model Navigator	Select 2-D space dimension Select COMSOL MultiPhysics|Fluid Dynamics|Incompressible Navier-Stokes|Steady-state Add. OK
Draw Menu	Select Draw objects. Create a rectangle with height 1 and width 2 with the bottom left corner at the origin
Physics Menu: Boundary conditions (F7)	Select boundary 1 Condition type: Inflow/Outflow. Enter $u_0 = 1$ Select boundary 4 Condition type: Normal flow/pressure Enter $P_0 = 0$ Leave boundary conditions for boundaries 2 and 3 at the default no-slip condition. OK
Physics Menu: Subdomain settings (F8)	Select subdomain 1. $\rho = 1$; $\mu = 1/Re$. OK
Options Menu: Constants	Select name. Enter Re Select expression. Enter 1. OK
Mesh	Click on the toolbar triangle for the default mesh Refine the mesh twice
Solve Menu: Solver Parameters (F11)	Select General|Linear system solve|GMRES Select Advanced|Null-space function|sparse Click on the solve (=) tool on the toolbar

This will allow us to increase the mesh resolution in 3-D flows by reducing the size of the geometry and to investigate the cumulative effect grooves in series. There are two methods for solving this problem. The first defines a driving pressure difference for the flow by setting the inlet and outlet pressures while the second uses a further equation to define the flow rate across the inlet boundary which is then balanced using an extra degree of freedom representing the pressure drop across the system. It is this second method which we will use as it gives us more control over the system parameters. We thus make the following changes to our model set up, shown in Table 2.

Finally, following instructions in Table 3, re-run the same simulation for a series of Reynolds numbers within the range $1 \leq Re \leq 15$ in increments of 2 using a parametric solver to analyse the effect of increasing inertia on the problem. A good test of accuracy is then to compare the profiles (for velocity and pressure) with an analytic solution. This is shown in Figure 4.

3. Newtonian flow in 3-D

Now that the methodology has been evaluated for 2-D flows we can move to a 3-D flow. The first problem is solved in a similar fashion to the simple nonperiodic 2-D flow using the geometry shown in Figure 5(a).

Table 2. Periodic boundary conditions for the inlet and outlet.

Model Navigator	Select PDE modes|Classical PDEs|Weak form point|Stationary analysis Select dependant variable name and change to *dp* Add. OK
Physics Menu	ns mode: Select Periodic conditions Periodic|Boundary Conditions Select source, boundary 1, expression *u* Select destination, boundary 4, expression *u* Select Source vertices, 1 and 2 Select Destination vertices, 3 and 4 Run through the same procedure for expression *v*. Apply OK
Physics Menu: Point settings (F5)	wp mode: Deactivate points 1–3. Select point 4 and set the weak term equal to *dp*_test*(flowrate_target-flowrate). Apply OK
Options Menu	Select Integration Coupling variables|Point variables In the source tab select point 4 and define a term *dp* with the expression *dp*. Deselect the global destination and select boundary 1 in the destination tab. Apply. OK
Options Menu	Select Integration Coupling variables Boundary|variables In the source tab select boundary 1 and define a term flowrate with the expression — $(u * nx + v * ny)$. Deselect the global destination and select point 4 in the destination tab. Apply OK
Physics Menu: Boundary conditions (F7)	wp mode. Select boundary 1 Condition type: Outflow/Pressure. Enter $P_0 = dp$ Select boundary 4 Condition type: Outflow/Pressure Enter $P_0 = 0$
Options Menu: Constants	Select name. Enter flowrate_target Select expression. Enter 1. OK
Solve Menu: Solver Parameters (F11)	Select Advanced|Type of scaling|none Click on the solve (=) tool on the toolbar

Table 3. Parametric solver.

Solve Menu: Solver Parameters (F11)	Select General|Parametric nonlinear Select Name of parameter|*Re* Select List of parameter values|1:2:15. Apply. OK Click on the solve (=) tool on the toolbar

A similar methodology is used as as with the simple 2-D laminar flow case, with the exception of setting a Poiseuille flow inlet boundary using the parameterization variables s_1 and s_2. The mesh parameters used were manually optimized as follows to place a more refined mesh at the areas of interest (near the groove). This results in 70000 elements and 317000 Dof which takes on the order of hours to solve. The solution is shown in

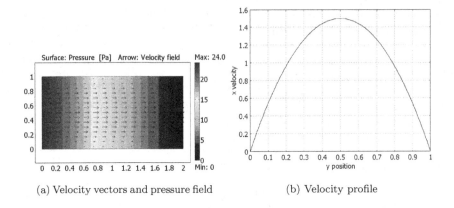

(a) Velocity vectors and pressure field   (b) Velocity profile

Figure 3.   (a) Pressure field along the 2-D channel with arrows indicating flow and
(b) the corresponding cross channel profiles for the x component of velocity at eight
points down the channel.

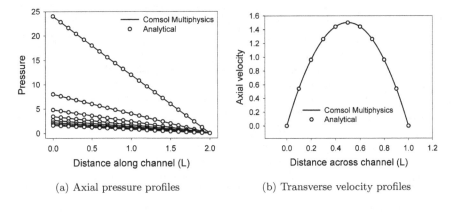

(a) Axial pressure profiles   (b) Transverse velocity profiles

Figure 4.   COMSOL Multiphysics solution (line) and analytic solution (points) for (a)
the collapsed pressure field along the 2-D channel and (b) the corresponding cross channel
profiles for the x component of velocity.

Figure 6. We can see from the streamlines that the inlet and outlet boundaries have been set far enough away from the area of interest (near the
groove) that they do not affect the area of interest. As before it is possible
to use the parametric solver to steadily increase Re. Once these have been
solved, it is possible to export the FEM structure and extract more values
of interest using the MATLAB code below. The following code creates the
image shown in Figure 7(a) to visualize the flow into the groove while

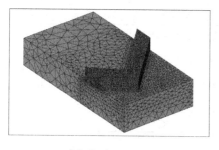

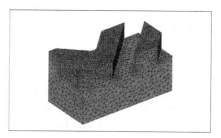

(a) single groove   (b) periodic geometry

Figure 5.   (a) Single groove geometry and (b) periodic geometry.

Table 4.   Geometry creation in 3-D.

Model Navigator	Select 3-D space dimension Select COMSOL MultiPhysics|Fluid Dynamics|Incompressible Navier-Stokes|Steady-state Add. OK
Draw Menu	Select working plane settings and create a new work plane with xy, $z = 0$. Create a 2-D herringbone in this plane by drawing lines at the following points (0,0.48 → 0,0.73), (0,0.73 → 0.33,0.4), (0.33,0.4 → 1,1.07), (1,1.07 → 1,0.82), (1,0.82 → 0.33,0.15), (0.33,0.15 → 0,0.48) before selecting these lines followed by the coerce to solid button. Select the shape using Draw Mode|extrude, and extrude it 0.16 in the z direction. Select 3-D work plane move this shape 0.35 in the z direction. Following this procedure create another five lines at points (0,0 → 0,0.78), (0,0.78 → 0.33,0.45), (0.33,0.45 → 1,1.12), (1,1.12 → 1,0), (1,0 → 0,0), extrude 0.15 and move 0.2 in the z direction. Create another shape similar to the second, but extrude it 0.2 and place it directly below the other shape. In the 2-D work plane create another five lines at (0,0.78 → 0.33,0.45), (0.33,0.45 → 1,1.12), (1,1.12 → 1,1.4), (1,1.4 → 0,1.4), (0,1.4 → 0,0.78), extrude 0.35. In the 3-D work plane create a rectangle of dimension $x = 1$, $y = 0.5$, $z = 0.35$ with the corner axis base point at position $x = 0$, $y = 1.4$, $z = 0$. Select all five shapes and use draw mode → create composite object, to create a single object and click on delete interior boundaries

similar post-processing can be used to evaluate the effectiveness of the groove at moving fluid cross-channel as a function of Re and groove height (further simulations with different groove heights required), shown in Figure 7(b).

Table 5.   Boundary settings.

Physics Menu: Boundary conditions (F7)	Select boundary 1 Condition type: Inflow/Outflow. Enter $v_0 = 9/4 * U_{mean} * 16 * s_1 * (1 - s_1) * s_2 * (1 - s_2)$ Select boundary 4 Condition type: Normal flow/pressure. Enter $P_0 = 0$ Leave boundary conditions for boundaries 2 and 3 at the default no-slip condition. OK
Options Menu: Constants	Select name. Enter U_{mean} Select expression. Enter 1. OK

Table 6.   Mesh settings.

Mesh	Select Mesh parameters|subdomain Select Domain 2|0.05, domain 3|0.026 and domain 4|0.08 Apply. OK
Solve Menu: Solver Parameters (F11)	Select General|Linear system solve|GMRES Select Advanced|Null-space function|sparse Click on the solve (=) tool on the toolbar

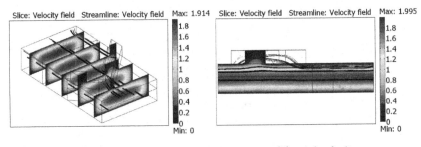

(a) tangential velocity   (b) axial velocity

Figure 6.   (a) Cross channel velocity field and streamlines and (b) the corresponding down channel velocity profile and streamlines, illustrating outlet affects downstream. The increased value for the maximum velocity in the right image when compared to the left image is due to the development of the flow away from a parabolic profile due to the influence of the grooves as the fluid moves downstream.

```
[xx,yy]=meshgrid(0:0.02:0.33,0.48:0.005:0.73); yy=yy-xx;
[xx1,yy1]=meshgrid(0.34:0.02:1,0.15:0.005:0.4); yy1=yy1+(xx1-0.33);
xx=[xx,xx1]; yy=[yy,yy1]; zz=zeros(51,51); zz=zz+0.35;
xx=reshape(xx,51,51); yy=reshape(yy,51,51);
xxx=[xx(:)';yy(:)';zz(:)'];
w=postinterp(fem,'w','Re',xxx,'solnum',1); ww=reshape(w,51,51);
surf(xx,yy,ww) axis([0 1 0.15 1.07 -0.25 0.3]) xlabel('Width
(L)','Fontsize',14) ylabel('Distance from inlet (L)','FontSize',14)
colorbar
```

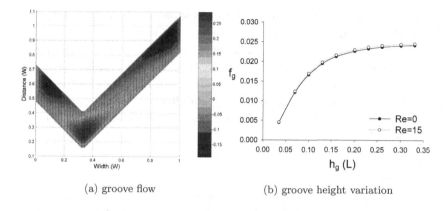

(a) groove flow   (b) groove height variation

Figure 7.   (a) Image created using MATLAB script showing the velocity flowing into and out of the groove and (b) Effect of groove height, h_g, on the integrated cross channel flow within the groove for $Re = 0$ and $Re = 15$. After a groove height of approx $0.20L$ there seems little increase in fluid movement with increasing height.

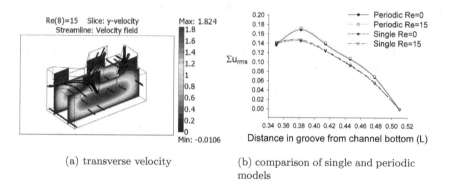

(a) transverse velocity   (b) comparison of single and periodic models

Figure 8.   (a) Cross channel velocity field and streamlines for $Re = 15$ and (b) Values of root mean squared x velocity for single and periodic groove systems at various horizontal cross sections of the groove, $Re = 0$ and 15.

Finally we can model a set of infinite same sign grooves. For this we need to model a length of channel equal to $0.5L$, reducing the domain and allowing increased mesh refinement per unit area. Using techniques used previously we can create the geometry shown in Figure 5(b), before we apply periodic boundary conditions and a weak point mode to as for the previous 2-D case. This time we define the flowrate as $-(u*nx + v*ny + w*nz)$ to take into account the third dimension and select automatic scaling in the advanced solver parameters. The solution is shown in Figure 8(a) and further post-processing can compare the single to infinite groove case, such as that shown in Figure 8(b). An indication of the mesh error can be acquired using Richardson extrapolation [4] for increasing mesh refinements. The periodic solutions were found to be more erroneous than that for the single grooves. This is probably due to reduced stability resulting from the extra equation and degree of freedom. Further analysis of the convective mixing can be found in [5].

References

[1] A. D. Stroock, S. K. W. Derlinger, A. Ajdari, I. Mezic, H. A. Stone and G. M. Whitesides, Chaotic mixer for microchannels, *Science* **295**(5555) (2002) 647–651.

[2] J. Aubin, D. F. Fletcher and C. Xuereb, Design of micromixers using CFD modelling, *Chemical Engineering Science* **60** (2005) 2503–2516.

[3] J. P. Bennett and C. H. Wiggins, A computational study of mixing microchannel fows (2003), Cond-mat/0307482.

[4] http://www.grc.nasa.gov/www/wind/valid/tutorial/spatconv.html, accessed 13 April 2005.

[5] D. G. Hassell and W. B. Zimmerman, Investigation of the convective motion through a staggered herringbone micromixer at low Reynolds number conditions, *Chemical Engineering Science*, accepted (2006).

Chapter Eleven

ELECTROKINETIC FLOW

W.B.J. ZIMMERMAN and J. M. MACINNES

Department of Chemical and Process Engineering, University of Sheffield,
Newcastle Street, Sheffield S1 3JD United Kingdom
E-mail: w.zimmerman@shef.ac.uk

This chapter explores the multiphysics modeling appropriate for electrokinetic flow
in microchannel networks. The modelling of networks of long channels that merge
at a Y-junction incorporates simultaneous algebraic constraints for the long channels (0-D) as coupled boundary conditions to the 2-D Y-junction. These are implemented as extended multiphysics using coupling variables.

1. Introduction

The purpose of this chapter is to demonstrate the facility of setting up a
model for electrokinetic flow in COMSOL Multiphysics. A cutting edge
application for electrokinetic flow is microfluidics, wherein small quantities of chemicals (nanoliters) are transported "just-in-time" for complicated
switching and sequencing in a network of microchannels to achieve high re-
producibility of chemical reactions and compositional changes by tight control. Moving fluids by physicochemical phenomena is especially important
since it involves fast response times and no moving mechanical parts that
can become damaged. There is a strong overlap between microfluidics and
micromechanical machines (MEMs). For instance, moving macromolecules
adjacent to walls and side channels for use as soft actuators is considered
microfluidics, but these are also molecular machines, but at a scale too
small to be considered conventional moving parts.

In order to set up even our simplest electrokinetic model, however, multiphysics is essential — coupling electric potential, chemical transport, and
momentum transport (Navier-Stokes). Furthermore, a first approach introduces some coupling through boundary conditions to approximate the
electrochemical boundary layer motion. Although this coupling is linear, it used to be the case that to get an acceptable model in COMSOL

Multiphysics, weak boundary constraints were required. However, COM-SOL Multiphysics now deals with linear Dirichlet boundary conditions coupling field variables across application modes without recourse to user "flagged" weak constraints. General boundary and auxiliary conditions are tractable, as discussed in Chapter two, by COMSOL Multiphysics using its Lagrange multiplier technique. Nevertheless, previous versions of Femlab required the user to specify any boundary constraints that were coupled across application modes using a special weak boundary mode entered in the Model Navigator Multiphysics facility. Careful inspection of the m-file script generated for the model developed here shows that COMSOL Multiphysics automatically detected the need for weak boundary constraints linearly coupling velocity and electric fields, and implemented the necessary Lagrange multipliers.

If the user wishes implement derivative boundary constraints (non-normal), nontangential boundary constraints, or nonlinear boundary constraints, there is still a GUI facility for flagging up these constraints to COMSOL Multiphysics so that they are properly included in the Jacobian matrices. Since this case study no longer requires weak boundary constraints, rather than review them here, we refer the reader to the COMSOL Multiphysics User's Guide [1].

2. Electrokinetic flow

2.1. *Background*

Electrokinetic flow is produced by the interaction of an electric field and charged (ion) species in a liquid. Two distinct interactions are present: the electric force on the liquid in the double layer region adjacent to wall surfaces where there is a net charge and the movement of individual ions in the bulk of the flow (outside the double layer region) where there is generally no net charge. The double layer may be taken as infinitesimal for channel sizes of interest (say greater than about 1 μm) and its effect on the flow is then equivalent (MacInnes *et al.* [3], MacInnes [2]) to application of the boundary conditions for velocity, u_i, electric field, ϕ, and mass fraction of a relevant chemical species, Y:

$$u_i = \zeta \frac{\partial \phi}{\partial x_i} \quad n_i \frac{\partial \phi}{\partial x_i} = 0 \quad n_i \frac{\partial Y}{\partial x_i} = 0, \tag{1}$$

where n_i is the unit normal vector to the wall surface.

The system of equations that must be solved comprises the momentum equation, the continuity of mass equation, the charge continuity equation and a species equation. A simplest case may be expressed in nondimensional form by

Momentum transport and continuity (Navier-Stokes):

$$\frac{\partial u_i}{\partial t} + u_j \frac{\partial u_i}{\partial x_j} = -\frac{\partial p}{\partial x_i} + \frac{\partial}{\partial x_j} \left(\frac{1}{Re} \frac{\partial u_i}{\partial x_j} \right), \tag{2}$$

$$\frac{\partial u_j}{\partial x_j} = 0. $$

Species transport including electrophoresis:

$$\frac{\partial Y}{\partial t} + u_j \frac{\partial Y}{\partial x_j} = \frac{\partial}{\partial x_j} \left(\frac{1}{Pe} \frac{\partial Y}{\partial x_j} + \beta z Y \frac{\partial \phi}{\partial x_j} \right). \tag{3}$$

Charge balance:

$$\frac{\partial}{\partial x_j} \left(\sigma \frac{\partial \phi}{\partial x_j} \right) = 0. \tag{4}$$

The electric field satisfying (4) must also satisfy Gauss' law (c.f. equation (1) of Chapter 7), which becomes an equation determining charge density as a function of position in the flow. In the typical conditions of electrokinetic flow, the charge density may be taken as negligible for purposes of both charge conservation (4) and the momentum balance (2). The electrical conductivity and zeta potential may depend on concentration of species Y and linear relations are assumed here: $\sigma = 1 + \sigma_r(1 - Y)$ and $\zeta = -1 - \zeta_r(1 - Y)$, where subscript "$r$" indicates the ratio of the property in the two pure solutions involved in the flows considered.

Boundary conditions at the flow inlets are that electric potential, pressure and species concentration must be specified, and at flow outlets electric potential and pressure must be specified. Species concentration is not known at the outlet boundaries and an approximation regarding species diffusion, the only term that connects the species field within the domain to the species distribution on the outflow boundary, is required. As usual, the species diffusion is neglected at the outflow boundary, i.e. a Neumann boundary condition just on the diffusion part of the flux term Γ is used.

The electric field is taken as quasi-steady, that is the electric field adjusts practically instantly to changes in the velocity and concentration. The above equations represent a generic problem providing a test of the numerical implementation which when verified may allow computation of any particular electrokinetic flow conditions. For the test implementation, suggested coefficient values are $1/Re = 30$, $1/Pe = 0.03$, $\zeta_r = 1$ (no variation in wall zeta potential), $z = 0$ (no charge on species Y) and $\sigma_r = 1$ (no variation in electrical conductivity and hence zero charge density throughout the bulk liquid).

2.2. *Problem set up*

The basic problem one can solve is the propagation of a concentration front along a channel. Initially, a sharp front is placed at mid channel and the evolution of the front is then computed in time. The test problem is two-dimensional and the channel width can be taken as 1 unit, with the length equal to 6 units.

There are a number of distinct steps in problem complexity. (1) With the parameter values suggested above, the electric field will be uniform and in the direction of the channel. The concentration will move with a uniform flow with the front thickening from diffusion. (2) Setting $\zeta_r \neq 1$ gives a nonuniform wall zeta potential with walls exposed to full concentration of the computed species having zeta potential ζ_r and those exposed to zero concentration $\zeta = -1$, giving variation of slip boundary velocity through the first of boundary conditions 9.4. The electric field remains uniform and in the channel direction, but the velocity field will be altered. The concentration front will be modified from the pure diffusion case by the nonuniform velocity field. (3) Setting $z = \pm 1$ and $\beta = 1$ will introduce electrophoresis. The computed species will translate in the channel direction in addition to being moved by the liquid velocity. (4) Finally, setting $\sigma_r \neq 1$ introduces nonuniform electrical conductivity. This leads to changes in the electric field associated with changes in concentration (Y) so the electric field is no longer uniform or, where concentration gradient is not everywhere in the direction of the channel, in the channel direction.

2.3. *COMSOL Multiphysics implementation*

There are application modes for conductive media, convection and diffusion, and the Navier-Stokes equations. To have best knowledge of what the computation entails, we start with the Navier-Stokes equations and add two general modes for (3) and (4).

Load up COMSOL Multiphysics and in the Model Navigator, check the Multiphysics tab, set up following the instructions in Table 1.

Figure 2 shows all the information rolled up into one plot for the final time $t = 1$. By this time, all streamlines are parallel and velocity vectors uniform, i.e. a flat profile. The spreading of concentration and speeding

Table 1. Basic model of electrokinetic flow in a channel. Filename: EKchannel.mph.

Model Navigator	Select 2-D dimension Click Multiphysics button Select COMSOL Multiphysics|Fluid Dynamics|Incompressible Navier Stokes. Add Select COMSOL Multiphysics|PDE Modes|General Mode Dependent variable name: Y. mode name: species. Add Select COMSOL Multiphysics|PDE Modes|General Mode Dependent variable name: phi. mode name: potential. Add OK
Draw Menu	Specify rectangle with width 3 and height 0.5 with corner at the origin
Options Menu| Constants	$Pec = 30$; $Re = 0.03$; zel = 1; betael = 1; zetar = 1; sigr = 1

Physics Menu: Subdomain settings (select subdomain 1 and use Multiphysics menu to switch modes)	Mode	Γ	F	da
	ns	$\rho g\ 1$; $\eta g\ 1/Re$; $F_x = 0$; $F_y = 0$		
	species	$-(1/Pec) * Y_x -$ betael*zel*Y*phi$x$ $-(1/Pec) * Y_y -$ betael*zel*Y*phi$y$	$-u*Y_x - v*$ Y_y	1
	Init tab: $Y(t_0) = 1$			
	potential	$-$ sig*phi$x$ $-$ sig*phi$y$	0	0

Options Menu| Expressions| Global	Name: zeta Definition: $-(Y + zetar*(1 - Y))$ Name: sig Definition: $Y + sigr*(1 - Y)$

Physics Menu: Boundary settings	ns mode			
	bnd 1	bnd 2	bnd 3	bnd 4
	Outflow/ pressure $p = 0$	$u =$ zeta*phi$x$ $v =$ zeta*phi$y$	$u =$ zeta*phi$x$ $v =$ zeta*phi$y$	Outflow /pressure $p = 0$
	species mode			
	bnd 1	bnd 2	bnd 3	bnd 4
	Dirichlet $G = 0$; $R =$ $-Y$	Neumann $G = 0$	Neumann $G = 0$	Neumann $G = $ betael*zel* Y*phi$x$
	potential mode			
	bnd 1	bnd 2	bnd 3	bnd 4
	Dirichlet $G = 0$; $R =$ $3 -$ phi	Neumann $G = 0$	Neumann $G = 0$	Dirichlet $G = 0$; $R = -$ phi

Mesh	Select the standard mesh and then refine once
Solve Menu	Set the time dependent solver Click on the solve (=) tool on the toolbar

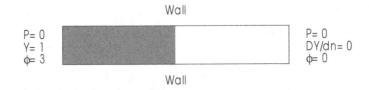

Figure 1.   Problem definition in a nutshell.

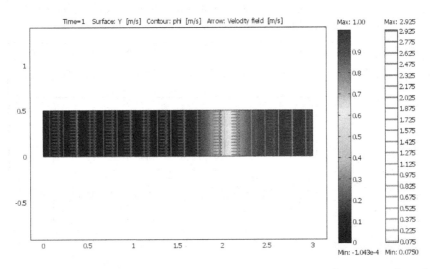

Figure 2.   Combined concentration (Y, color), electric potential (phi, contour), and velocity vector (u, v, arrow) plot. Coupling on the boundary of species electrophoresis/diffusion with electric field drags the fluid along.

up of the flow are all driven by the electric field, which is now apparently uniform in magnitude. A few cross plots (see Figure 3) show that the steady state electric field relaxes its transients within the first output time and remains constant thereafter (phi is linear for all times after $t = 0.1$). As expected, electrokinetic flow is dragged along by its boundary layer coupling to the electric field.

2.4. *Links to physical boundaries*

Current microchannel devices may consist of many distinct channel segments joined at several junctions. Future ones may well comprise hundreds of segments joined at a similar number of junctions. Detailed computation of the flow in such a system is unlikely to be feasible for some time to come and, indeed, is probably not desirable. Rather, an approach in which a particular junction of interest or perhaps an evolving mixing zone such as that considered in MacInnes *et al.* [4] and in MacInnes [7] is probably appropriate. The approach emerges from the fact that electrokinetic flows in microchannel networks virtually always are characterized by very low Reynolds number, $Re \ll 1$. In channel segments of uniform section and liquid and wall properties, the flow is developed along the entire segment length except for a region within about one channel width of junctions or other disturbances to uniformity. If the segment is many channel widths in length, it is a good approximation to neglect the junction effects and one can write linear relations between pressure and electric potential differences and the liquid volume flow rate, Q, and the charge flow rate, I:

$$-\frac{ReR^2 A}{f\Delta s}\Delta p + \frac{\zeta A}{\Delta s}\Delta \phi = Q \quad \text{and} \quad -\frac{\sigma A}{\Delta s}\Delta \phi = I. \tag{5}$$

These equations are coupled to the detailed flow solution through the liquid and charge flow rates. We will consider the specific example of an electrokinetic switching at a "Y" junction in the arrangement shown in Figure 4. By changing voltages at reservoirs A, B and C in an alternating pattern, "slugs" of the liquid fed in at A interspersed with the liquid fed in at B will be formed in the channel leading to C. No property nonuniformity will be present so the zeta potential and conductivity are uniform over each channel segment. We wish only to compute the flow in the vicinity of the junction where slug formation takes place.

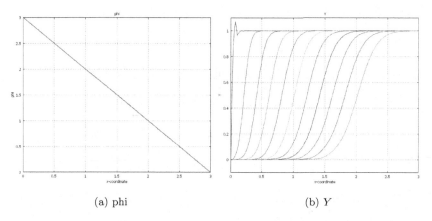

(a) phi   (b) Y

Figure 3. Histories of electric potential (phi) and species concentration Y along boundary 2 (bottom wall).

The "Y" junction domain is computed in detail with algebraic relations linking to the known boundary conditions at reservoirs A, B and C to provide the boundary conditions at a, b and c. In the case of nonuniform composition in channel segment C, a one-dimensional pie must be solved for the species composition to provide the link to downstream conditions.

The boundary conditions at the junction flow boundaries a, b and c come from the network flow equations linking junction boundary conditions to the known conditions of the reservoirs connected up to the network. In the simple case considered here (where $R = A = 1$), the linking equations for boundary a are

$$-\frac{Re}{f\Delta s_a}(p_A - p_a) + \frac{\zeta}{\Delta s_a}(\phi_A - \phi_a) = Q_a \quad \text{and} \quad -\frac{\sigma A}{\Delta s_a}(\phi_A - \phi_a) = I_a,$$

(6)

where Q_a and I_a are the flow rates at boundary a given in terms of computed conditions on that boundary by

$$Q_a = \int_A u_i n_i dA \quad \text{and} \quad I_a = -\int_A \sigma \frac{\partial \phi}{\partial x_i} n_i dA.$$

(7)

Precisely similar relations result at boundaries b and c.

Example: Y-junction electrokinetic valve

The physics of this problem is identical to the microchannel flow in the previous section, equations (1)–(4), but the geometry as shown in Figure 4 is different.

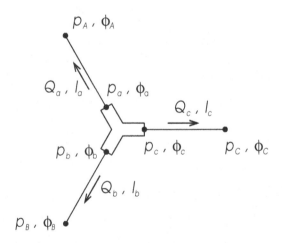

Figure 4.   The simple network considered.

We used to recommend starting this example from the FEMLAB MAT-file for the previous section, and to edit the geometry in Draw mode. However, COMSOL Multiphysics erases your model entries if you delete your geometry, so the only thing you save in starting from EKchannel.mph are your constants and expressions. Still, labor saved are key strokes you can use in the future. I must admit that I "tricked" COMSOL Multiphysics into accepting my new geometry, mesh, and xmesh into retaining my sub-domain settings by exporting and importing fem structures to Matlab and swapping around. However, the boundary settings are locked this way, and unlocking them required monkeying around with the equation settings (boundary equations) under the Physics menu. You might have a more straight forward exercise in just following the steps in Table 2.

The geometry may be formed from a composite of three rectangular solids rotated to the correct relative positions and orientations. Here, the alternative approach is used in which the line command to draw segments approximating the exact geometry, coercing the closed curve to a solid, and then editing the location of the vertices to achieve the desired symmetry. The vertices of the rectangular "spokes" taxed our recall of high school analytic geometry. Once the vertices of the equilateral triangle are placed at $\{(1, -0.5), (1, 0.5), (0.133977, 0)\}$, the six vertices of the three rectangular sections are determined from requiring that the slopes be appropriate — $\{+1, -1, 0\}$, The last is the easiest, with vertices at $(3, -0.5)$ and $(3, 0.5)$ for a channel of length two. All other channels should have the same length at least initially. So, for example, the lowest vertex is found by satisfying the distance formula and the slope constraint simultaneously:

$$\sqrt{(x_1 + 1)^2 + (y_1 + 1)^2} = 2\,,$$
$$\frac{y_1 + 0.5}{x_1 - 1} = 1\,. \tag{8}$$

There are two solutions to this quadratic system. The one we are after is $(x_1, y_1) = (-0.414214, 1.91421)$. We coded this system of equations in MATLAB since the algebra is straightforward, though tedious.

Before solving the time evolution, we are going to try a new trick to improve convergence. The biggest problem that was making the previous simulation slow was the rapid change in the velocity and electric fields required in the first few instants from the initial condition (no field, no flow) to practically pseudosteady flow and field. The viscous and dielectric response times are much faster than the diffusive and convective time scales, so this problem is inherently stiff. In the computational modeling of the

Table 2. Basic model of electrokinetic flow in a Y-junctions.   Filename: EK_Yjunction.mph.

Model Navigator	Open EKchannel.mph
Draw Menu	Erase previous geometry. Specify objects: line x: 1 3 3 1 −0.414214 −1.28024 0.133977 −1.28024 −0.414214 1 y: −0.5 −0.5 0.5 0.5 1.91421 1.41421 0 −1.41421 −1.91421 −0.5 OK Select Coerce to|solid from the pull down menu
Options Menu| Constants	$Pec = 30$; $Re = 0.03$; zel $= 1$; betael $= 1$; zetar $= 1$; sigr $= 1$; phi1 $= 5$; phi2 $= 5$

	Mode	Γ	F	da
Physics Menu: Subdomain settings (select subdomain 1 and use Multiphysics menu to switch modes)	ns	ρg 1; ηg 1$/Re$; $F_x = 0$; $F_y = 0$		
	species	$-(1/Pec)*Yx -$ betael*zel*Y*phi$x$ $-(1/Pec)*Yy -$ betael*zel*Y*phi$y$	$-u*Yx-v*$ Yy	1
	Init tab: $Y(t_0) = 1$			
	potential	$-$ sig*phi$x$ $-$ sig*phi$y$	0	0

Options Menu| Expressions| Global	Name: zeta Definition: $-(Y + $ zetar*$(1 − Y))$ Name: sig Definition: $Y + $ sigr*$(1 − Y)$

Physics Menu: Boundary settings	ns mode

bnd 1, 3	bnd 2, 4, 5, 6, 7, 8	bnd 4
Outflow/ pressure $p = 0$	$u = $ zeta*phi$x$ $v = $ zeta*phi$y$	Outflow/ pressure $p = 0$

species mode

bnd 1	bnd 3	bnd 2, 4, 5, 6, 7, 8	bnd 9
Dirichlet $G = 0$; $R = $ $1 - Y$	Dirichlet $G = 0$; $R = -Y$	Neumann $G = 0$	Neumann $G = $ betael* zel*Y*phi$x$

potential mode

bnd 1	bnd 3	bnd 2	bnd 4
Dirichlet $G = 0$; $R = $ phi1 − phi	Dirichlet $G = 0$; $R = $ phi3 − phi	Neumann $G = 0$	Dirichlet $G = 0$; $R = -$ phi

Mesh	Select the standard mesh and then refine once
Solve Menu	Set the time dependent solver Click on the solve (=) tool on the toolbar

Navier-Stokes equations for *incompressible* flow, this problem is encountered for the pressure. Since in the incompressible approximation, sound speed is infinite, the pressure field adjusts instantaneously to changes in velocity. Computationally, time stepping over such widely different time scales leads to problems with stiffness and slow convergence, requiring miniscule time steps. The SIMPLE algorithm (Patankar [5]) circumvented this pitfall by staging the time stepping of the velocity with rapid solution to the pressure field consistent with mass conservation and the current velocity field by solving a separate elliptic equation for the pressure — the Lighthill Poisson equation. The difference is that the corrections to the last pressure field are not found — small changes on the order of the time step — but rather the pressure can be wholly different from that at the previous time step. Instantaneous changes in the pressure that depend on the field everywhere in the domain are thus catered for, and the Navier-Stokes simulation is no longer stiff. The COMSOL Multiphysics Navier-Stokes application mode has this fast Poisson solver for pressure built in. But our electrostatic potential mode does not. So even though in principle ϕ should change instantaneously to changes in applied voltages, which should change the slip velocity virtually instantaneously, the ns mode will respond on the incompressible time scale, but the electric field needs to be relaxed. So to mimic the SIMPLE algorithm, we need to implement a fast elliptic solver for flow and electric field while freezing the concentration profile. Once the velocity field and electric field have been established, we can release the mass transport. Our no electrokinetic relaxation time in the potential mode is necessary for the model to advance steadily with only small changes to u, v, p and phi at each time step. The fast elliptic step overcomes the rapid relaxation time needed for abrupt changes in the electric field.

Staging the solutions to the potential mode and ns mode, since the basic model has no back action from the species concentration Y, is just a computational labor saving device at worst. Why compute these four variables at each time step in progressing Y if the work is redundant? By now, you must be familiar with the drill of using the Solver Manager to select the variables to be solved for and those for output, and using the restart solver to implement the staging process. The resulting complete solution looks like Figure 5 after all three stages. Note that species electrophoresis (caused by different mobilities) would lead to an inherent back coupling which would invalidate our staging procedure.

Animate and watch the streams merge and flow out. We can see that the final velocity and potential fields are unchanged from those set up in the

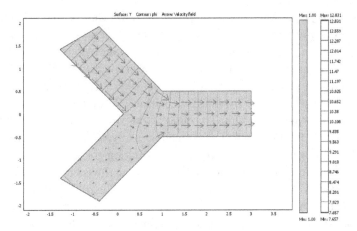

Figure 5. Stationary solution for flow field and potential lines from the initial conditions ($Y = 1$ everywhere) and boundary potentials imposed, such that the pseudosteady flow and electric fields are achieved instantaneously after imposition of the boundary potentials. The species transport can then be integrated with a frozen flow profile.

initial "fast elliptic solver" step. This would not be true if the conductive properties of the fluid, the zeta potential at the wall boundaries or viscosity were concentration dependent. The local concentration profile would then globally affect the potential field, thus modifying the velocity field, which in turn disperses the fluid differently. Such highly coupled electrokinetic flows are well suited to numerical analysis by a modest change to this coding — adding the concentration coupling to the conductivity according to a mixture rule. Zimmerman and Homsy treat the mixture rule for concentration dependency of viscosity, for instance, in the instability of viscous fingering in porous media [3].

2.5. *Y-junction switcher: An application of linkages through coupling variables*

As described in relation to Figure 4, by including relatively long channels to the reservoirs of the component species where electric potentials are applied, the flow and the electric potential can be described by algebraic relations for the inlet and outlet dependencies. We implement these physical linkages between nodes a and A, for instance, using the appropriately named coupling variables. Originally, we set up a pseudo 0-D geometry (geom 2 as in Chapters four and seven), but on advice from COMSOL consultants, realized that the conceptual 0-D domain was not needed — it

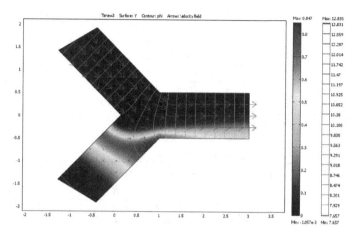

Figure 6. $t = 3$ solution for the species concentration Y showing merging of streams with $Y = 0$ and $Y = 1$.

can all be done in coupling variables. A pair of algebraic equations for I_i and Q_i are to be solved simultaneously. Fortunately, the equations can be solved by back substitution since the matrices are triangular — I_i can be found directly and Q_i after substitution. If these algebraic systems were fully coupled, then the linear equations could be solved symbolically and the expressions would be far more complicated.

Using the model MPH-file of the last section, set up the following twelve coupling variables, you can set up these network linkages by following Table 3. It should be noted that the extended multiphysics variables are of two types: boundary integration coupling variables and global expressions. In the first implementation of this model, all the extended multiphysics variables were coupling variables. The ease of specifying global destination of coupling variables makes it possible for half the work to be done as expressions. The additional expressions are all for quantities that enter the boundary conditions. Since these expressions could be directly entered in the boundary conditions, their usage is merely for convenience. The expression clarify that the computed variables are the already identified volumetric flow rates Qa, Qb, Qc and currents Ia, Ib, Ic. The flow of information in this model can be characterized as follows — (1) Assume that the Y-junction has inlet and outlet potentials ϕ_i and pressures p_i at the connections to the long channels where potentials and pressures are fixed at the other ends. Find these quantities and use them to solve for the required fluxes I_i and Q_i of current and velocity that are

Table 3.   Electrokinetic Y-junction attached to long channels (coupling variables). Filename: EK_Yjunction_merge.mph.

Model Navigator	Open EK_Yjunction.mph
Options Menu| Constants	$Pec = 30$; $Re = 0.03$; zel = betael = zeta1 = zetar = sigr = $f = 1$ PHIA = 24; PHIB = 12; PHIC = 0; PA = PB = PC = 0; dsa = 10; dsb = 10; dsc = 6;

Physics Menu: Boundary settings	ns mode			
	bnd 1	bnd 3	bnd 2, 4, 5, 6, 7, 8	bnd 4
	Outflow/ pressure $p = pb$	Outflow/ pressure $p = pa$	$u =$ zeta*phi$x$ $v =$ zeta*phi$y$	Outflow/ pressure $p = pc$

species mode: set bnd 3 to $R = 1 - Y$
potential mode

	bnd 1	bnd 3	bnd 2	bnd 4
	Dirichlet $G = 0$; $R =$ phib $-$ phi	Dirichlet $G = 0$; $R =$ phia $-$ phi	Neumann $G = 0$	Dirichlet $G = 0$; $R =$ phic $-$ phi

Options Menu| Integration coupling variables|Boundary variables	Select bnd 3. Enter name Qa. Expression: $nx*u + ny*v$ Select bnd 1. Enter name Qb. Expression: $nx*u + ny*v$ Select bnd 9. Enter name Qc. Expression: $nx*u + ny*v$ Select bnd 3. Enter name phia Expression: phi Select bnd 1. Enter name phia. Expression: phi Select bnd 9. Enter name phic. Expression: phi Keep global destination. OK
Options Menu| Expressions| Global	Add the following: Name: Ia. Expression: sigr*(PHIA $-$ phia)/dsa Name: Ib. Expression: sigr*(PHIB $-$ phib)/dsb Name: Ic. Expression: sigr*(PHIC $-$ phic)/dsc Name: pa Expression: PA $+ f$*dsa*(Qa $-$ zeta1*(PHIA $-$ phia)/ dsa)/Re Name: pb Expression: PB $+ f$*dsb*(Qb $-$ zeta1*(PHIB $-$ phib) /dsb)/Re Name: pc Expression: PC $+ f$*dsc*(Qc $-$ zeta1*(PHIC $-$ phic)/ dsc)/Re OK
Solve Menu	Set the time dependent solver Click on the solve (=) tool on the toolbar

consistent with (6). These fluxes are used as BCs for the Y-junction across the same inlet/outlet boundaries. This completes the linkage cycle and creates a self-consistent model to be solved simultaneously for internal variables and external linkages.

Replacing the complete equations with the zero-dimensional equations relating current and material flow rates to pressure difference and potential difference across a channel segment is an excellent approximation if

(1) the channel is long compared to its width
(2) channel section shape is uniform
(3) liquid properties are uniform.

Under these restrictions the potential and pressure will be uniform at any section along the channel. Also, since conductivity is uniform current flux will be uniform over any channel section (and indeed anywhere in the channel segment). However, velocity will only be uniform when the pressure difference across the channel is zero. In general, and perhaps often in the anomalous cases of concern when making such a computational analysis, this will not be the case. Thus, it is important that the domain boundaries are such that the inlets and outlets are in developed flow regions of the connecting channels. Then, from the above, pressure (not velocity) and current flux (or potential) found from the coupling equations can be imposed uniformly at the boundaries with excellent approximation.

There is an underlying assumption in the above formulation. Use of a uniform velocity at each flow boundary is only possible if pressure gradient can be neglected. In "pure" electrokinetic flow, that is where conductivity, zeta potential, viscosity are each uniform, the approximation of uniform velocity at the flow boundaries is excellent. The total pressure in each reservoir must also be the same (Cummings *et al.* [6]). However, when liquid properties are not uniform or a differential of dynamic pressure between reservoirs is present, pressure gradients arise within the network and the assumption implicit that velocity is uniform at each boundary is not appropriate.

From the above, there are two possible general modelling treatments:

(1) To to determine I and Q from the flow boundaries and use relation (6) for the uniform pressure p and potential ϕ at the boundary. That pressure or potential are uniform at each boundary follows rigorously when the boundary is at a position where the flow is

developed, that is sufficiently far (say, a channel width) from a disturbance region such as a junction.

(2) To to determine ϕ and Q from the flow boundaries and use relation (6) for the uniform pressure p and current I at the boundary.

Here, we coded linkages of type (2). We did code the other type of linkages (type 1) which have boundary conditions that are Dirichlet-based or expressing the potentials and pressures across the boundary faces, rather than the flux based boundary conditions based here. The potential/pressure based methodology failed miserably (or was that we were miserable that it failed?) The explanation that we offer is that even though we did not specify any weak boundary constraints, COMSOL Multiphysics automatically detected them for us, so we are bound by the limitations of weak boundary constraints. There are two limitations that may apply here ([1], p. 492):

- Pointwise and weak constraints on the same set of variables on adjacent boundaries do not work. This means that if you must constrain all boundaries on a solid and want to use a weak constraint on one boundary segment, you must use the weak constraint on the entire boundary of the solid, if the boundary is connected.
- Discontinuous constraints result in (theoretically) infinite Lagrange multipliers. In practice, you get large oscillations.

2.6. *Examples preparing the microfluidic Y-junction switcher*

Now we are prepared for implementing the microfluidic Y-junction switcher in two steps:

A. Merging two input streams

All as before, but uniform properties ($\zeta_r = \sigma_r = 1$) with an uncharged species ($z = 0$). $\Delta s_a = \Delta s_b = 10$, $\Delta s_c = 6$ and $p_A = p_B = p_C$. Set the voltages to $\phi_A = 24$, $\phi_B = 12$, $\phi_C = 0$.

B. Alternation between voltages

$\phi_A = 24$, $\phi_B = 12$, $\phi_C = 0$ and $\phi_A = 12$, $\phi_B = 24$, $\phi_C = 0$ with a period of around 6 time units. This is a square wave signal for A and a complementary one for B.

Table 4. Boundary fluxes across the three open boundaries.

	$\bar{I} = \int \hat{n} \cdot \nabla\phi ds$	$\bar{\phi} = \int \phi ds$	$\bar{u} = \int u ds$	$\bar{v} = \int v ds$	$\bar{p} = \int p ds$
bnd 3(a)	1.1184	12.816	0.83909	−0.8391	19.537
bnd 1(b)	0.13518	10.649	0.10208	0.1021	2.6554
bnd 9(c)	−1.2535	7.5213	1.2499	0	−0.64633
Σ	0.00008				

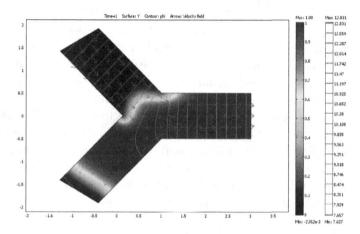

Figure 7. Developed flow of species $Y = 0$ along upper leg with inhibited flow of species $Y = 1$ in the lower leg for $t = 1$. Hardly anything changes from $t = 3$ onwards within the domain. The concentration profile is pseudosteady.

A. Merging two input streams

We are all set up for merging two input streams. The constant potentials are set for nodes A, B, and C according to the values of three constants PHIA, PHIB, PHIC. Use Add/Edit Constants to introduce these three values as in Table 3.

As before, we need to stage our solution to set up the pseudosteady velocity and potential fields initially, then turn on the species transport.

The result is shown in Figure 7. The fully developed flow of species $Y = 0$ along upper leg with inhibited flow of species $Y = 1$ in the lower leg for $t = 1$. Hardly anything changes from time $t = 3$ onwards within the domain. The concentration profile is near its steady distribution. It is prudent to check the consistency of the calculation of the velocity and potential solutions. Using Boundary Integration under Post Menu, we find the values in Table 4.

The conservation of charge is satisfied to 10^{-4}. The conservation of mass does not hold so well. You can verify that $\sqrt{2} \times (0.83909 + 0.10201) = 1.33 \neq 1.25$. This discrepancy suggests that the velocity flow field is not spatially well resolved at this level of meshing. To improve the result, it is likely that greater mesh density is required in the "Y" vertex which clearly has discontinuity in velocity from the upper leg to the lower leg.

The algebraic relations hold roughly for the electric potential, pressure, current, and velocity, for instance: $u_a \approx \frac{I_c}{\sqrt{2}} = 0.790828$. Some are spot on $-\frac{0 - \phi_c}{\Delta s_C} = \frac{0 - 7.5213}{6} = -1.2535$. In general, the electric potential conditions hold exceedingly well, but the velocity field/pressure fields do not hold as well. Again, this is a strong indicator that the solution is not mesh resolved.

Plot (see Figure 7) and Animate the solution. The animation shows clearly how the front evolves to set up a fully developed upper flow, bypassing the lower leg. The intermediate voltage in the "b-leg" of the Y-junction tends to hold back the flow of species $Y = 1$ from the merging into the main flow along the (a)–(c) open switch — only slow diffusion out and in, along with some modest convection, occur. With concentration dependent conductivity (and viscosity [2]), it is possible to counteract the diffusion to some extent which sharpens some fronts.

B. *Alternation between voltages*

Now for this application to be a Y-junction switch, we need to be able to replace the constant voltages ϕ_a and ϕ_b with $\phi_a(t)$ and $\phi_b(t)$, respectively. The signals could be arbitrary, however, in practice they are discontinuous level adjustments, which can be idealized as a square wave. A suitable choice is the alternation between values ϕ_{a0} and ϕ_{b0} (PHIA and PHIB), with the signals 90° out of phase. We coded the square wave in coupling variables in two different ways — (1) using logical functions to turn a sine wave into a square wave; (2) Using a truncation of the Fourier sine series for a square wave. Neither work particularly well since they both introduce very rapid time scales. Since the electrostatic model already introduces an infinitesimally small time scale, this model becomes very stiff with infinitely step switching of voltages. These attempts are chronicled in the predecessor book. Here, we present the best strategy we found — fast elliptic solution using the Solve For variables facility of the Solver Manager.

A *MATLAB wrapper for individual half periods*

Since we found that programming the square wave as signals in the coupling variables did not work, this section builds on what were learnt.

The smoother the signal, the greater likelihood that coding the time dependence succeeds in an efficient time integration. The discontinuity in the ideal square wave is the enemy of convergence. For our third attempt, we recognize that we have already used an excellent strategy to overcome the effects of the discontinuity in the initial conditions — staging the fast elliptic step without the species transport, and then turning on the transient solution with the species now mobile. We could simply implement the square wave by successively manually swapping the values of the constants PHIA and PHIB, restarting the stationary nonlinear solver to find the fast elliptic switch of the flow and potential fields, then let the species transport continue under the new flow conditions by restarting with this initial condition and the transient solver. The MATLAB code we wrote merely puts a loop around this operation to continue as long as specified. Since we have put loops around a number of COMSOL Multiphysics model m-files generated from the GUI, this is not a new technique. However, the coding warrants a look for a crafty way of patching together the array of solution vectors (fem.sol.u) and the array of solution times (fem.sol.tlist) so that the COMSOL Multiphysics functions can be used to display the results. Using the manual technique in the COMSOL Multiphysics GUI described above, the user would only have the most recent half-period available for animation. The m-file script outline below should be run in MATLAB, and thereafter the fem structure can be animated by using COMSOL Multiphysics commands, for instance. Using the femsol() command does allow importing the concatenated fem.sol structure into the COMSOL Multiphysics GUI.

```
load stagedY.mat fem
%%%%%%%%%%%%%%%%%%%%%%%WBJZ contants %%%%%%%%%%%%%%%%%%
tau=3; cycles=1; phia=24; phib=12;
% Constants
fem.const={...
    'Pec',    30,    'Re',    0.03,    'zel',    1,    'betael', 1,...
    'zetal',  1,     'zetar', 1,       'sigr',   1,    'f',      1,...
    'PHIA',   phia,  'PHIB',  phib,    'PHIC',   0,    'PA',     0,...
    'PB',     0,     'PC',    0,       'dsa',    10,   'dsb',    10,...
    'dsc',    6};
% Solve fast elliptic problem
fem=multiphysics(fem);
fem.xmesh=meshextend(fem);
fem.sol=femnlin(fem, ...
'nullfun','flnullorth', ...
'solcomp',{'lmv','u','phi','p','lmu','v'}, ...
'outcomp',{'lmv','u','phi','Y','p','lmu','v'}, ...
```

```
'ntol',1E-6,'report','off');
% Solve the time stepping problem
fem.sol=femtime(fem, ...
'init',fem.sol, 'nullfun','flnullorth', ...
'solcomp',{'Y'}, ...
'outcomp',{'lmv','phi','u','Y','p','lmu','v'},...
'tlist',[0:0.1:tau], ...
'atol',{'u','0.01','v','0.01','p','Inf','Y','0.001',...
'phi','0.001','lmu','0.001','lmv','0.001'},
... 'tout', 'tlist','report','off');
%%%%%%%%%%%%%%%%%%%%%%%%WBJZ%%%%%%%%%%%%%%%%%%%%%%%%%%%%%%%%%%%%%%%%%%
fem0=fem; fem1=fem;

for k=1:2*cycles-1 swap=phia; phia=phib; phib=swap; fem.const={...
     'Pec',      30,   'Re',   0.03,  'zel',    1,   'betael', 1,...
     'zetal',    1,    'zetar',  1,   'sigr',   1,   'f',      1,...
     'PHIA',   phia,  'PHIB',  phib, 'PHIC',   0,   'PA',     0,...
     'PB',      0,    'PC',     0,   'dsa',   10,   'dsb',   10,...
     'dsc',     6};
fem.sol=femnlin(fem, ...
'nullfun','flnullorth', ...
'solcomp',{'lmv','u','phi','p','lmu','v'}, ...
'outcomp',{'lmv','u','phi','Y','p','lmu','v'}, ...
'ntol',1E-6,'report','off');
fem.sol=femtime(fem, ...
'init',fem.sol, 'nullfun','flnullorth', ...
'solcomp',{'Y'}, ...
'outcomp',{'lmv','phi','u','Y','p','lmu','v'}, ...
'tlist',[(k*tau):0.1:((k+1)*tau)], ...
'atol',{'u','0.01','v','0.01','p','Inf','Y','0.001',...
'phi','0.001','lmu','0.001','lmv','0.001'}, ...
'tout','tlist','report','off');
u=[fem1.sol.u, fem.sol.u];
ut=[fem1.sol.ut, fem.sol.ut];
tlist=[fem1.sol.tlist, fem.sol.tlist];
fem.sol=femsol({u,ut},'tlist',tlist);
fem0=fem; fem1=fem;
clear u ut tlist;
end save Ystaging.mat fem
```

Note that the variables generated include lmu and lmv. This is the syntax for the Lagrange multipliers which are the degrees of freedom of the weak boundary constraint automatically detected and generated.

The model m-file commands are found from running the first half period with the initial fast elliptic step without species, then a time dependent solution with species transport, then swapping the phia and phib values, then repeating the fast elliptic step and time dependent restarts. The loop is then placed around the second set of solutions. The final part is to

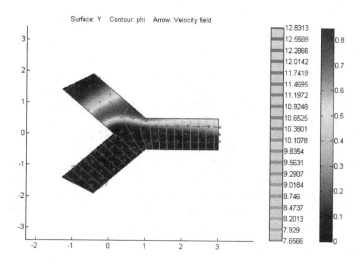

Figure 8. Lower flow pattern in the second half-period.

append the fem.sol structure with the current set of solution vectors and tlist. Now run the animation to appreciate the speed and electrokinetic switching in action. Figure 8 shows the configuration in the "lower" pseudosteady state (second half period). The striking feature of the animation is how reproducible each cycle is — diffusion does not smooth anything out cumulatively.

3. Summary

This chapter explored the multiphysics modeling appropriate for electrokinetic flow in microchannel networks. In setting up our case study, we discovered how to use COMSOL Multiphysics automatically detects and implements weak boundary constraints for coupled boundary conditions that incorporate tangential derivatives. Our model demonstrates the ability of COMSOL Multiphysics to treat general network models by extended multiphysics through coupling variables.

Acknowledgments

We are indebted to Johan Sundqvist and Niklas Rom of COMSOL for consultation on the handling of weak boundary constraints in this chapter.

References

[1] COMSOL Multiphysics User's Guide v. 3.2 (2005), p. 491ff.

[2] J. M. MacInnes, Computation of reacting electrokinetic flow in microchannel geometries, *Chem. Eng. Sci.* **57**(21) (2002) 4539–4558.

[3] W. B. Zimmerman and G. M. Homsy, Nonlinear viscous fingering in miscible displacement with anisotropic dispersion, *Phys. Fluids A* **3**(8) (1991) 1859.

[4] J. M. MacInnes, X. Du and R. W. K. Allen, Prediction of electrokinetic and pressure flow in a microchannel T-junction, *Phys. Fluids* (2003), in press.

[5] S. V. Patankar, *Numerical Heat Transfer and Fluid Flow* (Hemisphere Publishing Corporation, New York, 1980).

[6] E. B. Cummings, S. K. Griffiths, R. H. Nilson and P. H. Paul, Conditions for similitude between the fluid velocity and electric field in electroosmotic flow, *Analytical Chemistry* **72**(11) (2000) 2526–2532.

[7] J. M. MacInnes, Modelling pressure and electrokinetic flow in complex microfluidic devices, *Microfluidics: History, Theory, and Applications*, CISM Lecture Series, Vol. 466, ed. W. B. J. Zimmerman (Springer-Verlag-Wien, 2006).

Chapter Twelve

PLASMA SIMULATIONS VIA
THE FOKKER-PLANCK EQUATION

ALI SHAJII and DANIEL SMITH

MKS Instruments, 90 Industrial Way, Wilmington, MA, USA, 01887

The area of plasma simulations has historically posed a great challenge to the computational community. The simulations have been done at a number of different approximation levels. These include a wide range of attempts to solve the full kinetic problem, all the way to fluid-like approximations and the related drift-diffusion equations. In recent years, the Fokker-Plank treatment of the electrons in the plasma, coupled with the fluid treatment of the ions has shown very promising results. In this chapter we demonstrate the solution of the Fokker-Planck equation in general, followed by an example of the use of this equation in relation to obtaining the electron density profile in a capacitive-like discharge. The hope is that the simplified model motivates the reader to follow a similar process in modeling a full plasma discharge.

1. Introduction

In a number of different industries plasma processing constitutes a critical step in the overall "process" cycle [2][3]. Historically, plasma modeling has been done in a number of different approximations, and associated levels of accuracy in capturing the relevant physics of the problem [3][5][7]. The various levels include:

- Full kinetic models (i.e. Boltzmann equation for multi-species) [4],
- Particle simulations using Monte-Carlo techniques [3],
- The Fokker-Planck approximation [1][7],
- Multi-fluid models (and the so called drift-diffusion variation of such techniques) [3].

Plasma modeling and simulation is a notoriously difficult endeavor for a number of reasons. For one, the most straight forward models which involve using the multi-fluid equations do not capture many of the relevant plasma physics. Furthermore, the "hydrodynamic" coefficients are highly dependent on the specific problem under consideration and are not as easily measured as those of pure gases and fluids [6]. On the other extreme, the full kinetic models involving the Boltzmann equation are computationally very expensive and difficult to set up.

An intermediate ground between the full kinetic and fluid models is that of either using the Fokker-Planck (FP) approximation or Monte Carlo (MC) particle simulations. These options allow for a good balance between the required computational complexity and capturing the most important physical details of the plasma.

The main goal of this chapter is to show the overall capability of COMSOL Multiphysics in solving the FP equation. The main emphasis is on a simple example, in order to give an overall view of the problem at hand. Specifically, in Section 2 we give a simple but intuitive description of the FP and how it relates to particle simulation via Brownian motion. This leads to the main contribution of the chapter, which is the modeling of electron dynamics in an external electric field. Finally, in Section 4 we present a detailed discussion of the COMSOL Multiphysics implementation of the model.

2. 1D FP and the Langevin equation

The most straight forward derivation of the FP is related to the so called Langevin equation [1]. Consider a "Brownian" particle in a bath of liquid. Such a particle, if small enough in size, will experience two effective types of forces. One such a force is due to the friction between the particle and the fluid medium, which will tend to slow down the mean particle speed. The second force is due to the random collision of the particle with the fluid "molecules." The motion of the Brownian particle is thus governed by [1]:

$$\dot{v} = -\gamma v + \Gamma(t), \tag{1}$$

where γ is the damping coefficient, and the stoichastic term $\Gamma(t)$ represents the continuous collision of the particle with those of the background fluid. In this simple example, as is commonly the case, we have assumed the friction force is linearly dependent on the particle velocity. Also, in accord to the stochastic approximation, the Langevin force satisfies the following [1]:

$$\langle \Gamma(t) \rangle = 0,$$

$$\langle \Gamma(t)\Gamma(t') \rangle = q\delta(t - t'),$$

where $q = 2\gamma kT/m$ [1], and $\langle \rangle$ denotes the ensemble averaging. Here, T is the fluid temperature, k is the Boltzmann constant, and m is the mass of the brownian particle. Also, note that the force Γ can easily be implemented using the MATLAB command `randn` (see Section 4 below). A MC technique is easily implemented to solve (1) for a suitable number of particles and initial conditions. This number may be in excess of 1 million in order to obtain an accurate statistical measure of the velocity distribution of the particles at time t. The simplest initial condition is $v(t = 0) = v_0$.

In summary, the Langevin equation describes the motion of a Brownian particle within a background fluid medium. Without the stochastic force, of course, the path of the particles could be trivially computed as $v = v_0 e^{-\gamma t}$. However, due to Γ, the result of solving Eq. (1) for many particles, is a normally distributed curve with the width of this distribution determined by q [1].

The FP equation provides an alternative to the MC technique. The Langevin equation, with the stochastic force, can be shown to be equivalent to the following PDE:

$$\frac{\partial w}{\partial t} = \frac{\partial}{\partial v}\left(\frac{q}{2}\frac{\partial w}{\partial v}\right) + \gamma\left(w + v\frac{\partial w}{\partial v}\right), \tag{2}$$

with initial and boundary conditions

$$w(t = 0) = \delta(v - v_0) \qquad w(v \to \pm\infty) = 0, \tag{3}$$

where w is the probability distribution of the particles in velocity space, that is, the probability of finding a particle, in an infinite ensemble, with a velocity v is given by wdv. A simple implementation of initial condition (3) is shown in Figure 1.

Solving this PDE is equivalent to running a Monte-Carlo simulation for an infinite number of particles, in obtaining the probability distribution function at any $t > 0$. Figure 2 shows a comparison of the two methods. The MC implementation involved solving 20000 ODEs with initial condition $v_0 = 0$.

Even though, in this simple example the computational time for solving the PDE is the same as that of the ODE, such is not the case for highly nonlinear damping forces and particle motion in 3-D space. When the system of ODE's for the particle tracing is nonlinear or an external force value needs to be queried from a separate FEA simulation, the computational time for the MC method can become prohibitedly expensive.

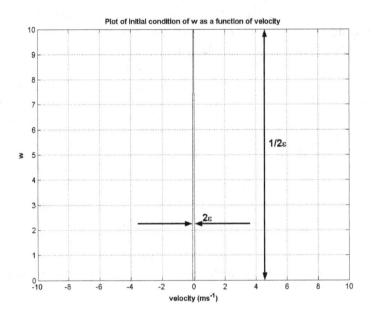

Figure 1.   Plot of initial condition for w.

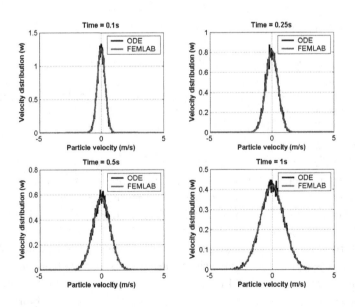

Figure 2.   Plot of Fokker-Planck versus Monte Carlo at various timesteps.

As an example, consider the modified Langevin equation:

$$\dot{v} = -\gamma v^3 + \Gamma(t)\,,\tag{4}$$

and the associated FP equation [1]:

$$\frac{\partial w}{\partial t} = \frac{\partial}{\partial v}\left(D_2 \frac{\partial w}{\partial v}\right) + \gamma\left(w\frac{\partial D_1}{\partial v} + D_1 \frac{\partial w}{\partial v}\right)\,,\tag{5}$$

where $D_1 = -\gamma v^3$ and $D_2 = q/2$. The distribution function for this case is plotted in Figure 3. Even with 1 million particles, the probability distribution from the ODE's is numerically noisy, whereas FP results are smooth. The statistical noise shown here is a classic problem with MC simulations with finite number of particles (i.e. statistical post-processing of the data is a delicate and critical step).

Next, with the material in this section as a general motivating background, we proceed to a more complex analysis using the FP equation.

3. Electron dynamics in an external field

Here, we consider the dynamics of an electron stream in an external electric field generated by a set of capacitive plates. One plate is held at a constant

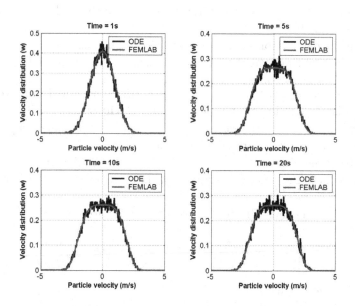

Figure 3. Plot of Fokker-Planck versus Monte Carlo at various timesteps for nonlinear Langevin equation.

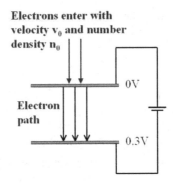

Figure 4. Diagram of Simulation Domain.

voltage, while the other side is grounded. The electrons are assumed to be introduced from the grounded plate with a specified velocity and are accelerated towards the annode. Furthermore, the region between the plates is taken to be filled with a stagnant fluid of temperature T. Figure 4 depicts the geometry under consideration.

The main assumptions of the model are:

- Ion generation due to electron collision with the background media is negligible,
- Electron-electron interactions are also negligible,
- The problem is treated in only one spatial dimension (1-D space).

A FP equation must now be solved in order to accurately resolve the electron distribution in both physical (y) and velocity space (v_y). In this example we also calculate the electron density, electron flux and electrostatic field between the set of parallel plate capacitors.

Since in general, the FP equation requires 1 more spatial dimension than the electrostatic problem, we require solving a 1-D electrostatics problem and a 2-D FP equation.

The governing equations for the electron distribution are

$$-\frac{\gamma kT}{m}\frac{\partial^2 w}{\partial v_y^2} = v_y - \frac{\partial w}{\partial y} - \left(\frac{F_y}{m}\right)\frac{\partial w}{\partial v_y} + \gamma\left(w + v_y\frac{\partial w}{\partial v_y}\right), \qquad (6)$$

and for the electric field

$$\nabla \cdot \mathbf{E} = \frac{\rho}{\epsilon_0}, \qquad (7)$$

which in terms of a scalar potential and electron density becomes

$$\frac{\partial^2 V}{\partial y^2} = \frac{e n_e(y)}{\epsilon_0},$$ (8)

where

$$n_e(y) = \int_{-\infty}^{\infty} w \, dv_y,$$ (9)

and

$$F_y(y) = e \frac{\partial V}{\partial y}.$$ (10)

Note that both n_e and F_y are functions of y. The boundary and initial conditions for the problem are:

$$w(y = 0, v_y) = \delta(v_y - v_0),$$ (11)

$$w(y, v_y = \pm\infty) = 0,$$ (12)

$$V(y = 0) = 0,$$ (13)

$$V(y = 1) = 0.3,$$ (14)

where m is the electron mass, k is Boltzmann's constant, T is the background fluid temperature, e is the electron charge, ϵ_0 is the permittivity of free space, and as before γ is the damping constant. Also, here the probability of finding an electron in an infinite ensemble with velocity v_y and at location y is $w \, dv \, dy$.

Equations (6) and (8) are two coupled PDE's of different spacial dimensions. Because COMSOL Multiphysics allows an arbitrary number of geometries of different spacial dimensions to be coupled together, we can now solve the equations simultaneously. The electron density is introduced into the electrostatic problem by way of a projection coupling variable, and the force on the particle is introduced into the FP problem by way of an extrusion coupling variable.

To summarize, the electron stream enters the "capacitive discharge" area with velocity v_0, and then subjected to three types of forces: Γ, $-\gamma v$, and the ∇V. These forces cause the electron velocity and density profiles to change along the y-direction, hence modifying the electric potential

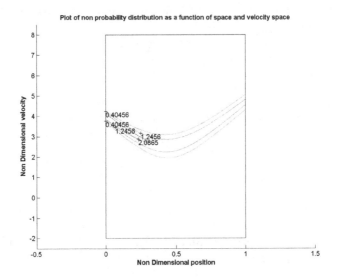

Figure 5.   Probability distribution.

V across the plates, as compared to the simple linear profile that would otherwise exist without the presence of the electrons.

Figure 5 shows the solution for w as a function of v_y and y. Note the initial slowing down of the electrons as they enter the computational domain. This nonlinear, nonmonotonic curve is due to the effective shielding of the cathode by the electrons as shown in Figure 6. Observe the peak electron density near the center of the domain. Figure 8 presents the electric potential across the capacitor which has a minimum at the location of the maximum of n_e.

Finally, the electron flux is given by:

$$\Phi = \int_{-\infty}^{\infty} w v_y dv_y \, ,$$

which due to mass flux conservation should remain constant in the y-direction. Figure 7 plots the electron flux as a function of y confirming the conservation of the mass flux in the simulation.

In the next section we present a detailed description of the COMSOL implementation of the model. Before proceeding, however, it is important to note that event though the above model does not constitute a full plasma simulation, it does bring out many of the important required features needed for such a simulation, that is, a more complete model would

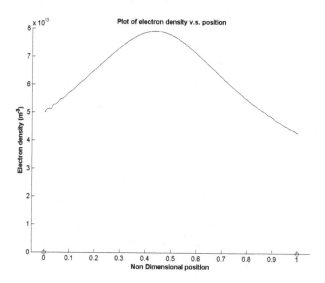

Figure 6.   Electron density.

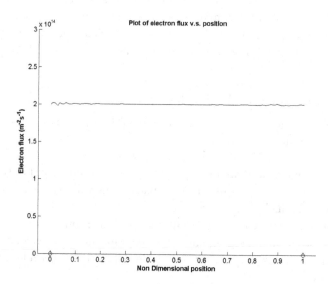

Figure 7.   Average velocity.

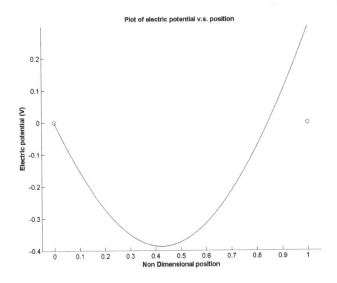

Figure 8.   Plot of electric potential versus position between parallel plates.

still constitute the FP equation, but now coupling would have to be made to the ion dynamics. Also, in many cases of interest the FP equation would have to be solved in 2-D space. The velocity dependence of the FP equation would then be handled in a very similar way as demonstrated here [7].

4. COMSOL Multiphysics implementation

This model uses some advanced features of COMSOL Multiphysics. Immediately obvious problems are:

- The topological domain in velocity space is infinite and only of the order of centimeters in physical space. This will lead to a geometry with extremely high aspect ratio.
- The two PDE's are of different spatial dimensions.
- A series of line integrals must be taken for all y over velocity space and then this value must be made available to the electrostatic problem.
- Similarly, the force resulting from the electrostatic problem must be made available to the Fokker-Planck domain at any point in physical and velocity space.

- There is strong coupling between the two physics fields, in that the electron density is a strong function of the external applied force and the external applied force is a strong function of the electron density.

Each of the above points presents a potential problem, but COMSOL Multiphysics has the tools to overcome them all. Bullet 1 is overcome by scaling the PDE's (6) and (8). This results in a geometry with a low aspect ratio which in turn results in a good mesh quality. The scaled form of the PDE's are:

$$-\frac{\partial^2 w}{\partial v_y^2} = \left(\frac{v_T}{v_C}\right) - v_y \frac{\partial w}{\partial y} \alpha \left(\frac{dV}{dy}\right) - \frac{\partial w}{\partial v_y} + w + v_y \frac{\partial w}{\partial - v_y}, \quad (15)$$

and

$$\nabla^2 V = -\beta \int_{-\infty}^{\infty} w \, dv_y, \quad (16)$$

where

$$v_T = \sqrt{\frac{kT}{m}},$$

$$v_C = \gamma d,$$

$$\alpha = \frac{eV_0}{mv_T \gamma d},$$

$$\beta = \frac{ed^2 n_0}{V_0 \epsilon_0},$$

and m is the mass of an electron, k is Boltzmann's constant, T is the background temperature, e is the charge on an electron, ϵ_0 is the permittivity of free space, d is the distance between the parallel plates, V_0 is the voltage at the lower plate, γ is the damping constant, and n_0 is the electron density at $y = 0$.

Using the model navigator, we create two geometries, Geom 1 of spacial dimension 2, with independent variables y, v_y and z and a General Form PDE with dependent variable w. Add a second geometry, Geom 2 with independent variables x, y and z and a General Form PDE with dependent variable V. In Geom 1, draw a rectangle from $y = 0$ to $y = 1$ and $v_y = -2$ to $v_y = 8$. Define the constants

Table 1.   Table of constants.

Name	Expression
m	9.1e−31
k	1.38e−23
T	300
e	1.6e−19
e_0	8.84e−12
d	0.001
V_0	0.3
gamma	1/1e−7
n_0	3e13
v_T	sqrt $(k*T/m)$
v_C	d*gamma
alpha	$e*V_0/(m*v_T$*gamma*$d)$
beta	$e*d*d*n_0/(V_0*e_0)$

Define the following scalar expressions in Geom 1,

Table 2.   Scalar Expressions, Geom 1.

Name	Expression
a	2
smooth	0.1
$vy0$	0.2
w_0	$(1/(2*vy0))$*(flc2hs$(v_y+vy0-a$, smooth)-flc2hs$(v_y-vy0-a$, smooth))

In subdomain settings set: and in boundary settings, set $w = 0$ at

Table 3.   Subdomain settings, Geom 1.

Γ	0	$-wvy$
F	$-(v_T/v_C)*v_y*w_y+w+v_y*wvy-$ alpha*phix_extr*wvy	
Init	w_0	

$v_y = -2$ and $v_y = 8$ and at $y = 0$ set $w = w_0$. At $y = 1$, specify a dirichlet boundary condition with both r and g equal to zero. In Geom 2 draw a line from $x = 0$ to $x = 1$ and enter

Table 4.   Subdomain settings, Geom 2.

Γ	$-Vx$
F	$-$ beta*ne
Init	V_0*x

For boundary conditions in Geom 1 use: In Geom 2, set $V = 0$ at $x = 0$

Table 5. Boundary settings, Geom 1.

Location	Type	Coefficients
$y = 0$	Dirichlet	$g = 0$, $r = w_0 - w$
$y = 1$	Dirichlet	$g = 0$, $r = 0$
$v_y = -1$	Dirichlet	$g = 0$, $r = -w$
$v_y = 10$	Dirichlet	$g = 0$, $r = -w$

and $V = V_0$ at $x = 1$. Now we need to couple the physics of the two geometries together. In Geom 1, define projection coupling variables

Table 6. Projection Coupling Variables, Geom 1.

Name	Expression
ne	w
flux	$v_y * w$

For Source Transformation, select General Transformation and specify x: y and y: v_y. In the destination tab, select Geometry: Geom 2, Level: Subdomain, Subdomain Selection: 1 and Destination Transformation x: x. In Geom 2, define an extrusion coupling variable as below. Select General Transformation and for Source Transformation x: x. For Destination, select Subdomain 1 in Geom 1 and specify the Destination transformation x: y. A fine mesh is needed at $y = 0$ in Geom 1 and also near $x = 0$ in Geom 2. Select the nonlinear solver and Solution Form: weak and click Solve.

Table 7. Extrusion Coupling Variables, Geom 2.

Name	Expression
phi_extr	V_x

5. Appendix — Weak solution form

The weak solution form must be used in this model. This is because there will be a substantial contribution to the Jacobian from the projection coupling variables. In order for COMSOL to take into account this coupling the Weak Solution Form must be used. Compare the two spy plots below (See Chapter 1). The first figure shows the sparsity of the Jacobian when

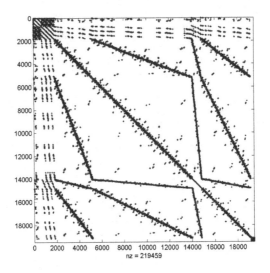

Figure 9. Sparsity plot of Jacobian using general solution form.

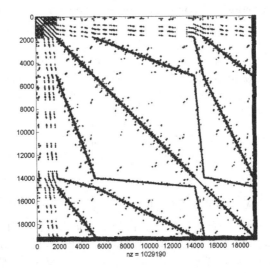

Figure 10. Sparsity plot of Jacobian using weak solution form.

the General form is used. The number of nonzeros is 219459. In the second figure, the Weak form is used and the number of nonzeros is 1092160. There are two strips of the matrix that are full corresponding to the projection coupling variable. In order to solve the model when the cross coupling between the two geometries is the dominant effect, the extra terms in the Jacobian are essential, otherwise Newton's method will not find a solution.

6. About the authors

Dr. Ali Shajii is currently the Director of Corporate Advanced Technology at MKS Instruments, a leading provider of materials delivery and reactive gas (plasma) equipment in the semiconductor industry. He received his PhD in Engineering Physics and Applied Math from MIT in 1994. His main areas of expertise are in dynamics of reacting and nonreacting flows, electrodynamics, and control theory of complex and highly nonlinear systems. He currently holds 18 US patents, and has over 30 journal publications in the aforementioned areas.

Daniel Smith is a staff scientist at MKS Instruments in the Corporate Advanced Technology group. He has a masters degree in applied mathematics from the University of St Andrews in 2002 and a Masters in Numerical computing from the University of Manchester. His main areas of expertise are viscous compressible flow, multiphase flow, computational electromagnetics and multiphysics problems.

References

[1] H. Risken, *The Fokker-Planck Equation* (Springer, 1996).
[2] J. R. Roth, *Industrial Plasma Engineering* (IOP Publishing, Ltd., 1995).
[3] M. A. Lieberman and A. J. Lichtenberg, *Principles of Plasma Discharges and Materials Processing* (Wiley, 1994).
[4] S. Harris, *An Introduction to the theory of the Boltzmann Equation* (Dover, 2004).
[5] S. Ichimaru, *Basic Principles of Plasma Physics — A Statistical Approach* (Addison-Wesley, 1980).
[6] J. P. Hansen and I. R. McDonald, *Theory of Simple Liquids* (Academic Press, 2003).
[7] V. I. Kolobov, Fokker-Planck modeling of electron kinetics in plasmas and semiconductors, *Computational Materials Science* **28** (2003) 302–320.

Chapter Thirteen

CREVICE CORROSION OF STEEL
UNDER A DISBONDED COATING

F. M. SONG and N. SRIDHAR

*Mechanical and Materials Engineering, Southwest Research Institute, 6220,
Culebra Road, San Antonio, Texas 78238, USA
E-mail: fsong@swri.org*

Crevice corrosion is one of the primary forms of corrosion that reduce the service life of buried pipelines. The corrosion mechanisms are not fully clear and
need to be understood further. Two coupled Laplace's equations were solved for a
two-dimensional (2-D) crevice geometry to simulate this corrosion. The nonlinear
boundary condition is a challenge in the numerical solution to this problem. FEM-
LAB, a finite element code, was used and the problem was solved successfully.
Projection and integration coupling variables and a trial method to search for a
potential at no cathodic protection (CP) condition were used. The computational
profiles of dissolved oxygen concentration, steel potential and current flow in the
crevice were calculated.

1. Background

1.1. *Motivation*

Of the roughly US\$276 billion reported as the cost of corrosion in the United
States in 2004, US\$7 billion was attributed to the corrosion of oil and gas
pipelines [1]. Buried in soil the pipelines are protected from corrosion by
both polymer coatings and CP. CP is an electrical means to prevent or
mitigate corrosion by lowering the electrochemical potential of the pipeline
steel in contact with soil ground water. Even though the pipelines are protected, the residual corrosion is still significant. Such corrosion is especially
significant in the crevices between the coating and steel where groundwater
can accumulate and CP may not be present. This work provides an example of studying the corrosion mechanisms through mathematical modeling.
More information about the modeling can be found elsewhere [2]–[5].

1.2. *Uniqueness and challenges of the modeling*

The challenges to solve the problem include the complex nonlinear boundary condition at the steel surface and the complex model geometry to simulate realistic coated pipeline corrosion. Projection and integration coupling variables and other techniques were used to solve the problem. Different mesh sizes were used at different boundaries to solve the problem more efficiently.

1.3. *Description of the model*

The model geometry in Figure 1 schematically depicts the crevice corrosion. It is a symmetrical half. The other half on the left to boundary B1 is not shown. In Figure 1, the holiday* mouth and the disbonded coating (B4) divide the geometry into two regions. The upper region represents soil solution and the lower region is the crevice where steel corrosion occurs. The boundaries (B5) are assumed to be where a reference electrode is located to measure the pipe steel potential. Along with and outside of this boundary (B5) the potential versus reference electrode

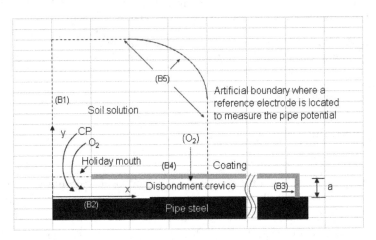

Figure 1.   Schematic diagram of the pipeline crevice corrosion geometry. B1 is symmetry boundary; B2 is steel surface, a flux boundary; B3 and B4 are impermeable coating, zero flux boundary; B5 is measurement boundary, uniform potential and uniform oxygen concentration.

---

*Holiday is a technical term meaning a defect in the coating and here it is assumed to be a rectangular crack.

Table 1. Geometric dimensions.

Parameter		Value
Half length of the holiday mouth		0.001 m
Gap between coating and the steel surface		0.0005 m
Crevice length		0.1 m
Height of upper domain		0.0995 m
Width of upper domain		0.01 m

and the oxygen concentration are assumed to be uniform. The coordinates used for the modeling, x and y, are labeled. The dimensions of the geometry are given in Table 1.

Dissolved oxygen in soil solution can enter into the crevice through both the holiday mouth and through the coating. Since oxygen has very low permeability through the coating, only the oxygen transport through the holiday mouth is considered. CP current enters into the crevice through the holiday mouth only.

At steady state, the model governing equations are mass conservation for dissolved oxygen in terms of oxygen concentration (c_{O_2}):

$$\frac{\partial^2 c_{O_2}}{\partial x^2} + \frac{\partial^2 c_{O_2}}{\partial y^2} = 0, \tag{1}$$

and, under the assumption of uniform ionic composition inside and outside the crevice, charge conservation in terms of steel potential versus a reference electrode anywhere in the solution (ψ):

$$\frac{\partial^2 \psi}{\partial x^2} + \frac{\partial^2 \psi}{\partial y^2} = 0. \tag{2}$$

The boundary conditions at B5 have been described. The fluxes of oxygen concentration and potential are both zero at B1 because it is a symmetry boundary. The fluxes are zero at B3 and B4 because of insulation of the coating to current and to oxygen diffusion. B2 is a flux boundary to be defined below.

At the pipe steel surface, both anodic and cathodic reactions occur. The anodic half-cell reaction is iron oxidation or corrosion:

$$Fe \rightarrow Fe^{2+} + 2e^-. \tag{3}$$

The corrosion current density expressed by the Tafel equation is:

$$i_{\text{corr}} = i^0_{Fe\_ref} 10^{(\psi_s - E^{\text{eq}}_{Fe\_ref})/b_{Fe}}, \tag{4}$$

where ψ_s is ψ at the steel surface. $i^0_{Fe\_ref}$ and $E^{\text{eq}}_{Fe\_ref}$ are respectively the exchange current density and equilibrium potential of Equation (3) at a same reference condition. b_{Fe} is the Tafel slope of iron oxidation.

For nonacidic soil solution, the primary cathodic half-cell reactions are water reduction:

$$H_2O + e^- \rightarrow H + OH^- , \tag{5}$$

and oxygen reduction:

$$O_2 + 2H_2O + 4e^- \rightarrow 4OH^- . \tag{6}$$

Their Tafel equations are respectively:

$$i_{H_2O} = -i^0_{H_2O\_ref} 10^{(-\psi_s + E^{eq}_{H_2O\_ref})/b_{H_2O}} , \tag{7}$$

and

$$i_{O_2} = -i^0_{O_2\_ref} \frac{c_{O_2}s}{c_{O_2\_ref}} 10^{(-\psi_s + E^{eq}_{O_2\_ref})/b_{O_2}} , \tag{8}$$

where $i^0_{H_2O\_ref}$ and $E^{eq}_{H_2O\_ref}$ are the exchange current density and equilibrium potential of Equation (5) at a same reference condition. b_{H_2O} is the Tafel slope of water reduction. $i^0_{O_2\_ref}$ and $E^{eq}_{O_2\_ref}$ are exchange current density and equilibrium potential at a same reference condition with an oxygen concentration of $c_{O_2\_ref}$. c_{O_2s} is oxygen concentration at the steel surface. b_{O_2} is the Tafel slope of oxygen reduction. Note that the reference conditions for Equations (3), (5) and (6) are independent of each other.

At B2, the oxygen diffusion flux equals its cathodic reduction rate:

$$-D_{O_2} \frac{dc_{O_2}}{dy} \bigg|_{y=0} = \frac{i_{O_2}}{n_{O_2}} F , \tag{9}$$

where D_{O_2} is oxygen diffusivity in ground water. n_{O_2} is the number of electrons transferred in the elemental reaction of Equation (6) and F is Faraday's constant. Following ohm's law, the following equation is obtained:

$$\frac{d\psi}{\rho dy} \bigg|_{y=0} = i_{\text{net}} , \tag{10}$$

where ρ is ground water resistivity and $i_{\text{net}} = i_{\text{corr}} + i_{O_2} + i_{H_2O}$.

2. FEMLAB modeling procedures and solution

2.1. *Variables and constants*

Table 2 gives the Application modes and Geometry drawing instructions for this model. Table 3 gives the Constants data entry necessary for the

Table 2. Model set up — Application modes and drawing.

(1) Start **FEMLAB**

(2) In the **Model Navigator,** select **2-D** from the **Space dimension** list

(3) In **FEMLAB** under **Application Modes,** select **PDE General Form, Stationary analysis** from the **PDE Modes** directory

(4) Enter c_{O_2} phi in the **Dependent variables** edit field in the **Model Navigator**

(5) Click the **Multiphysics** button and click the **Add** button to add the application mode to the model

(6) Click **OK**

Drawing

(1) Select **Axes/Grid Setting** in the **Options** menu

(2) Clear **Axis equal** and enter axis setting in the following table

(3) Click **Grid**, clear **Auto** and enter the grid setting in the table

(4) Click **OK**

AXIS		GRID	
x min	$-5e-2$	x spacing	$5e-2$
x max	0.15	Extra x	$1e-3, 6e-3, 0.01$
y min	$-5e-3$	y spacing	$5e-3$
y max	0.015	Extra y	$5e-4, 5.01e-4, 6e-3$

(1) Click the **Rectangle/Square** from the Draw toolbar

(2) Draw a rectangle with lower left corner at $(0, 0)$ and upper right corner at $(0.1, 5e-4)$. Caution to distinguish $5e-4$ and $5.01e-4$

(3) Click the **Line** from the Draw toolbar

(4) Draw the line from $(0, 5e-4)$ to $(0, 0.01)$ to $(6e-3, 0.01)$

(5) Click the **2nd Degree Bezier Curve** in the Draw toolbar and drag to $(0.01, 0.01)$ to $(0.01, 6e-3)$

(6) Click **Line** in the Draw toolbar and drag to $(0.01, 5.01e-4)$ to $(1e-3, 5e-4)$

(7) Right click in the same position

(8) Click the **Create Composite Objects** button in the Draw toolbar

(9) Enter R1+CO1 in the **Set Formula** edit field

(10) Click **OK**

(11) Finish the geometry by clicking **Zoom Extends** in the top toolbar

Note that the two lines, which share the joint point $(1e-3, 5e-4)$ but are separated slightly respectively connecting to $(0.1, 5e-4)$ and $(1e-3, 5.01e-5)$, are supposed to be one overlapping boundary. This artificial small separation makes the zero flux across this boundary easily defined.

Table 3. Constants table.

Name	Value	Definition	Unit
F	96485	Faraday's constant	/
$E^{eq}_{Fe\_ref}$	−0.908	Fe versus Fe^{2+} equilibrium potential	V versus $Cu/CuSO_4$
$i^0_{Fe\_ref}$	2e−4	Fe versus Fe^{2+} exchange current density	A/m²
b_{Fe}	0.04	Iron oxidation Tafel slope	V/decade
$E^{eq}_{H_2O\_ref}$	−0.860	H_2O versus H_2 equilibrium potential	V versus $Cu/CuSO_4$
$i^0_{H_2O\_ref}$	0.002	H_2O versus H_2 exchange current density	A/m²
b_{H_2O}	0.12	Water reduction Tafel slope	V/decade
n_{O_2}	4	No. of electron transfer	/
$c_{O_2\_ref}$	1.24	O_2 reference concentration	mol/m³
$E^{eq}_{O_2\_ref}$	0.369	O_2 versus OH^- equilibrium potential	V versus $Cu/CuSO_4$
$i^0_{O_2\_ref}$	4e−9	O_2 versus OH^- exchange current density	A/m²
b_{O_2}	0.12	Oxygen reduction Tafel slope	V/decade
D_{O_2}	1.96e−9	Oxygen diffusivity in water	m²/s
c_{O_20}	0.26	Bulk oxygen concentration	mol/m³
ro	25	Soil solution resistivity	ohm.m
gap	5e−4	The gap btw coating and steel	m
phi0	−0.9	Measured potential in bulk	V versus $Cu/CuSO_4$

model set up. Under the Options menu follow the steps below to set up the contants, expressions, and coupling varibles:

(1) Select **Constants** in the **Options** menu and define the constants according to the table below (the Definition and Unit columns should be ignored. They are written here for reference only)

(2) Click **OK**

(3) Select **Boundary Expressions** from **Expressions** in the **Options** menu

(4) Click 2 in the **Boundary selection** under **Boundary Expressions**

(5) Enter the expressions according to Table 4

(6) Click **OK**

(7) Select **Boundary Variables** from **Integration Coupling Variables** in the **Options** menu

(8) Click **Source**, select 4 under **Boundary selection**

Table 4. Expressions entry.

Name	Expression
i_{corr}	$i^0_{Fe\_ref} * 10^\wedge((\text{phi}-E^{\text{eq}}_{Fe\_ref})/b_{Fe})$
i_{H_2O}	$-i^0_{H_2O\_ref} * 10^\wedge(-(\text{phi}-E^{\text{eq}}_{H_2O\_ref})/b_{H_2O})$
i_{O_2}	$-i^0_{O_2\_ref} * (c_{O_2}/c_{O_2\_ref}) * 10^\wedge(-(\text{phi}-E^{\text{eq}}_{O_2\_ref})/b_{O_2})$
i_{net}	$i_{\text{corr}} + i_{H_2O} + i_{O_2}$

(9) Enter Intphiy in **Name** and phi$y$/ro in **Expression**

(10) Clear **Global destination**

(11) Click **Destination**, select **Point** in **Level** and select **4** under **Point selection**

(12) Click **OK**

(13) Select **Subdomain Variables** from **Projection Coupling Variables** in the **Options** menu

(14) Click **Source**, select **1** under **Subdomain selection**

(15) Enter c_{O_2}av in **Name** and c_{O_2}/gap in **Expression**

(16) Select **Global transformation**

(17) Click **Destination**, select **Boundary** in **Level** and select **2** under **Boundary selection**

(18) Click **OK**

2.2. *Meshing*

(1) Click **Mesh Parameters** in **Mesh** menu

(2) Enter 2e−4 in **Maximum element size** in **Global**

(3) Click **Boundary**, select **2** under **Boundary selection**

(4) Enter 2e−5 in **Maximum element size**

(5) Click **OK**.

2.3. *PDE system set up*

Follow the instructions in Table 5 to set up the boundary conditions (boundary settings) and PDE system (subdomain settings) under the Physics Menu.

2.4. *Solution*

The no CP condition is solved first. With this condition known, the effect of CP on the crevice corrosion can be studied. No CP means zero current flowing into the geometry/system. Without generation and consumption of current in the upper region, the integration variable Intphiy must be

Table 5.　PDE System definition.

Subdomain Settings

(1)　Click **Subdomain Settings** in **Physics** menu

(2)　Select **1** and **2** in the **Subdomain selection**

(3)　Click the **F** tab and change the source terms zero

(4)　Click the **Init** tab and under **Initial Value** enter $c_{O_2 0}/10$ for $c_{O_2}(t_0)$ and phi0 for **phi(t_0)**

(5)　Click **OK**

Boundary Conditions

(1)　Select **Boundary Settings** in **Physics** menu

(2)　Enter boundary coefficients according to the following table:

BOUNDARY	5, 8, 10	2	1, 3, 6–7, 9
Type	Dirichlet	Neumann	Neumann
r	$-c_{O_2} + c_{O_2 0}$	/	/
	$-$ phi + phi0	/	/
g	0	$i_{O_2}/n_{O_2}/F/D_{O_2}$	0
	0	$-$ro*inet	0

Note: In the case of no CP, the integration of the potential flux along (or total current across) the holiday mouth or Boundary B5 in Figure 1 is zero. In Section 2.5, parametric solver (by varying phi0) and the integration coupling variable (Intphiy) are used.

0.　Parametric solver is used and by varying phi0, a plot of Intphiy versus phi0 allows for the determination of phi0 at no CP condition. In the following, the screening of the no CP condition and the effect of CP are comprehensively modeled at one time.

(1)　Click **Solve Parameters** in **Solve** menu

(2)　Select **Parametric nonlinear** under **Solver**

(3)　Under **General** enter phi0 in the **Name of parameter** tab

(4)　Enter $-0.9{:}0.01{:}-0.85$ $-0.84{:}0.001{:}-0.83$ -0.82 -0.81 in the **List of parameter values** tab

(5)　Click **OK**

(6)　Click **Solver Manager** in the **Solve** menu

(7)　Select the **Initial value expression** radio button under the **Initial Value** tab

(8)　Click **Solve** to solve the problem

2.5. *Post-processing*

First, plot Intphiy versus phi0 to determine the no CP condition.

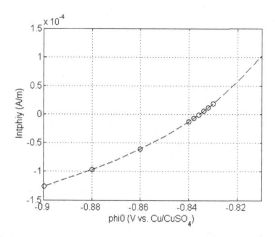

Figure 2. Current across the holiday mouth per unit width (perpendicular to the geometry) versus potential at Boundary B5 in Figure 1.

(1) Select **Domain Plot Parameters** in **Postprocessing** menu
(2) Click **Point** and enter Intphiy in the **Expression**
(3) Select **4** under **Point selection**
(4) Click **Apply**

From this operation, a new figure (Figure 2) is created where the phi0 corresponding to Intphiy = 0 is estimated to be −0.836 V. This is the potential when CP is absent in the system.

At this no CP potential, the average oxygen concentration across the gap is plotted versus the ratio of distance over gap (Figure 3). In the same figure, the oxygen concentrations at the inner surface of the coating and at the steel surface are plotted to show their difference. The plotting procedure is given below.

(1) Click **General** and select **Line/Extrusion plot** under **Plot type**
(2) Select **−0.836** under **Solutions to use**
(3) Click **Line/Extrusion** in the top menu of the current window
(4) Select **Line plot** under **Plot type**
(5) Enter c_{O_2}av in **Expression** under **y-axis data**
(6) Select **2** under **boundary selection**
(7) Select **Expression** under **x-axis data**
(8) Click **Expression**
(9) In the new **X-Axis Data** window, enter x/gap in **Expression** under **x-axis data**

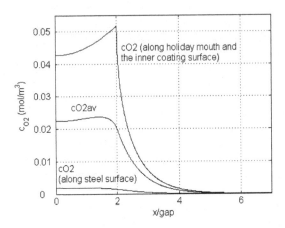

Figure 3.   Oxygen concentration profile along the crevice. c_{O_2} av is oxygen concentration averaged across the gap between steel surface and coating ($0 \leq y \leq gap$).

(10) Click **OK** and the new window disappears
(11) Click **Apply** in the original window

The above procedure completes the plot of the average oxygen concentration. To continue plotting the concentration at the steel surface and that along the mouth and the inner surface of the coating, follow:

(1) Click **General** again in the top menu of the current window
(2) Select **Keep current plot** near the bottom of the current window
(3) Click **Line/Extrusion** in the top menu of the current window
(4) In the **Expression** under **y-axis data**, replace c_{O_2}av by c_{O_2}
(5) Click **Apply**
(6) Select **4** and **7** under **Boundary selection**
(7) Click **Apply**

Since oxygen concentration becomes zero quickly from the holiday mouth, the left portion of the diagram is expanded by the following procedure.

(1) Click the new figure
(2) Select **Zoom Window** in the tool bar
(3) Select the portion of the figure to be expanded

To plot the effect of the CP or the potential phi0 on the steel corrosion current density (Figure 4) and potential (Figure 5), follow the procedure below:

(1) Click **General** again in the top menu of the current window
(2) Clear **Keep current plot** near the bottom of the current window
(3) Select **New figure** in **Plot in**
(4) Select −**0.9**, −**0.85** and −**0.836** under **Solutions to use**
(5) Click **Line/Extrusion** in the top menu of the current window
(6) Select **phi** in **Predefined quantities** under *y*-**axis data**
(7) Select **2** under **Boundary selection**
(8) Click **Apply**
(9) Replace **phi** with icorr in **Expression** under *y*-**axis data**
(10) Click **OK**

To plot the current flow in the geometry, follow:

(1) Select **Plot Parameters** under **Post-processing**
(2) Click **General** and select **Arrow** under **Plot type**
(3) Select −**0.85** in **Parameter value** under **Solution to use**
(4) Click **Arrow** in the top menu of the current window
(5) Select **grad (phi)** in **Predefined quantities** under **Subdomain** under **Arrow data**
(6) Enter 50 and 100 respectively in *x* **points** and *y* **points** under **Number of points** under **Arrow positioning**
(7) Click **OK**

To expand the region near the holiday mouth, follow the procedures 23–25. The diagram is shown in Figure 6.

3. Discussion of the results

Figure 2 is a plot of the current passing across the holiday mouth per unit width (perpendicular to the geometry) versus potential at Boundary B5 of Figure 1. The current is negative at low potentials and increases with increasing potential. It passes zero at phi0 about −0.836 V and becomes positive at higher potentials. The negative current means cathodic current and indicates cathodic protection. The positive current means anodic current and indicates enhanced anodic dissolution or corrosion. The current of zero indicates no CP, a condition that the crevice corrosion occurs naturally without interference by external polarizations.

Figure 3 shows the oxygen concentration profiles respectively along the steel surface, its average across the crevice gap ($0 \leq y \leq gap$) in the x direction and the concentration along the crevice mouth and the inner surface of the coating. Clearly, oxygen disappears after, only, a few gap

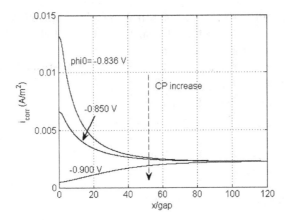

Figure 4. The corrosion current density profile along the steel surface for three potentials at Boundary B5. The effect of CP on the corrosion current density profile is shown.

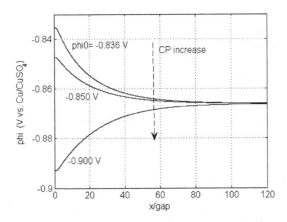

Figure 5. The steel potential profile along the steel surface for three potentials at Boundary B5. The effect of CP on the steel potential profile is shown.

sizes into the crevice $(x/a < 6)$ due to oxygen diffusion limitation. This leads to a very small oxygen concentration at the steel surface compared to the other two concentrations. The oxygen concentration decreases into the crevice because of the impermeable coating through which oxygen cannot pass.

Figures 4 and 5 show respectively the effect of CP on the corrosion current density and potential. For steel, the current density unit A/m^2 is equivalent to a corrosion rate of 1.2 mm/year. As CP increases or

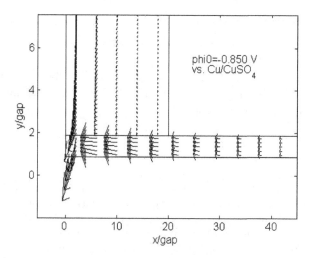

Figure 6.   Current flow in the geometry near the holiday mouth.   The potential at Boundary B5 is -0.850 V versus Cu/CuSO$_4$.

the potential at Boundary B5 of Figure 1 decreases, the corrosion current density and potential decrease. For potentials of phi0 being -0.836 V and -0.850 V, the corrosion rate decreases into the crevice because CP is insufficient to consume all oxygen diffusing into the crevice and hence, the oxygen concentration still exists at the steel surface which is higher near the holiday region than inside the crevice.   For the potential of phi0 at -0.9 V, CP is more than sufficient to consume all oxygen diffusing in the crevice. The additional CP further cathodically polarizes the steel surface and slightly decreases the anodic dissolution rate or corrosion rate.

Figure 6 shows the current flow in the geometry near the holiday region. That the current flows from the upper region or soil into the crevice indicates presence of CP which decreases the corrosion rate. That the current also flows from inside of the crevice to the holiday region indicates insufficient CP that cannot consume all oxygen diffusing into the crevice. The higher oxygen concentration near the holiday and lower concentration inside the crevice results in a differential oxygen concentration cell which creates a local anode inside the crevice and a local cathode near the holiday region.   The current flows from the anode to the cathode.   Due to this cell (which would be strongest when there is no CP), the steel corrosion rate inside the crevice is enhanced while near the holiday region it is relieved.

Acknowledgments

This work is a part of the research project: "An Approach to Determining Reassessment Intervals Through Corrosion," sponsored by the US Department of Transportation under contract: DTRS56-04-T-0002 (James Merritt, Program Manager). PRCI, SoCal, Valero Energy and Gaz de France provided in-kind support.

References

[1] http://www.corrosioncost.com/home.html.
[2] F. M. Song and N. Sridhar, A two-dimensional model for steel corrosion under a disbonded coating due to oxygen with or without cathodic protection, *Corrosion*, submitted.
[3] F. M. Song, D. A. Jones and D. W. Kirk, *Corrosion* **60**(2) (2005) 145–154.
[4] F. M. Song, D. W. Kirk, J. W. Graydon and D. E. Cormack, *Corrosion* **59**(1) (2003) 42–49.
[5] F. M. Song, D. W. Kirk, J. W. Graydon and D. E. Cormack, *Corrosion* **58**(12) (2002) 1015–1024.

Chapter Fourteen

NUMERICAL SIMULATION OF A
MAGNETOHYDRODYNAMIC DC MICRODEVICE

JAIME H. LOZANO PARADA and WILLIAM B.J. ZIMMERMAN

Department of Chemical and Process Engineering, University of Sheffield,
Newcastle Street, Sheffield S1 3JD United Kingdom
E-mail: cpp02jhl@sheffield.ac.uk

1. Introduction

Magnetohydrodynamic theory (MHD) studies the interaction of conducting fluids with electromagnetic fields. It plays a central role in many areas of physics such as thermonuclear reactions [1], solar and space plasmas [2], and rocket engines [3]. There is a growing interest in applying MHD for the control of flows at the microscale for lab-on-chip applications [4]–[6]. MHD micropumps are driven by Lorentz forces defined, in direction and magnitude, by the vectorial product of the magnetic field B and electric field E. The important feature of these pumps is the possibility of local fluidic control, enabling to direct precisely fluid flow through defined paths in a microchannel network with no need for mechanical devices [7]. This local fluidic control through the Lorentz force allows versatility of control such as bi-directional pumping, acceleration, slowing down or even reversing of the flow. In contrast to electrokinetic pumps using high axial voltages [8], MHD micropumps use low transversal electric fields. Low Joule heating therefore makes them appropriate for driving biological flows that are sensitive to high temperatures and voltages. Simple electronic set-ups can control sequences of independent micropumps inside complex microfluidic systems. Flow rates are controlled by the strength of the electromagnetic field.

It seems that to date, there are no publications on the simulation of MHD micropumps. In the present work we will present some results concerning the simulation of a micropump based on the Galerkin finite element method (FEM) that is implemented in the commercial software COMSOL

Multiphysics 3.2. The numerical procedure used here is the pressure correction algorithm called SIMPLE, which first guesses a pressure field and solves for the velocity field determined from the Navier–Stokes equations for incompressible flows. These velocities do not necessarily satisfy the Poisson-type continuity equation, so corrections to the pressure field lead to corrections in the velocities, which in turn restore the mass balance. Simultaneous to determining the velocity field, the equation for electric potential is solved. This produces the Lorentz forces that are fed back into the N-S equations and integrated as volumetric body forces. The coupling between Lorentz forces and the velocity field continues until convergence is achieved using Newton iteration.

2. Theory

In order to study the behaviour of an electrically neutral, conducting fluid under the influence of cross electric and magnetic fields in a microchannel reactor (MR) we formulate a body of MHD equations for an incompressible Newtonian flow with isotropic properties.

In the laboratory scale, the governing equations of MHD can be decoupled into three main problems [9]: A problem of fluid mechanics based on continuous equations for mass and momentum conservation, a problem of electrodynamics based on the Maxwell equations and a thermal problem based on the energy conservation equation. However, de-coupling between fluid mechanics and electrodynamics is not possible in problems defined by boundary conditions incorporating pressure. But it is easily realizable when the flow velocity is sufficiently low that the pressure distribution is, in first approximation hydrostatic, then it is possible to define a purely magnetohydrostatic problem. In this, the effect of the magnetic field upon the velocity field is appreciable, but instead the effect of the velocity flow on the convection of the magnetic field lines is negligible (this is the well-know weak coupling) unless the inlet velocity is increased in such a way the magnetic Reynolds number R_m is greater than one. At $R_m \approx 1$, convection of magnetic field lines is of the order of 1% of the static magnetic field B_0, but at $R_m \approx 10$ the perturbed magnetic field lines will be of approximately the same magnitude as the unperturbed B_0, meaning that the convection of the field lines is substantial. In our case, the mean flow velocity should be around $u_0 \approx 10^4$ m/s as to produce convection of the B lines, which is impossible for our conditions. Thus the fluid flow convection of the magnetic field lines is negligible.

We derive the set of equations for both a MHD DC micropump and a MHD DC micro valve based on the model proposed in Ref. 10. But instead of using the computer code PHOENICS that is based on finite volume methods along with external FORTRAN subroutines attached to the N-S equations to account for the Lorentz body forces, in our simulations we use the computer package COMSOL Multiphysics 3.2 which is based on the finite element method. The advantage of using it is that it is not necessary to write external subroutines since generic expressions for sources and body forces are already built-in, and so the solving algorithms are optimized with this into consideration. As a result, we could obtain faster convergence by Newton iteration.

The following are the scaling factors with which the problem is stated in nondimensional form:

$$\rho^* = \frac{\rho}{\rho_0}; \quad t^* = \frac{t\sigma B_0^2}{\rho_0}; \quad x^* = \frac{x}{L}; \quad y^* = \frac{y}{R}; \quad u^* = \frac{u}{u_0}; \quad B^* = \frac{B}{B_0},$$

$$p^* = \frac{p}{\sigma u_0 B_0^2 R}; \quad J^* = \frac{J}{\sigma u_0 B_0}; \quad Ha^2 = \frac{\sigma B_0^2 R^2}{\eta}; \quad \phi^* = \frac{\phi}{u_0 B_0 R},$$

(1)

where ρ (m^{-3}) is the flow density, σ (siemens/m) is the electrical conductivity, u_0 (m/s) and B_0 (Tesla) are the velocity and the magnetic field strength. Ha is the Hartmann number and Ha^2 can be seen as the ratio of magnetic to normal viscous forces, p (Pa) the pressure and R is the width of the channel, η (Kg/m·s) is the dynamic viscosity.

With the above assumptions and non-dimensional numbers, the system of MHD equations in steady state can be formulated thus: The continuity equation is given by

$$\nabla \cdot \mathbf{u}^* = 0,$$

(2)

where $\mathbf{u}$ (m/s) is the divergence-free bulk velocity field subject to nonslip boundary conditions on the walls.

The net electromagnetic force acting upon an assemble of charged particles is the Lorentz force, which enters the Navier-Stokes equations for incompressible media as a body force (3)

$$\frac{Re}{Ha^2}\mathbf{u}^* \cdot \nabla \mathbf{u}^* - \frac{1}{Ha^2}\nabla^2 \mathbf{u}^* = -\nabla p^* + \mathbf{J}^* \times \mathbf{B}^*,$$

(3)

where $\mathbf{J}$ (Ampere/m^2) is the current density given by the Ohm's law.

$$\mathbf{J}^* = -\nabla \phi^* + \mathbf{u}^* \times \mathbf{B}^*,$$

(4)

where ϕ (Volts) is the scalar potential given by (5). The first term in the right hand side of (4) is the externally applied electric field, and the second term, the electric field induced by the magnetic field. The cross product $\mathbf{J}^* \times \mathbf{B}^*$ in (3) is the Lorentz force. The electric potential is obtained from:

$$\nabla^2 \phi^* = \nabla \cdot (\mathbf{u}^* \times \mathbf{B}^*).\tag{5}$$

The external electric field is subject to insulating boundary conditions ($\nabla\phi \cdot \mathbf{n} = 0$) on the walls. The electric and magnetic field are described by the Maxwell equations. Integration of equation (4) with respect to z gives the mean current density

$$\mathbf{j}_{0y} = \frac{1}{h} \int_{-h/2}^{h/2} \mathbf{j}_z dz = \sigma(E_0 - B_0 u_0).\tag{6}$$

From a careful analysis of (6) we can draw some very interesting physical conclusions that have technological connotations whether the mean current is zero, positive or negative. Let's suppose that we insert a pair of electrodes on the lateral walls (parallel to $\mathbf{B}$) of the microchannel and short-circuit them, then the induced electric field $E_0 = 0$ and it follows from (6) that $j_{0y}^{<}0$ for all positive values of B_0. Now, we introduce some resistance in the external circuit and gradually increase its value, then the mean current will be in the negative y-direction, decreasing in magnitude as the circuit's resistance increases, when the resistance is high enough (infinite resistance) the circuit will open, which means that $j_0 = 0$. When $j_0 < 0$ the action of the mean Lorentz force ($j_0 \times B_0$) is that of a "brake." This braking action is maximum when the current is maximum, this is when the external resistance is zero.

For a finite external resistance, electrical power can be extracted from the fluid and the system behaves as a magnetohydrodynamic generator. It is very important to notice that when $j_0 = 0$ the mean flow velocity is precisely determined by the relation $E_0 = u_0 B_0$, so that by measuring the induced field, the fluid flux may be determined. Let us suppose that instead of a resistance we set a power source in the external circuit. Then a current is driven through the fluid in the positive y-direction and the Lorentz force acts in the direction of the flow, so the external power accelerates the flow. This is the principle of an electromagnetic pump.

To solve equations (2)–(5), several approaches are possible with COMSOL Multiphysics 3.2. The simplest one is to make use of those modes already available in the present version, so we choose the **Incompressible**

Navier-Stokes mode for momentum balance and the **Conductive media dc** for potential.

3. Geometry

Our MHD system is confined within a MR with rectangular cross section. The width and height are around 300 and 100 microns, respectively. The length is around ten fold the width. The domain is divided into three portions (left, centre, right) to account for the local variation of the magnetic field along the axis. The magnetic field is constant in the central zone and exponentially decaying in the left and right zones. The computational domain is meshed bearing in mind that the high aspect ratio may pose serious problems to flux conservation through internal borders, so scaling of spatial coordinates to equivalent lengths and matching the respective meshes has been done carefully. The flow configuration, and the magnetic and electric fields are shown in Figure 1. Parallel to the x–y plane we place a pair of magnets (it can be an electromagnet as in the case of an AC micropump) producing a magnetic field in the positive z-direction, the fluid flows in the positive x-direction and when it passes through the central region

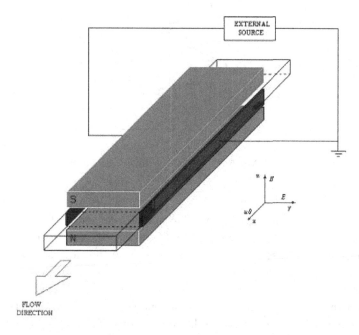

Figure 1. Microchannel geometry.

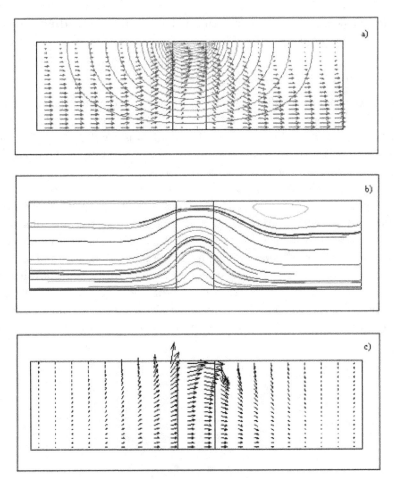

Figure 2. (a) Velocity profiles and contours of electric potential; (b) flow lines; (c) Lorentz force vectors for $N = 5 \times 10^3$; $Ha = 4$; $Re = 3.2 \times 10^{-3}$.

is distorted by the magnetic field. The induced current and electric field lie in the y-direction. The configuration shown in Figure 1 represent both the pump or valve cases, depending on the sign of the electrode potential. The fluid is fully developed laminar flow in the upstream region (left side in Figure 2). For simplicity we assume that the channel height is much less than the width or length, so the current density in the magnetic field direction is much less than in the other dimensions. Thus, this problem can be described with a 2-D model.

3.1. Boundary conditions

Table 1. Boundary conditions.

Inlet	$u^*(-1, y) = u_0 \left(1 - \dfrac{y}{R}\right)^2$
	$v^*(-1, y) = 0$
Wall	$u^*(x, R) = v^*(x, R) = 0$
	$\phi^*(-1, R) = \phi^*(1, R) = 0$
Electrode	$\phi^*(x, R) = V_0$ for $\|x\| \leq \dfrac{a}{2}$ (7)
Insulating Walls	$\dfrac{\partial \phi^*}{\partial \mathbf{n}}(x, R) = 0$ for $\|x\| > \dfrac{a}{2}$
Symmetry Axis	$\dfrac{\partial u^*}{\partial y}(x, 0) = 0$
	$\phi^*(x, 0) = 0$
Outlet	$p^*(1, y) = 0$

4. COMSOL Multiphysics 3.2 implementation

Start up **COMSOL Multiphysics 3.2** and enter the **Model Navigator**.
Table 2 gives the instructions about how to set up the model.

5. Results

The variation of the interaction parameter (Re/Ha^2) and the Hartmann
number accounts for the variation of the Reynolds number and the magnetic field strength. Several cases were modelled for both the pump and
valve configurations. In all cases non-conducting walls outside the electrode
region are considered.

To demonstrate the concept of pumping and braking with cross electromagnetic fields by the introduction of a magnetic field and an electric field
in the transverse direction by placing a pair of electrodes on the walls of
the microchannel, plots of velocity and pressure profiles are presented for
each case.

5.1. MHD DC micropump

Figure 2 shows a flow pattern and contours of electric potential (a), flow
lines (b) and Lorentz vectors (c) of the MHD microchannel. Since there

Table 2.   MHD microdevice model.

Model Navigator	Select **2-D** from the pull-down menu Select **Fluid Dynamics|Incompressible Navier-Stokes Add**. Click **Multiphysics** button Select **Electromagnetics** and **Conductive media DC** OK
Draw Menu	Draw a rectangle from $(-1, 0)$ to $(-0.11, 0.5)$ Draw a rectangle from $(-0.11, 0)$ to $(0.11, 0.5)$ Draw a rectangle from $(0.11, 0)$ to $(1, 0.5)$ Select **Axes/Grid Settings** from menu **Options** Set $x_{\min} = -1.2$; $x_{\max} = 1.2$; $y_{\min} = -0.5$; $y_{\max} = 1$
Options Menu| Constants	$Re = 2e{-}3$; $Ha = 10$; $u_0 = 1$, $B_0 = 1$; $a = 1/9$; $N = 500$; $V_0 = -10$
Options Menu|Scalar Expressions|Global	$BL = B_0 * \exp(9 * (x + a))$ $BR = B_0 * \exp(9 * (a - x))$
Physics Menu: Boundary settings	ns mode Inlet. Set BC: Inflow/Outflow velocity Set $u_0 = 1 - (y/0.5)\char`^2$ and $v_0 = 0$ For the Axis boundary condition: Slip/Symmetry For the upper wall select boundary condition: No slip For the electrode select boundary condition: No slip For the output boundary condition select: Outflow/Pressure **DC** mode For the Inflow Boundary (in ns mode) select boundary condition: $V = 0$ For the Axis select boundary condition: $V = 0$ For the upper walls select boundary condition: Electric Insulation For the electrode select boundary condition: Electric Potential $= V_0$
Physics Menu| Subdomain Settings	ns mode. Subdomain 1 $\rho = 1/N$; $\eta = 1/Ha\char`^2$; $F_x = -BL * V_y - BL\char`^2 * u$; $F_y = 0$ Subdomain 2 $\rho = 1/N$; $\eta = 1/Ha\char`^2$; $F_x = -B0 * V_y - B_0\char`^2 * u$; $F_y = 0$ Subdomain 3 $\rho = 1/N$; $\eta = 1/Ha\char`^2$; $F_x = -BR * V_y - BR\char`^2 * u$; $F_y = 0$ DC mode Subdomain 1 $\sigma = 1.0$ and $Q = 9.0 * v * BL + BL * v_x - BL * u_y$ Subdomain 2 Set $\sigma = 1.0$ and $Q = v_x - u_y$ Subdomain 3 $\sigma = 1.0$ and $Q = -9.0 * v * BR + BR * v_x - BR * u_y$
Solve Menu	Click on the solve $(=)$ tool on the toolbar

are no electrodes in the upstream and downstream regions and the walls are insulators, the electromotive forces will tend to oppose the electrostatic field in the y-direction. The first ones dominate where the magnetic field is strong, while outside this region the magnetic flux density decays quickly and the electrostatic forces dominate. This closes the current lines so the system will behave like a brake. It can be seen in Figure 2 when approaching the central region from upstream the velocity profile flattens as the flow enters the constant magnetic field region (centre). The velocity profile develops in a M-shaped form since the Lorentz forces are more intense close to the electrode and the resistance to the axial component of velocity is weaker there than in the centre. In this case the distortion of the developed parabolic profile when the fluid enters the electrode region is dramatic. In the electrode region, the flow is accelerated developing a concave profile. Downstream the velocity maximum moves toward the centre where the M-shape profile is more notorious. As the flow goes further down and the electromotive forces decrease exponentially the maximum of the velocity profile shifts towards the centre until the flow recovers its parabolic profile again. The contours of electrostatic potential diffuse outwards from a maximum at the electrode and zero along the axis. Figure 2(b) shows the corresponding flow lines. As the flow is pushed towards the walls by the Lorentz forces, the flow lines concentrate there, and then the flow accelerates to preserve continuity. Figure 2(c) shows vectors of Lorentz force, which are given by the cross product of the current density $\mathbf{J}$ and the magnetic field strength $\mathbf{B}$. The magnetic field is external and is produced by whether a permanent magnet or an electromagnet. The magnetic Reynolds number is small given the present conditions, hence weak coupling is assumed, which means the magnetic field is not influenced by the fluid flow. The current density vector consists of two components, one due to the electrostatic field produced by the electrode pair and the other due to the electromotive field $\mathbf{U} \times \mathbf{B}$. In the upstream and downstream regions the electrostatic forces are dominant and the electric field vectors point toward the electrode (Figure 3(b)), hence in the upstream region the axial current component is in the positive x-direction and the opposite in the downstream region (Figure 3(a)). In Figure 3(c) the induced magnetic field current is shown. It is obvious that these components point in the negative y-direction since they are the cross product of the flow velocity with the magnetic field, and so, it is only possible for them to point in that direction given the magnetic field orientation we have chosen.

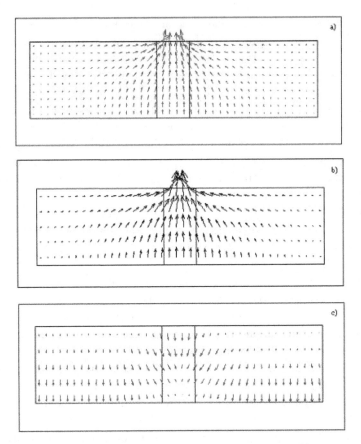

Figure 3.   (a) Current vectors; (b) Current due to the electric field; (c) Current induced by the magnetic field.

The influence of the Lorentz forces on the flow can be seen more clearly in Figure 4, which shows the pressure along the axis and along the wall close to the electrode for $Ha = 10$ and $N = 500$. As the fluid flow approaches the magnetic field region the pressure along the axis increases gradually, hence is evident that the channel becomes a pump. Along the wall pressure increase is slower, but as the electrode's centre is reached the pressure surmounts that of the axis. After the flow passes through the electrode the pressure both at the axis and the wall fall smoothly in the downstream region to meet the boundary condition at the outlet. As we will see later, reversing of electrode polarity inverts the pressure drop and the device becomes an electromagnetic valve.

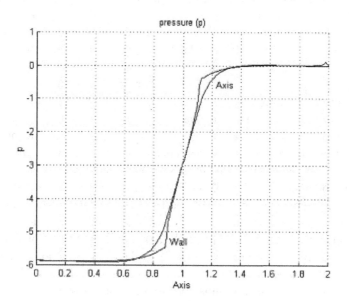

Figure 4.   Pressure along the axis and the wall.

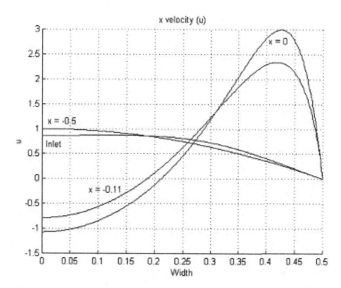

Figure 5.   Spatial evolution of the axial velocity profile for different axial positions (inlet to outlet direction).

Figure 5 shows the evolution of the axial velocity profile in the MHD micropump (for several positions along the domain) from the upstream region to the downstream region for the case $N = 500$ and $Ha = 10$.

The inlet boundary conditions define the parabolic profile into the channel. As the flow approaches the central region the Lorentz forces become dominant close the walls and the profile elongates along the flow direction. This results in an increase in the axial velocity close to the walls. This is a flow pattern very similar to electro-osmotic flow; however, the key difference is in that this profile was obtained only with cross electric and magnetic fields instead of large axial electric fields. Notice that to produce a net pumping effect the electrode is set at -10 Volts only!

Figure 6 shows the effect of the Hartmann number for $Re = 2 \times 10^{-3}$, on the axial velocity profile at the centre of the magnetic field dominated region. For small Ha the flow is not greatly affected by the electromotive forces, hence the velocity profile will have a similar shape as for $Ha = 0$.

As Ha increases the Lorentz forces overtake the viscous forces and tend to push the flow towards the walls in the direction of the flow, so close to the walls the Lorentz forces are unbalanced and strong enough as to elongate the velocity profile in the direction of flow.

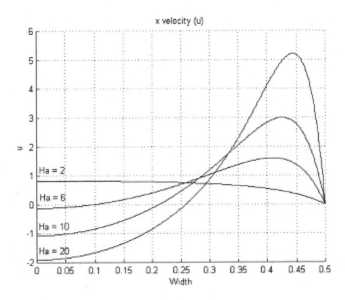

Figure 6. Effect of Hartmann number on the axial velocity profile at the centre.

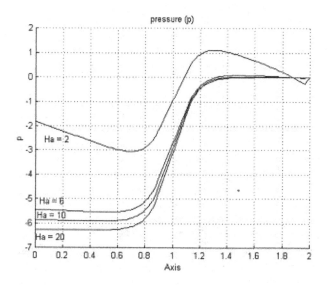

Figure 7.   Pressure profiles along the axis for several Hartmann numbers.

Once the fluid passes the central region and due to its small inertia, the flow is forced into a recirculation pattern, as can be observed close to the axis for $Ha = 6$. We believe it happens to preserve continuity.

The pumping effect produced by the electromagnetic forces in the central region is clearly seen, as said before, in Figure 7 where plots of pressure profiles along the axis are shown. Here a six-fold increase in pressure compared to the $Ha = 0$ case is observed. Also is to be noticed that an increase in Ha beyond six-fold does not change dramatically the pressure drop. This results convenient since high Ha numbers may by difficult to achieve experimentally in microchannel geometries due to limitations in the material's capacity to withstand high magnetic fields, so a moderate Ha number is a desirable figure.

5.2. *MHD DC microvalve*

In this section we investigate through the model on what happens when the polarity of the electrode is inverted. In this case, the electric field produces an electric current that enhances the current induced by the magnetic field, so the Lorentz force will be reinforced in the negative x-direction and the system will act as a brake (valve). Once the electrode's polarity is inverted, the induced electric field is enhanced by the imposed electric field. Figure 8(a) shows how the upstream parabolic velocity profile deforms when

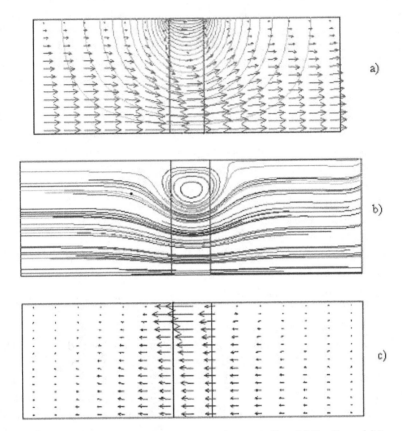

Figure 8.   (a) Contours of potential and axial velocity profiles; (b) flow lines; (c) Lorentz force vectors.

it enters the region dominated by the Lorentz forces. Close to the wall in the central region, the flow is reversed. Once the flow leaves the electrode and enters the downstream region, it recovers its parabolic shape again. Figure 8(b) shows the corresponding flow lines. The flow is pulled away from the walls and a recirculation vortex appears. The fluid accelerates to preserve continuity.

Figure 9 shows that when the electrode's polarity is reversed, the current vectors do as well. In the valve configuration all the current vectors point in the same direction augmenting each other. This means that the Lorentz forces are reinforced in the negative x-direction to slow down the flow.

Figure 10 shows the spatial evolution of the axial velocity profile for different axial positions. It is to be noticed that as the flow enters the

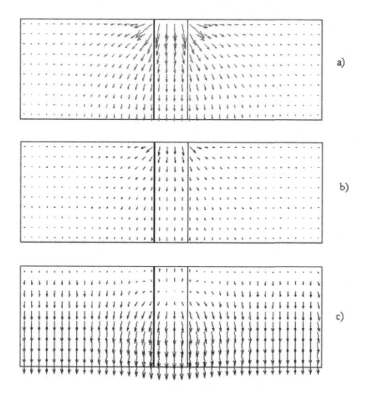

Figure 9. (a) Current density vectors; (b) current due to electric field; (c) current induced by the magnetic field.

magnetic field dominated region a recirculation vortex is formed close to the electrode. A slightly increase in velocity to preserve continuity is observed.

As in the case of the pump configuration, the evidence of braking effect can be readily seen in the pressure graph of Figure 11.

Along the axis the pressure decays smoothly from the inlet boundary in the upstream region and as the central region is approached it does it abruptly. Once the flow enters the downstream region, the pressure continues decaying softly again. Close to the walls the flow decays smoothly in the upstream region but starts its fall further down and it does it more abruptly than at the axis, but it starts more quickly its smooth fall in the downstream region. Figure 11 is very important since it demonstrate the slowing down effect of the electromagnetic field by causing the observed pressure drop.

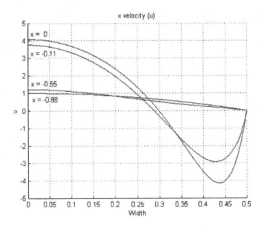

Figure 10.   Axial velocity profile for several axial positions.

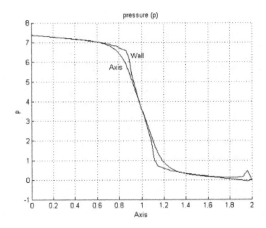

Figure 11.   Pressure along the axis and close to the walls.

6. Conclusions

In the present work, the numerical solution of a magnetohydrodynamic
model developed for a micro reactor has been implemented in COMSOL
Multiphysics 3.2 to investigate the interaction of cross electric and magnetic fields with a conductive flow. The pumping and breaking effects of
electromagnetic fields was analyzed for several Hartmann numbers. Data
of the influence of the Reynolds number on the pumping or braking effect
is not included since for $2 \times 10^{-3} < Re < 1$ which is typically the range
of interest in micro geometries, the Re number does not affect the flow

considerably, and its effects can only be observed in the recovery distance of the parabolic profile in the downstream region. It is worth noting that when the electrode is polarized negatively (pump configuration), the flow is deformed into an M-shape profile (typical of electrokinetic flows) and when the electrode voltage is reversed (valve configuration), the flow is deformed into a W-shape profile. It is demonstrated that conducting fluids can be driven by Lorentz forces produced by cross magnetic and electric fields in microchannel geometries and that the Joule heating and voltage requirement in the cross configurations are much less than in conventional electrokinetic flows, which enables these devices to be used for driving bioflows that are susceptible to deterioration by high temperatures and voltages. As the reader may have noticed, in these devices, as the Hartmann number increases the flow develops into recirculation vortexes that could improve the mixing of chemical reagents. Of course, in the present work we have just dealt with flow computation but mixing is something that MHD micro devices would perform with enhanced performance [11]. That is left for future work.

References

[1] P. Helander, Magnetohydrodynamics (MHD), *The 40th Culham Plasma Physics Summer School*, UKAEA Fusion (2003).

[2] P. Cargill, Space plasma physics: I + II, *The 40th Culham Plasma Physics Summer School*, UKAEA Fusion (2003).

[3] F. F. Chen, *Introduction to Plasma Physics and Controlled Fusion*, Vol. 1, 2nd edn. (Plenum Press, New York, 1984).

[4] J. Jang and S. S. Le, Theoretical and experimental study of MHD (magnetohydrodynamic) micropump, *Sensors and Actuators* **80** (2000) 84–89.

[5] A. Manz and G. Jenkins, A miniaturized glow discharge applied for optical emission detection in aqueous analytes, *J. Micromech. Microeng.* **12** (2002) N19–N22.

[6] J. Zhong, M. Yi and H. H. Bau, Magnetohydrodynamic (MHD) pump fabricated with ceramic tapes, *Sensors and Actuators A* **96** (2002) 59–66.

[7] H. H. Bau, J. Zhu, S. Qian and Y. Xiang, A magneto-hydrodynamically controlled fluidic network, *Sensors and Actuators B* **88** (2003) 205–216.

[8] J. M. MacInnes, Computation of reacting electrokinetic flow in microchannel geometries, *Chem. Eng. Sci.* **57** (2002) 4539–4558.

[9] R. Moreau, *Magnetohydrodynamics* (Kluwer Academic Publishers, Dordrecht, 1990).

[10] M. Hughes, K. A. Pericleous and M. Cross, The numerical modelling of DC electromagnetic pump and brake flow, *Appl. Math. Modelling* **19** (1995).

[11] H. H. Bau, J. Zhong and M. Yi, A minute magneto hydrodynamic (MHD) mixer, *Sensors and Actuators B* **79** (2001) 207–215.

Appendix

VECTOR CALCULUS FUNDAMENTALS IN COMSOL MULTIPHYSICS WITH MATLAB

W.B.J. ZIMMERMAN[1] and J.M. REES[2]

[1] *Department of Chemical and Process Engineering, University of Sheffield, Newcastle Street, Sheffield S1 3JD United Kingdom*

[2] *Department of Applied Mathematics, University of Sheffield, Hicks Building, Sheffield*

Vector calculus underpins partial differential equations and their numerical approximation. Modelers must have a good working knowledge of the basics of vector calculus to use finite element methods effectively. Perhaps because undergraduate engineers are not confronted with realistic applications of vector calculus, but rather learn it as a mathematical discipline, their ability to apply vector calculus in engineering modeling is limited. In this appendix, all the basics of vector calculus are introduced with reference to COMSOL MULTIPHYSICS WITH MATLAB utility and implementation. So the other way of reading this appendix is as a primer for COMSOL MULTIPHYSICS WITH MATLAB basics with regard to multivariable differential calculus. When we wrote this appendix, we debated whether or not to augment Chapter One (basics of numerical analysis) with the material directly, as numerical approximation of derivatives is fundamental to the solution of PDEs — a COMSOL MULTIPHYSICS "primitive" operation. Indeed, in learning spectral methods for solving PDEs, the fundamental theorem is the "derivative theorem" — how to use the spectral transform method to approximate derivatives. So by analogy, the fundamental utility of FEM is numerical differentiation. The debate was lost in that Chapter One aims to solve basic problems with COMSOL MULTIPHYSICS straightaway. Approximating derivatives, no matter how useful, is still an intermediate step in modeling, rarely the objective itself.

The only MATLAB basics we consider essential that are not used in making vector calculus the point of this appendix are eigenvalue analysis and logical expressions. These are sprinkled throughout the textbook anyway. It should be noted that nothing we use in this chapter cannot also be done in COMSOL Script. In general, the only Matlab command in this book that we use that is not in COMSOL Script is fminsearch.

1. Review of vectors

1.1. *Representation of vectors*

Since FEMLAB deals with scalar, vector, and matrix quantities, if only as input coefficients, a brief review of the representation of vectors (as a special case of MATLAB's matrix data type) is in order here. Scalar quantities can be represented by a single number, but vector quantities have magnitude and direction. Given a righthanded coordinate system as shown in Figure A1, any vector **a** is expressible in the form

$$
\begin{aligned}
\mathbf{a} &= a_1\mathbf{i} + a_2\mathbf{j} + a_3\mathbf{k}\,, \\
\mathbf{a} &= (a_1, a_2, a_3)\,,
\end{aligned}
\tag{1}
$$

where **i**, **j** and **k** are unit vector in the coordinate directions, a_1, a_2, a_3 are the components of **a** relative to this set of axes. They are the projections of **a** on to the unit vectors **i**, **j** and **k**. For a point P with coordinates (x, y, z), the position vector of P relative to the origin of the coordinate system, O, is

$$
\begin{aligned}
\mathbf{r} &= x\mathbf{i} + y\mathbf{j} + z\mathbf{k} \\
&= (x, y, z)\,.
\end{aligned}
\tag{2}
$$

MATLAB represents vectors in component form as either column (countervariant) or row (covariant) vectors:

```
>> a = [1; 2; 3];                      % column vector
>> a = [1   2   3];                    % row vector
```

In the row vector, the white space (any number of contiguous spaces) serves as the delimiter. The column vector is delimited by semicolons, or alternatively, by newlines:

```
>> a = [1
   2
   3];
```

1.2. *Scalar products, matrix multiplication, unit vectors, and vector products*

Typically, scalar products (or dot products) are defined by

$$
\mathbf{a} \cdot \mathbf{b} = |a||b| \cos\theta = a_1 b_1 + a_2 b_2 + a_3 b_3 = \sum_{i=1}^{3} a_i b_i\,,
\tag{3}
$$

where θ is the angle between the vectors **a** and **b**. To achieve the same result in MATLAB, we use the * operator

```
>> a = [1; 2; 3];
>> b = [-3 2 -1];
>> b*a
ans  =
   -2
```

This is a special case of a row vector (1×3 matrix) multiplying a column vector (3×1 matrix). As the first dimension of the latter and the second dimension of the former are the same, these matrices are compatible and can be multiplied according to the general rule for matrix multiplication

$$(AB)_{ik} = \sum_{j=1}^{n} A_{ij} B_{jk} . \tag{4}$$

If A is an $m \times n$ matrix and B is an $n \times l$ matrix, then AB is an $m \times l$ matrix. If the common size is not respected, then the matrices are incompatible and the product is not defined. MATLAB can compute scalar products as the special case of matrix multiplication, but care must be taken to respect compatibility of the vectors. For instance,

```
>> a*b
ans =
   -3     2    -1
   -6     4    -2
   -9     6    -3
```

What happened? Simply, **a** is a 3×1 matrix multiplying a 1×3 matrix, **b**. The product, **ab**, is a 3×3 matrix, viz.

$$(ab)_{ik} = \sum_{j=1}^{n} a_{ij} b_{jk} = a_{i1} b_{1k} . \tag{5}$$

In the case of vectors, the matrix $(ab)_{ik}$ is called the dyadic product of **a** and **b**, or a dyad. It is a special case of the matrix outer product, where the scalar product is also termed the inner product.

The scalar product of two row vectors or two column vectors can be computed in MATLAB using the transpose operator ', which is a unary operator and deceptively easy to mistake as a single quote of a character string, for instance.

```
ans =
    -2
```
but
```
>> a*b'
ans =
    -3    2   -1
    -6    4   -2
    -9    6   -3
```

still yields the dyad. Care must still be taken to respect the matrix compatibility. If a and b were row vectors, which combination, $b' * a$ or $a * b'$ yields the inner and outer products? MATLAB provides a special function dot for this purpose that blurs the distinction about compatibility:

```
>> help dot
 DOT  Vector dot product.
    C = DOT(A,B) returns the scalar product of the vectors A and B.
    A and B must be vectors of the same length.  When A and B are both
    column vectors, DOT(A,B) is the same as A'*B.
    DOT(A,B), for N-D arrays A and B, returns the scalar product
    along the first non-singleton dimension of A and B. A and B must
    have the same size.
    DOT(A,B,DIM) returns the scalar product of A and B in the
    dimension DIM.
    See also CROSS.
```

Example.

```
>> dot(a,b)
ans =
    -2
>> dot([1; 2; 3],[-3 2 -1])
ans =
    -2
```

It simply does not matter with dot which combination of row/column vectors is used.

Vector magnitude

The norm or magnitude of a vector is found by the formula

$$\|\mathbf{a}\| = \sqrt{\mathbf{a} \cdot \mathbf{a}} = \left(\sum_{j=1}^{n} a_j^2 \right)^{1/2}. \tag{6}$$

MATLAB will compute the norm of a vector with the formula

```
ans =
    3.7417
or with the built-in command norm
>> norm(a,2)
ans =
    3.7417
```

where sqrt( ) is the built-in square root function.

Unit vector

A unit vector is a vector whose norm is one. Unit vectors can be constructed by normalization, i.e.

$$\hat{a} = \frac{a}{\|a\|}. \tag{7}$$

For example,

```
>> ahat=a/norm(a,2)
ahat =
    0.2673
    0.5345
    0.8018
```

The division above is scalar division, which divides each element of the vector by the scalar. COMSOL Multiphysics routinely refer to built-in geometric unit vectors to describe boundaries. nx, ny, nz are reserved names for the unit vectors for the outward pointing normal to a boundary. Less commonly used are tx, ty which are the components of the tangential vector to a bounding curve. There are two orthogonal tangential directions to the general 2-D surface and therefore an infinity of tangential vectors.

Cross product

The vector or cross product is defined

$$a \times b = |a||b| \sin \theta \hat{n} = \sum_{i=1}^{3} \varepsilon_{ijk} a_j b_k \hat{e}_i, \tag{8}$$

where ε_{ijk} is the permutation tensor, which takes the value $+1$ when indices ijk are a positive permutation of 123, -1 if they are a negative permutation of 123, and zero otherwise. $\hat{n}$ is the unit normal vector to the plane containing a and b. $\hat{e}_i$ is the unit vector in the ith coordinate direction.

MATLAB provides a special function to compute cross products

```
>> help cross
CROSS  Vector cross product.
   C = CROSS(A,B) returns the cross product of the vectors
   A and B.  That is, C = A x B.  A and B must be 3 element
   vectors.
   C = CROSS(A,B) returns the cross product of A and B along the
   first dimension of length 3.
   C = CROSS(A,B,DIM), where A and B are N-D arrays, returns the cross
   product of vectors in the dimension DIM of A and B. A and B must
   have the same size, and both SIZE(A,DIM) and SIZE(B,DIM) must be 3.
   See also DOT.
For example,
>> cross(a,b)
ans =

   -8
   -8
    8
>> cross(b,a)
ans =

    8
    8
   -8
```

We see that the order of factors in a cross product switches the sign of the cross product, akin to changing the sense of the unit normal $\hat{n}$.

1.3. *Arrays: Simple arrays, cell arrays, and structures*

Array manipulation is essential to data extraction from FEMLAB. FEMLAB has organized models conveniently (for its developers and programmers) around fem structures for multiphysics and xfem structures for extended multiphysics. Pruning structures and cell arrays to extract meaningful information is a useful way of interrogating FEMLAB models (and solutions).

Simple arrays

Arrays have dimensions $(m \times n \times l \ldots)$. A matrix is a two-dimensional array. Each dimension has a length. So two very important commands are size( ) and length( ).

```
>> a = [ 1 2 3 4; 5 6 7 8];
>> size(a)
ans =
```

```
        2     4
```

Size of an array is itself a row vector of length equal to the array dimensions.

```
>> length(a(1,:))
ans =
    4
```

The colon (:) placeholder in the second argument of a specifies the entire range of the second dimension, in this case elements 1:4, i.e. 1 thru 4.

```
>> length(a(:,3))
ans =
    2
>> a(1,2:4)
ans =
    2     3     4
```

In fact, the colon refers to subarrays of a lower dimension. $a(1,:)$ is the first row; $a(:,3)$ is the third column of a. $a(1,2:4)$ gives a subarray of elements 2 thru 4 of row 1. In higher dimensions, the subarrays extracted are more complicated. For instance

```
>>b=ones(2,2,2)
b(:,:,1) =
    1     1
    1     1
b(:,:,2) =
    1     1
    1     1
>> b(1,:)
ans =
    1     1     1     1
```

Here, the subarrays are matrices in the first two cases, but in the third case, the final two dimensions are rolled up into a single row vector.

FORTRAN programmers will probably feel more comfortable addressing single elements

```
>> a(1,3)
ans =
    3
```

rather than subarrays, perhaps by using looping structures.

Array construction

Arrays can be automatically generated using colon notation, viz.

```
>> a=[0: 0.1: 1]*pi
a =
  Columns 1 through 8
0    0.3142    0.6283    0.9425    1.2566    1.5708    1.8850    2.1991
Columns 9 through
11 2.5133    2.8274    3.1416
```

which produces eleven values equally spaced between 0 and π. So does

```
a=linspace(0,pi,11) a =
  Columns 1 through 8
0      0.3142      0.6283      0.9425      1.2566      1.5708      1.8850
2.1991
Columns 9 through
11 2.5133      2.8274      3.1416
```

linspace is a versatile command for automatic matrix generation, performing a role that is often done in looping constructs in older programming languages. logspace comes in handy as well.

```
>> help linspace
 LINSPACE Linearly spaced vector.
    LINSPACE(X1, X2) generates a row vector of 100 linearly
    equally spaced points between X1 and X2.
    LINSPACE(X1, X2, N) generates N points between X1 and X2.
    For N < 2, LINSPACE returns X2.
    See also LOGSPACE, :.
```

```
>> help logspace
LOGSPACE Logarithmically spaced vector.
    LOGSPACE(X1, X2) generates a row vector of 50 logarithmically
    equally spaced points between decades 10^X1 and 10^X2.  If X2
    is pi, then the points are between 10^X1 and pi.
    LOGSPACE(X1, X2, N) generates N points.
    For N < 2, LOGSPACE returns 10^X2.
    See also LINSPACE, :.
```

Four other common array generators are zeros, ones, rand, and for matrices, eye. zeros initializes an array with zeros; ones with ones, rand with uniformly distributed random numbers (randn with normal deviates) and eye with the identity matrix.

```
>> help zeros
 ZEROS   Zeros array.
   ZEROS(N) is an N-by-N matrix of zeros.
   ZEROS(M,N) or ZEROS([M,N]) is an M-by-N matrix of zeros.
   ZEROS(M,N,P,...) or ZEROS([M N P ...]) is an M-by-N-by-P-by-...
   array of zeros.
   ZEROS(SIZE(A)) is the same size as A and all zeros.

>> help ones
 ONES    Ones array.
   ONES(N) is an N-by-N matrix of ones.
   ONES(M,N) or ONES([M,N]) is an M-by-N matrix of ones.
   ONES(M,N,P,...) or ONES([M N P ...]) is an M-by-N-by-P-by-...
   array of ones.
   ONES(SIZE(A)) is the same size as A and all ones.

>> help rand
 RAND    Uniformly distributed random numbers.
   RAND(N) is an N-by-N matrix with random entries, chosen from
   a uniform distribution on the interval (0.0,1.0).
   RAND(M,N) and RAND([M,N]) are M-by-N matrices with random entries.
   RAND(M,N,P,...) or RAND([M,N,P,...]) generate random arrays.
   RAND with no arguments is a scalar whose value changes each time it
   is referenced.  RAND(SIZE(A)) is the same size as A.

>> help eye
 EYE Identity matrix.
   EYE(N) is the N-by-N identity matrix.
   EYE(M,N) or EYE([M,N]) is an M-by-N matrix with 1's on
   the diagonal and zeros elsewhere.

   EYE(SIZE(A)) is the same size as A.
```

By now you may have noticed that if you know the name of the command, Matlab will tell you the appropriate syntax. But how do you find out the names of commands? In COMSOL Script, just type help and a list of command types will be given. Subsequent help calls "help type" reveal commands. For instance, "help numerics" gives a list of numerics commands. Then "help daspk" will give you the syntax for calling daspk, a stiff DAE solver created solely for FEMLAB and its successors.

Scalar — Array math

Arithmetic of scalars acting on arrays threads across the array. For instance,

```
>> 3*a
ans =
 Columns 1 through 8
```

```
   0    0.9425    1.8850    2.8274    3.7699
4.7124    5.6549    6.5973
Columns 9 through
11 7.5398    8.4823 9.4248

>> 5+a
ans =
Columns 1 through 8
5.0000    5.3142    5.6283    5.9425    6.2566
6.5708    6.8850    7.1991
Columns 9 through 11
7.5133    7.8274    8.1416
```

Array — Array element-wise math

Arithmetic of arrays on arrays is a tricky area. If the arrays are compatible sizes, then dot-operators are applied element-wise:

```
>> b=linspace(1,11,11)
b =
     1    2    3    4    5    6    7    8    9    10    11
>> size(a)
ans =
     1    11
>> size(b)
ans =
     1    11
>> a.*b
ans =
Columns 1 through 8
   0    0.6283    1.8850    3.7699    6.2832
9.4248    13.1947    17.5929
Columns 9 through11
22.6195    28.2743            34.5575
```

Cell arrays and structures

You could write a whole chapter about cell arrays and structures. The important thing to note about both is that they are containers for hetero-geneous mixtures of data types — floating point numbers, complex numbers, matrices, character strings, other cell arrays and structures. The cell array refers to each cell within the array by an array index. Cell arrays are directly defined with braces surrounding the list of array elements. The cell command will create an empty shell which can be assigned individual elements or subarrays.

```
>> ca = { 'every', 'good', 'boy', 'does', 'find', 3+3i, [0 1; -2 2] }
ca =
Columns 1 through 6
'every'    'good'    'boy'    'does'    'find'    [3.0000+ 3.0000i]
Column 7
[2x2 double]
```

Referencing can be done by array index, with parenthesis, returns the cell element.

```
>> ca(3)
ans =
    'boy'
```

Referencing by braces and index number returns the *contents* of the cell element.

```
>> ca{3}
ans =
boy
```

Perhaps this is clear with regard to the matrix cell constituent,

```
>> ca(7)
ans =
    [2x2 double]
>> ca{7}
ans =
    0    1
   -2    2
```

Structures are referenced by fields, which are named rather than enumerated, much as in the C programming language. The greatest utility in using a structure as averse to a cell array is that if you choose to alter the structure, addition or elimination of fields does not change the ordering of fields in a meaningful way. Elimination or addition of cell array elements, however, changes the numbering of cells or leaves "holes" in the array. The most common structure encountered by COMSOL Multiphysics users is the fem structure, which is how COMSOL Multiphysics organizes the complete set of data for its multiphysics models and their solutions. Exporting as fem structure to the MATLAB workspace from COMSOL Multiphysics produces the following for our electrokinetic flow model.

```
>> fem

fem =

            version: [1x1 struct]
               appl: {[1x1 struct]  [1x1 struct]  [1x1 struct]}
               geom: [1x1 geom2]
               mesh: [1x1 femmesh]
              shape: {1x7 cell}
            gporder: {[4]  [2]}
            cporder: {[2]  [1]}
           simplify: 'off'
             border: 1
            outform: 'weak'
               form: 'weak'
                equ: [1x1 struct]
                bnd: [1x1 struct]
                pnt: [1x1 struct]
               expr: {'zeta'  '-zeta1*(Y+zetar*(1-Y))'  'sig'  'Y+sigr*(1-Y)'}
            elemcpl: {[1x1 struct]}
               draw: [1x1 struct]
              const: {1x34 cell}
          globalexpr: {1x12 cell}
              xmesh: [1x1 com.femlab.xmesh.Xmesh]
                sol: [1x1 femsol]
```

The list is of fields in the structure fem above shows the description of the field contents. Each field can be addressed with the "dot" notation:

```
>> fem.expr

ans =

    'zeta'    [1x22 char]    'sig'    'Y+sigr*(1-Y)'
```

fem.expr is a cell array with four cells; the cell array is small enough that its contents can be displayed. Since it is a cell array, the braces index reference will act on the contents of the cell element.

```
>> fem.expr{2}

ans =

-zeta1*(Y+zetar*(1-Y))
```

As we can see, the fem structure has cell arrays, other structures, characters, and numbers as its constituents. There is no reason why we cannot have cell arrays of fem structures, which is indeed the make up of the xfem structure used by COMSOL Multiphysics for extended multiphysics, with one fem structure for each logical geometry. We have frequently had need of the postinterp command which acts on fem structures or xfem structures to produce values of solution variables interpolated at points within the domain discretized by finite elements:

```
[is,pe]=postinterp(xfem,xx);
[u]=postinterp(xfem,'u1',is);
```

Passing the whole of the xfem structure to the postinterp function gives it access to the complete description of the model and solution, for which it may have to execute different branches of commands given that specific structure. As users of COMSOL Multiphysics, we need to know enough about the MATLAB data structure of a COMSOL Multiphysics model and solution to extract relevant data if we have particular postprocessing or modeling requirements that are not built into the COMSOL Multiphysics GUI.

2. Scalar and vector fields: MATLAB function representations

Physical properties of matter typically depend on position and sometimes time. At length scales observable to humans (by eye), most physical quantities are treated as a continuum — having values at every mathematical point. These quantities are called fields. Quantities such as temperature and pressure that represent a single value are termed scalar fields. A scalar field is a single number, e.g.

$$\phi = \phi(\mathbf{x}) = \phi(x, y, z) \,. \tag{9}$$

A vector field in 3-D requires three components:

$$\mathbf{F}(\mathbf{x}) = \begin{bmatrix} F_1(x, y, z) \\ F_2(x, y, z) \\ F_3(x, y, z) \end{bmatrix} \,. \tag{10}$$

Each component of $\mathbf{F}$ is itself a scalar function of position.

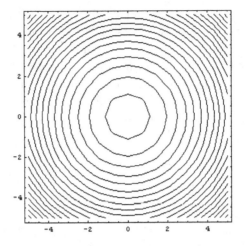

Figure 1.   Contours of $\phi = x^2 + y^2 = C$ for several different values of the constant C.

Example: $\phi = x^2 + y^2$

Figure 1 shows the contours of $\phi = x^2 + y^2 = C$ for several different values of the constant C. MATLAB has no data type to represent a field quantity. Rather, such quantities are represented as *functions*. There are three major ways of representing a function in MATLAB

(1) An inline function, defined only in the workspace

```
>> myfun = '1+log(r)';
>> myfuni=inline(myfun,'r')
myfuni =
     Inline function:
     myfuni(r) = 1+log(r)
>> a=feval(myfuni,1)
a =
     1
>> a=feval(myfuni,10)
a =
     3.3026
```

COMSOL Multiphysics has a new facility under the Options Menu — Functions for defining inline functions by formula.

(2) An m-file function, which is stored as a disk file and can be called from either the workspace or an m-file script. For instance the m-file

function temperat.m contains the following code and is stored in the MAT-LAB current directory.

```
function t=temperat(r)
%TEMPERAT evaluates T = 1 + ln r
%    T = temperat(r)
%
t=1+log(r);

>>a=temperat(10)
a =
      3.3026
```

(3) Interpolation functions. The values of the function at certain points are specified. Values at nearby points are estimated by assumption about how smooth the function is locally. Interpolation requires a series of MATLAB commands, but eventually results in a functional form. MATLAB has built-in functions for 1-D, 2-D, and 3-D data called interp1, interp2, and interp3, respectively. COMSOL Multiphysics has a new facility under the Options Menu — Functions for defining interpolation functions for data entered by hand or through data files in a given format. See p. 122ff of Chapter two for an example of entering the density profile of water as a function of temperature.

Typically COMSOL Multiphysics field entry for coefficients and boundary data is done by in-line forms expressing the predefined independent, dependent, and derived variables. For instance, in general form with a single dependent variable u and independent variable x, expressions such as

$$u + 5 * x + \sin(3 * pi * x) + 3 * u * ux,$$

can be entered. But MATLAB m-file functions (including interpolants) can be used just as readily. An important point is that COMSOL Multiphysics expects data entry as scalar components. If a vector or matrix is required, it is always through specification of scalar components, any of which can be (complex) functions. Another equally important point is that COMSOL Multiphysics expects that functions can deal with vectorial input, i.e. each argument to the function could be a vector. Therefore, you must be careful to explicitly use element-wise arithmetic operators such as .*, .êtc. in defining your m-file functions.

COMSOL Multiphysics represents its results in a FEM structure with the degrees of freedom specified in fem.sol for a mesh specified in fem.mesh (or fem.xmesh). COMSOL Multiphysics provides a special postinterp function to extract interpolated values from fem.sol for each dependent variable and derived variable. This book is littered with examples of using postinterp to represent functions. It can even be automated in an m-file function that calls the appropriate fem structure from a mat-file.

3. Differentiation in multivariable calculus

3.1. *The gradient of a scalar field*

If $\phi = \phi(x, y, z)$, then the vector

$$\nabla\phi = \frac{\partial\phi}{\partial x}\mathbf{i} + \frac{\partial\phi}{\partial y}\mathbf{j} + \frac{\partial\phi}{\partial z}\mathbf{k} = \begin{bmatrix} \dfrac{\partial\phi}{\partial x} \\[2mm] \dfrac{\partial\phi}{\partial y} \\[2mm] \dfrac{\partial\phi}{\partial z} \end{bmatrix}, \tag{11}$$

is called the *gradient* of the scalar field ϕ, and is denoted as well by grad ϕ. The gradient operator ∇ (the nabla character) is the vector operator

$$\nabla = \mathbf{i}\frac{\partial}{\partial x} + \mathbf{j}\frac{\partial}{\partial y} + \mathbf{k}\frac{\partial}{\partial z}, \tag{12}$$

in Cartesian coordinates in 3-D.

A COMSOL Multiphysics example

Suppose $\phi = x^2 + y^2$, then $\nabla\phi = (2x, 2y, 0)$.

But MATLAB does not directly deal with such symbolic calculations, however its symbolic toolbox does. COMSOL Multiphysics, however, routinely calculates the numerical approximation of the derivatives of a solution. So the gradient of a scalar field can be constructed by COMSOL Multiphysics "primitive" operations. How do we easily access this information? Here's the recipe in Table 1. It should be noted that since no PDE is actually being solved, Neumann BCs amount to a neutral or noncondition on the boundaries. Otherwise, only if the boundary data are compatible with the condition $0 = \text{phi} - x\hat{}2 - y\hat{}2$ is a solution possible.

Now export the fem structure to MATLAB (file menu). We will use postinterp to get the approximate numerical value, along with MATLAB bilinear regression.

Table 1.   Gradient model.

Model Navigator	2-D geom., PDE modes, general form (nonlin stat) independent variables: x, y dependent:phi
Options	Set Axes/Grid to $[-1, 1] \times [-1, 1]$
Draw	Rectangular domain $[-1, 1] \times [-1, 1]$
Boundary Mode/ Boundary Settings	Set all four domains to Neumann BCs
Subdomain Mode/Subdomain Settings	In domain 1, set $\Gamma = 0\ 0$; $da = 0$; $F = \text{phi} - x\hat{}2 - y\hat{}2$
Mesh mode	Remesh using default (418 nodes, 774 elements)
Solve	Use default settings (nonlinear solver)
Post-process	Switch to arrow mode (automatically set to vectors of grad ϕ.) See Figure A4

Table 2.   Numerical estimates of grad ϕ using the gradient model.

X	y	phi$x$	phi$y$
0.5	0.5	1.0000	1.0000
−0.25	0.75	−0.5000	1.5000
0.75	−0.5	1.5000	−1.0000
0.25	−0.75	0.5000	−1.5000

```
>>x=0.5;y=0.5;
[xx,yy]=meshgrid(-1:0.01:1,-1:0.01:1);
xxx=[xx(:)'; yy(:)'];
phix=postinterp(fem,'phix',xxx);
phiy=postinterp(fem,'phiy',xxx);
uu=reshape(phix,size(xx));
vv=reshape(phiy,size(xx));
u=interp2(xx,yy,uu,x,y);
v=interp2(xx,yy,vv,x,y);
[u,v]
```

```
ans = 1.0000    1.0000
```

All the reshape commands are to make sure that the data is in the correct format for interp2. Table 2 samples the gradient calculation at other points.

By any accounting method, the use of FEM for finding first derivatives is fairly accurate. The global error of $O(10^{-16})$ as reported in the convergence criteria leads to a minimum of four decimal places in the estimated gradients. Figure 2 graphs the numerical approximation to the gradient.

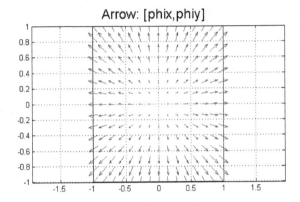

Figure 2.   Arrow plot of vectors of grad ϕ.

3.2. *The directional derivative*

The directional derivative of ϕ is the rate of change of ϕ (x, y, z) along a given direction. If $\hat{n}$ is the unit vector in that direction, then the directional derivative is given by

$$\frac{\partial \phi}{\partial n} = \hat{n} \cdot \nabla \phi. \tag{13}$$

The coordinate directions are the easiest to compute, e.g.

$$i \cdot \nabla \phi = \frac{\partial \phi}{\partial x}. \tag{14}$$

We used directional derivatives in the ECT models of Chapter seven to directly compute the normal derivatives of the electric potential (see chapter seven). Clearly, directional derivatives are intimately related to the concept of flux. The total flux across a material surface for a "linear" property (Fick's Law, Fourier's Law, etc.) is proportional to the integral of the normal derivative along that surface. The local flux is proportional to the normal derivative at a point.

At this point in most vector calculus texts, it is demonstrated that the direction in which the rate of change of ϕ is greatest is the direction of grad ϕs and that $|\text{grad}\phi|$ is the rate of change in that direction. We can show this at say the point $(x, y) = (0.25, -0.75)$ by stepping through the angles $\theta = [ws\pi]$ and plotting the (scalar value) $\frac{\partial \phi}{\partial n}$. MATLAB code that achieves this is written below.

```
>> theta=linspace(0, pi, 100);
dirder = zeros(size(theta));
```

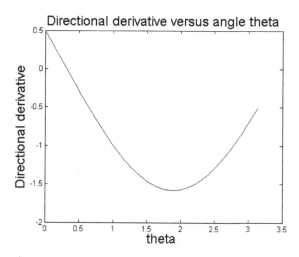

Figure 3.   Directional derivative versus direction (angle θ) in radians. Note the presence of a minimum in directional derivative — the direction of steepest descent, which corresponds to the gradient direction.

```
for k=1:length(theta)
dirder(k)=cos(theta(k))*u+sin(theta(k))*v;
end
plot(theta, dirder)
```

Of course I refuse to apologize for my FORTRAN-ish programming bias which is revealed in the looping structure above. Were I in a more MATLAB-ish mode, then judicious use of threading achieves the same results without the loop:

```
>>dirder = u*cos(theta)+v*sin(theta);
plot(theta,dirder)
```

cos and sin functions thread across each element of the vector theta, producing an output vector of the same length.

3.3. *Level sets/level surfaces*

Note that the directional derivative (dirder) crosses the x-axis, i.e. there is a direction for which the directional derivative is zero — no rate of change at all in that direction. It can be shown that the direction $\hat{n}$ for which $\frac{\partial \phi}{\partial n} = 0$ is perpendicular to the gradient direction. So in this direction, $\phi = $ constant locally. Tracing out the curve (in 2-D) or surface (in 3-D) of each constant identifies a family of curves (surfaces) called level sets of ϕ (see Chapter eight). In 2-D, level sets are also called contours. The terminology of the directional derivative is analogous to survey maps, where ϕ is the elevation of land. The contours all have the same height above sea level (level sets); the directional derivative $\hat{n} \cdot \nabla \phi$ is the rate of climb in the direction $\hat{n}$, and the gradient is in the direction of steepest climb (or descent) and the rate of climb is $|\text{grad}\phi|$. In fluid dynamics, the quantity that is most often represented by a contour plot is the streamfunction, with contours all being streamlines (particle paths in steady flow) tangent to the velocity field. In Chapter three, the buoyant convection example shows how to compute streamfunction (see equation (3) of Chapter 3).

3.4. *Derivatives of vector fields*

The vector differential operator ∇ may be applied to a vector field $\mathbf{F}(\mathbf{x})$ in two ways: (1) the scalar product $\nabla \cdot \mathbf{F}$ called the divergence, (2) the vector product $\nabla \times \mathbf{F}$, called the curl.

The divergence is given by

$$\text{div } F = \nabla \cdot F = \sum_{k=1}^{3} \frac{\partial F_k}{\partial x_k}$$

$$= \left(\frac{\partial}{\partial x}, \frac{\partial}{\partial y}, \frac{\partial}{\partial z} \right) \cdot (F_1, F_2, F_3)$$

$$= \frac{\partial F_1}{\partial x} + \frac{\partial F_2}{\partial y} + \frac{\partial F_3}{\partial z} \,. \tag{15}$$

The curl is given by

$$\text{curl } F = \nabla \times F = \sum_{i=1}^{3} \sum_{j=1}^{3} \sum_{k=1}^{3} \varepsilon_{ijk} \frac{\partial F_k}{\partial x_j} \hat{e}_i$$

$$= \begin{vmatrix} i & j & k \\ \frac{\partial}{\partial x} & \frac{\partial}{\partial y} & \frac{\partial}{\partial z} \\ F_1 & F_2 & F_3 \end{vmatrix}$$

$$= \left(\frac{\partial F_3}{\partial y} - \frac{\partial F_2}{\partial z}, \frac{\partial F_1}{\partial z} - \frac{\partial F_3}{\partial x}, \frac{\partial F_2}{\partial x} - \frac{\partial F_1}{\partial y} \right), \tag{16}$$

ε_{ijk} is the permutation tensor introduced earlier. Of course one can see readily that div $\mathbf{F}$ is a scalar, while curl $\mathbf{F}$ is a vector.

Table 3.   div and curl recipe.

Model Navigator	3-D geom., PDE modes, general form (nonlin stat) independent variables: x, y z; 3dependent: u_1, u_2, u_3
Options	Set Axes/Grid to $[0,1] \times [0,1] \times [0,1]$
Draw	Block BLK1 $= [0,1] \times [0,1] \times [0,1]$
Boundary Mode/ Boundary Settings	Set all four domains to Neumann BCs
Subdomain Mode/Subdomain Settings	set $\Gamma_x = 0\ 0\ 0$; $da_1 = 0\ 0\ 0$; $F_1 = u_1 - x\hat{}2$ set $\Gamma_y = 0\ 0\ 0$; $da_2 = 0\ 0\ 0$; $F_2 = u_2 - 3*x*y$ set $\Gamma_z = 0\ 0\ 0$; $da_3 = 0\ 0\ 0$; $F_3 = u_3 - x\hat{}3$
Mesh mode	Remesh using mesh scaling factor 3 (201 nodes, 719 elements)
Solve	Use default settings (nonlinear solver)
Post-process	(1) Color plot of $u_1x + u_2y + u_3z$ for the divergence (2) Arrow plot for the curl of $(u_3y - u_2z, u_1z - u_3x, u_2x - u_1y)$

The operator $\mathbf{F} \cdot \nabla$ is often seen in advection terms in heat or mass transport equations. Clearly, it is *not* the divergence, since

$$F \cdot \nabla = F_1 \frac{\partial}{\partial x} + F_2 \frac{\partial_2}{\partial y} + F_3 \frac{\partial}{\partial z}, \qquad (17)$$

which is still an operator, in contrast to (15), which is a scalar.

As we saw before, the numerical approximation of derivatives is a "primitive" of COMSOL Multiphysics, so we should be able to compute approximations to both div and curl.

A COMSOL Multiphysics example

Suppose $F = (x^2, 3x_y, x^3)$. Here's the recipe in Table 3.

Again, it should be noted that since no PDE is actually being solved, Neumann BCs amount to a neutral or noncondition on the boundaries. Otherwise, only if the boundary data are compatible with the conditions $0 = u_1 - x\hat{}2$, $0 = u_2 - 3*x*y$,

$0 = u_3 - x\hat{}3$ is a solution possible.

Symbolically, it is straightforward to compute

$$\nabla \cdot \mathbf{F} = 5x,$$
$$\nabla \times \mathbf{F} = (0, -3x^2, 3y).$$

So how good is the numerical approximation? Try the divergence:

```
>> xxx=[0.42; 0.57; 0.33];
postinterp(fem,'u1x+u2y+u3z',xxx)
ans =    2.1137
```

```
>> 5*0.42
ans =      2.1000
```

Clearly, for such a coarse mesh, half a percent error is not a bad result. Now for the curl.

```
>>xxx=[0.42;0.57;0.33];[postinterp(fem,'u3y-u2z',xxx);
postinterp(fem,' u1z-u3x',xxx);postinterp(fem,'u2x-u1y',xxx)]
ans =
    0.0043
   -0.5319
    1.7100
>> [0; -3*0.42^2; 3*0.57]
ans =
         0
   -0.5292
    1.7100
```

The worst error here is again half a percent. Figure 4 shows the numerical approximation by FEM to the divergence, which qualitatively shows isosurfaces consistent with $\nabla \cdot \mathbf{F} = 5x$. Figure 5 shows the arrow plot of curl F. Since most of us have little feel for three-dimensional vector plots, determining whether the plot is consistent with the closed form calculation is beyond our visual capacity for numeracy. Nevertheless, the FEM solution shows the very important feature of numerical solutions —

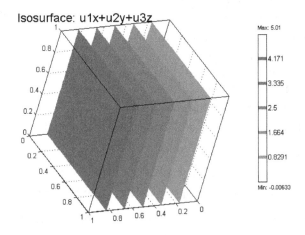

Figure 4.   Isosurfaces of divergence computed for the example $\mathbf{F} = (x^2, 3xy, x^3)$.

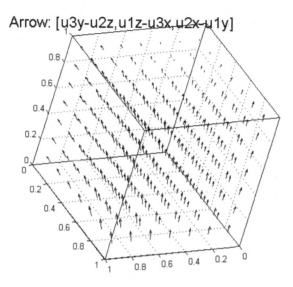

Arrow: [u3y-u2z,u1z-u3x,u2x-u1y]

Figure 5.   Arrow plot of curl computed for the example $\mathbf{F} = (x^2, 3xy, x^3)$.

visualization of solutions. Does anyone have a feel for the analytic solution $\nabla \times \mathbf{F} = (0, -3x^2, 3y)$ either?

3.5. *The Laplacian operator*

For the scalar field ϕ

$$\nabla \cdot \nabla \phi = \nabla \cdot \left(\frac{\partial \phi}{\partial x}, \frac{\partial \phi}{\partial y}, \frac{\partial \phi}{\partial z} \right)$$

$$= \frac{\partial}{\partial x} \left(\frac{\partial \phi}{\partial x} \right) + \frac{\partial}{\partial y} \left(\frac{\partial \phi}{\partial y} \right) + \frac{\partial}{\partial z} \left(\frac{\partial \phi}{\partial z} \right)$$

$$= \frac{\partial^2 \phi}{\partial x^2} + \frac{\partial^2 \phi}{\partial y^2} + \frac{\partial^2 \phi}{\partial z^2}. \tag{18}$$

By parallel with other nabla operators,

$$\text{div}(\text{grad}) = \nabla \cdot \nabla$$

$$= \left(\frac{\partial}{\partial x}, \frac{\partial}{\partial y}, \frac{\partial}{\partial z} \right) \cdot \left(\frac{\partial}{\partial x}, \frac{\partial}{\partial y}, \frac{\partial}{\partial z} \right)$$

$$= \frac{\partial^2}{\partial x^2} + \frac{\partial^2}{\partial y^2} + \frac{\partial^2}{\partial z^2}. \tag{19}$$

So for shorthand, the operator div(grad) is called the Laplacian and denoted ∇^2. Typically, the Laplacian is used in differential equations, rather than computed directly. For instance, Laplace's equation

$$\nabla^2 \phi = 0, \qquad (20)$$

is an example where the Laplacian is known (zero) but the function ϕ is to be found. FEMLAB routinely computes the first derivatives of a dependent variable, but not necessarily the second, directly. But as we already know how to compute both div and grad separately, computing div(grad) is a matter of using auxiliary dependent variables v_1, v_2, v_3 that are assigned values in the last example of $F_1 = v_1 - ux$; $F_2 = v_2 - uy$; $F_3 = v_3 - uz$ so that

$$\nabla^2 u = v_1 x + v_2 y + v_3 z. \qquad (21)$$

We should comment here that COMSOL Multiphysics directly estimates the second derivatives referred to symbolically as uxx, uyy, uzz, so computation of the second derivatives and the Laplace operator can be done simply by solving the equation $u = f(x, y, z)$ and using the symbolic derivatives in post-processing. There is a caveat — to compute second derivatives, one should use at least third order elements for accuracy.

3.6. *Scalar and vector potentials*

Quite a lot of space in vector calculus books is devoted to the topics of scalar and vector potentials.

A scalar potential ϕ for a vector field $\mathbf{F}$ is a scalar function for which $\nabla \phi = \mathbf{F}$. The textbooks show that this is only possible if, and only if,

$$\text{curl}\mathbf{F} = 0.$$

Similarly, a vector potential $\mathbf{A}$ for a vector field $\mathbf{F}$ is a vector function for which $\mathbf{F} = \text{curl}\mathbf{A}$. Again, the textbooks show that this is only possible, if and only if,

$$\text{div}\mathbf{F} = 0.$$

Scalar and vector potentials are useful for simplifying pde systems that are either irrotational or divergence free (solenoidal). In the case of fluid flow, either inviscid or completely viscous flow are simplified dramatically by such potentials. One might ask, can COMSOL Multiphysics help in the task of identifying these potentials? In the case of 2-D flows, we already saw that the streamfunction acts like a vector potential (3.3), so the answer is a qualified yes. For many years in both electrodynamics and hydrodynamics, the hunt for vector potentials or scalar potentials to simplify calculations was paramount — many analyses end is solving, even approximately, for such a potential. Yet whether sufficient symmetries exist in a given modeling situation to use scalar and vector potentials to simplify the calculations is now almost a moot point. General pde engines like COMSOL Multiphysics can compute numerical approximations to the primitive variables in the most general cases, limited only by their CPU requirements. So the virtue of finding such simplifications is a reduction of CPU usage, for which we must still pay the price of numerical differentiation to arrive at the primitive variables (using our grad and curl recipes) if detailed solutions are required.

It is perhaps a sobering note to end our Appendix on that general purpose numerical solvers like COMSOL Multiphysics limit the need for many of the complexities of vector calculus, since much of the higher theory was developed to treat intentionally idealized models. Nevertheless, the basics of vector calculus are necessary to understand what such pde engines do, and how they do it. Theory simultaneously becomes more important in some aspects — dealing with complexities that are still beyond computability, proposing physical models that are amenable to numerical computation — but also less necessary for "run-of-the-mill" applications. Theorists should be challenged that they must remain ahead of the game to still be relevant practioners due to the advent of general purpose solvers like COMSOL Multiphysics.

INDEX